Virus Diseases of Tropical and Subtropical Crops

CABI PLANT PROTECTION SERIES

Plant pests and diseases cause significant crop losses worldwide. They cost growers, governments and consumers billions annually and are a major threat to global food security: up to 40% of food grown is lost to plant pests and diseases before it can be consumed. The spread of pests and diseases around the world is also altered and sped up by international trade, travel and climate change, introducing further challenges to their control.

In order to understand and research ways to control and manage threats to plants, scientists need access to information that not only provides an overview and background to the field, but also keeps them up to date with the latest research findings. This series presents research-level information on important and current topics relating to plant protection from pests, diseases and weeds, with international coverage. Each book provides a synthesis of facts and future directions for researchers, upper-level students and policy makers.

Titles Available

1. *Disease Resistance in Wheat*
   Edited by Indu Sharma
2. *Phytophthora: A Global Perspective*
   Edited by Kurt Lamour
3. *Virus Diseases of Tropical and Subtropical Crops*
   Edited by Paula Tennant and Gustavo Fermin

Virus Diseases of Tropical and Subtropical Crops

Edited by

Paula Tennant

The University of the West Indies, Mona, Jamaica

and

Gustavo Fermin

Universidad de Los Andes, Mérida, Venezuela

CABI is a trading name of CAB International

CABI
Nosworthy Way
Wallingford
Oxfordshire OX10 8DE
UK

Tel: +44 (0)1491 832111
Fax: +44 (0)1491 833508
E-mail: info@cabi.org
Website: www.cabi.org

CABI
745 Atlantic Avenue
8th Floor
Boston, MA 02111
USA

Tel: +1 (0)617 682 9015
E-mail: cabi-nao@cabi.org

A catalogue record for this book is available from the British Library, London, UK.

Library of Congress Cataloging-in-Publication Data

Names: Tennant, Paula. | Fermin, Gustavo.
Title: Virus diseases of tropical and subtropical crops / editors, Paula Tennant, The University of the West Indies, Mona, Jamaica, Gustavo Fermin, Universidad de Los Andes. Mérida, Venezuela.
Description: Boston, MA : CABI, 2015.
Identifiers: LCCN 2015030126 | ISBN 9781780644264 (alk. paper)
Subjects: LCSH: Plant viruses--Tropics. | Virus diseases of plants--Tropics. | Tropical crops--Diseases and pests.
Classification: LCC SB736 .V574 2015 | DDC 632/.8--dc23 LC record available at http://lccn.loc.gov/2015030126

ISBN-13: 978 1 78064 426 4

Commissioning editor: Alex Hollingsworth
Editorial assistant: Emma McCann
Production editor: Tim Kapp

Typeset by SPi, Pondicherry, India
Printed and bound by Gutenberg Press Limited, Tarxien, Malta

Contents

_____

Contributors

Isabelle Abt, INRA-Cirad-SupAgro Montpellier, UMR 385 BGPI, Cirad TA A-54K, Campus International de Baillarguet, F-34398 Montpellier, France; Bayer S.A.S./Bayer CropScience, 16 rue Jean Marie Leclair - CS 90106, 69266 Lyon Cedex 09, France. E-mail: isabelle.abt@supagro.inra.fr

Olufemi Alabi, Department of Plant Pathology & Microbiology, Texas A&M AgriLife Research and Extension Center, Weslaco, TX 78596, USA. E-mail: alabi@tamu.edu

Ricardo Alcalá-Briseño, Research Scholar, Dept. of Plant Pathology, University of Florida, Gainesville, FL 32611. E-mail: ralcala@ufl.edu

Icolyn Amarakoon, Department of Basic Medical Sciences, Aqueduct Road, University of the West Indies, Mona, Kingston, Jamaica. E-mail: icolyn.amarakoon@uwimona.edu.jm

Abdolbaset Azizi, Department of Plant Pathology, Faculty of Agriculture, Tarbiat Modares University, Tehran, Iran. E-mail: abdolbasetazizi@gmail.com

Guy Blomme, BIOVERSITY International, c/o ILRI, Addis Ababa, Ethiopia. E-mail: g.blomme@cgiar.org

Aman Bonaventure Omondi, BIOVERSITY International, Bujumbura office, Burundi. E-mail: b.a.omondi@cgiar.org

Niyongere Célestin, Institut des Sciences Agronomiques du Burundi (ISABU), Avenue de la Cathédrale Régina Mundi, BP 795 Bujumbura, Burundi. E-mail: niyocelestin@gmail.com

V. Celia Chalam, National Bureau of Plant Genetic Resources (ICAR-NBPGR), Pusa, New Delhi - 110012, India. E-mail: mailcelia@gmail.com

Tsung-Chi Chen, Department of Biotechnology, Asia University, Wufeng, Taichung 41354, Taiwan; Department of Medical Research, China Medical University Hospital, China Medical University, Taichung 40402, Taiwan. E-mail: kikichenwolf@hotmail.com

Mohamad Chikh-Ali, Department of Plant, Soil and Entomological Sciences, University of Idaho, 875 Perimeter Drive MS 2339, Moscow, ID 83844-2339, USA. E-mail: mchikhali@uidaho.edu

Indranil Dasgupta, Department of Plant Molecular Biology, University of Delhi, South Campus, New Delhi-110021, India. E-mail: indasgup@south.du.ac.in

Angela Eni, Department of Biological Sciences, Covenant University, Ota, Nigeria. E-mail: angela.eni@covenantuniversity.edu.ng

Fulgencio Espejel, Plant-Virus Interaction Laboratory, Dept. of Genetic Engineering at Cinvestav-Unidad Irapuato, Km 9.6 Libramiento Norte Carretera Irapuato-León, Irapuato, Guanajuato 36821, Mexico. E-mail: fespejel@ira.cinvestav.mx

Vicente Febres, Horticultural Sciences Department, University of Florida, Florida, USA. E-mail: vjf@ufl.edu

Gustavo Fermin, Instituto Jardín Botánico de Mérida, Faculty of Sciences, Universidad de Los Andes, Mérida 5101, Mérida, Venezuela. E-mail: fermin@ula.ve

Latanya Fisher, Horticultural Sciences Department, University of Florida, Florida, USA. E-mail: lfisher@ufl.edu

Cherie Gambley, Department of Agriculture and Fisheries (DAF), Applethorpe Research Station, GPO Box 501, Stanthorpe, Queensland 4380, Australia. E-mail: cherie.gambley@daf.qld.gov.au

Augustine Gubba, Department of Plant Pathology, School of Agricultural, Earth and Environmental Sciences, University of KwaZulu-Natal, Private Bag X01, Scottsville 3209, Pietermaritzburg, South Africa. E-mail: gubbaa@ukzn.ac.za

Cindy-Leigh Hamilton, Biotechnology Centre, The University of the West Indies, Mona, Kingston, Jamaica. E-mail: cindyleigh87@gmail.com

Masarapu Hema, Department of Virology, Sri Venkateswara University, Tirupati – 517501, AP, India. E-mail: hemamasarapu70@gmail.com

Emmanuel Jacquot, INRA-Cirad-SupAgro Montpellier, UMR 385 BGPI, Cirad TA A-54K, Campus International de Baillarguet, F-34398 Montpellier, France. E-mail: emmanuel.jacquot@rennes.inra.fr

Fuh-Jyh Jan, Department of Plant Pathology, National Chung Hsing University, Taichung 40227, Taiwan; Agricultural Extension Center, National Chung Hsing University, Taichung 40227, Taiwan. E-mail: fjjan@nchu.edu.tw

Edward E. Kanju, International Institute of Tropical Agriculture, PO Box 34441, Dar es Salaam, Tanzania. E-mail: e.kanju@cgiar.org

Alexander V. Karasev, Department of Plant, Soil and Entomological Sciences, University of Idaho, 875 Perimeter Drive MS 2339, Moscow, ID 83844-2339, USA. E-mail: akarasev@uidaho.edu

P. Lava Kumar, International Institute of Tropical Agriculture, Oyo Road, PMB 5320, Ibadan, Nigeria. E-mail: l.kumar@cgiar.org

James P. Legg, International Institute of Tropical Agriculture, PO Box 34441, Dar es Salaam, Tanzania. E-mail: j.legg@cgiar.org

Sudeshna Mazumdar-Leighton, Department of Botany, Delhi University, Delhi-7, India. E-mail: smazumdar@botany.du.ac.in

Rabson M. Mulenga, Zambia Agriculture Research Institute, Mount Makulu Central Research Station, Private Bag 7, Chilanga, Lusaka, Zambia. E-mail: rabson2010@gmail.com

Basavaprabhu L. Patil, National Research Centre on Plant Biotechnology (ICAR-NRCPB), IARI, Pusa Campus, New Delhi - 110012, India. E-mail: blpatil2046@gmail.com

Melaine Randle, Biotechnology Centre, The University of the West Indies, Mona, Kingston, Jamaica. E-mail: melainerandle@gmail.com

Marcia Roye, Biotechnology Centre, The University of the West Indies, Mona, Kingston, Jamaica. E-mail: marcia.roye@uwimona.edu.jm

Laura Silva-Rosales, Plant-Virus Interaction Laboratory, Dept. of Genetic Engineering at Cinvestav-Unidad Irapuato, Km 9.6 Libramiento Norte Carretera Irapuato-León, Irapuato, Guanajuato 36821, Mexico. E-mail: lsilva@ira.cinvestav.mx

Benice J. Sivparsad, Department of Plant Pathology, School of Agricultural, Earth and Environmental Sciences, University of KwaZulu-Natal, Private Bag X01, Scottsville 3209, Pietermaritzburg, South Africa. E-mail: sivparsad@ukzn.ac.za

Pothur Sreenivasulu, Former Professor of Virology, Sri Venkateswara University, Tirupati–517501, AP, India. E-mail: pothursree@gmail.com

Paula Tennant, Department of Life Sciences, 4 Anguilla Close, The University of the West Indies, Mona Campus, Jamaica. E-mail: paula.tennant@uwimona.edu.jm

John Thomas, Queensland Alliance for Agriculture and Food Innovation, The University of Queensland, Ecosciences Precinct, Level 2C West, GPO Box 267, Brisbane, Queensland 4001, Australia. E-mail: j.thomas2@uq.edu.au

Jeanmarie Verchot, Department of Entomology and Plant Pathology, Oklahoma State University, Stillwater, OK 74074, USA. E-mail: verchot.lubicz@okstate.edu

Preface

Paula Tennant and Gustavo Fermin

————————————————

Plant viruses continue to be, probably more than ever, major contributors to severe yield and economic losses in crop production, particularly in tropical and subtropical regions. The tropics, regions that are limited by the Tropic of Cancer at approximately 23°30′ N and the Tropic of Capricorn at 23°30′ S, and subtropical regions that are immediately north and south of the tropic zone, present different ecological characteristics as well as ideal conditions for the perpetuation of host plants and virus vectors for most of the year. Moreover, the impact of plant viruses on sustainable crop production is influenced by dramatic climatic changes that are occurring throughout the world. Changes in host plants and insect vector populations resulting in increasing instability within virus–host ecosystems could affect the spread of plant viruses. Some of the threatening and economically important virus diseases in tropical and subtropical zones which affect global food production include tungro in rice, mosaic in sugarcane, soybean and cassava, tristeza in citrus, leaf curl in tomato and ringspot in papaya, among others. Of the estimated 3000 recognized viruses, slightly more than 1000 are plant viruses—most of which are geminiviruses and potyviruses, and together make up almost half of the most economically important plant viruses. Key factors shaping the emergence of new plant virus diseases, including the intensification of agricultural trade, changes in cropping systems and climate change, will add to our need to know more about plant viruses (inclusive of the unidentified ones) and the diseases they cause, as well as the role played by new virus variants—sometimes with increased virulence and/or widened host range.

Technological advances in, for example, diagnostic and agronomic practices have reduced the risk of virus disease epidemics. Additionally, advances in sequencing strategies, deeper knowledge of vector dynamics and the discovery of new viruses, and of virus biology and virus/vector, vector/host and virus/host interactions, will eventually facilitate the development of new disease resistant varieties and strategies that would mitigate the damage and losses inflicted by plant diseases of viral etiology. With technology, there is hope that useful genes for virus disease resistance can be transferred to agricultural food crops. Combined with ecological and epidemiological investigations, it is increasingly possible to effectively enter into an arms race against the most virulent and damaging plant viruses. Integrated disease management, employing both classical and modern technologies, must however be put into force. This will necessitate the field of engineered resistance moving beyond pathogen derived resistance (i.e. traditional transgenic crops) and into the realms of an understanding and the manipulation of virus recognition and response driven by major, dominant R resistance genes. Besides R genes, the use of recessive resistance genes, targeted manipulation of RNA interference pathways and plant hormone-mediated resistance can add to the arsenal of weapons against virus driven diseases. Plant pathology is a science with important ties to human welfare; accordingly, we should all be aware of the biological basis of diseases and their control, the legislation governing environmental protection, the release of genetically engineered organisms, along with trends aimed at securing land preservation for agricultural use, and genetic reserves.

The primary goal of this book is to provide readers with the latest information on important virus diseases of crops in tropical and subtropical countries and the envisaged directions into the future of plant virus control. The volume comprises 18 chapters. The first chapter covers general information on the impact of virus diseases, methods for estimating disease severity, recent data on host-virus interactions, serological and molecular techniques for the diagnosis of virus pathogens, and disease management strategies. Subsequent chapters examine selected virus diseases, inclusive of the hosts, vectors and viruses and their intricate relationships, symptoms development, virus transmission, host resistance and the underlying biochemical and genetic factors leading to disease, as well as their control using a variety of techniques including genetic modification. Each chapter is written by distinguished scientists who have made significant contributions in their respective fields. Each chapter represents in part the fruits of their original research, but also incorporates the work of others on the diseases and their respective etiological agents, for a comprehensive review. It is evident that the structure of the various virus epidemics covered in this book is different and complex, dependent on the nature of the virus or viruses, their relationship with their vectors, geographical location, among other factors such as political and social approaches that deal with food security. Much progress has been achieved, but there is much work remaining on the pursuit and development of knowledge-based disease management strategies, regulatory policies and synergies between all the stakeholders. Yet there are reasons to be optimistic.

Finally, we wish to express our appreciation to all the contributing authors. Without their expertise, commitment and investment of time, the book could never have been completed. We are especially grateful to members of the editorial board, Rachel Cutts, Joris Roulleau, Alex Hollingsworth and Emma McCann, for their assistance and guidance in shaping and compiling the book. Gratitude is extended to many of our colleagues for their advice and collaboration. Lastly, we are indebted to our families for their patience and understanding while performing this gratifying project.

1 Viruses Affecting Tropical and Subtropical Crops: Biology, Diversity, Management

Gustavo Fermin,[1]* Jeanmarie Verchot,[2] Abdolbaset Azizi[3] and Paula Tennant[4]

[1]*Instituto Jardín Botánico de Mérida, Faculty of Sciences, Universidad de Los Andes, Mérida, Venezuela;* [2]*Department of Entomology and Plant Pathology, Oklahoma State University, Stillwater, Oklahoma, USA;* [3]*Department of Plant Pathology, Faculty of Agriculture, Tarbiat Modares University, Tehran, Iran;* [4]*Department of Life Sciences, The University of the West Indies, Mona Campus, Jamaica*

1.1 Introduction

Viruses are the most abundant biological entities throughout marine and terrestrial ecosystems. They interact with all life forms, including archaea, bacteria and eukaryotic organisms and are present in natural or agricultural ecosystems, essentially wherever life forms can be found (Roossinck, 2010). The concept of a virus challenges the way we define life, especially since the recent discoveries of viruses that possess ribosomal genes. These discoveries include the surprisingly large viruses of the *Mimiviridae* (Claverie and Abergel, 2012; Yutin *et al.*, 2013), the Pandoraviruses that lack phylogenetic affinity with any known virus families (Philippe *et al.*, 2013) and Pithovirus sibericum that was recovered from Siberian permafrost after being entombed for more than 30,000 years (Legendre *et al.*, 2014). Apparently they co-occurred and even predated cellular forms on our planet, yet arguably they have no certain place in our current view of the tree of life (Brüssow, 2009; Koonin and Dolja, 2013; Thiel *et al.*, 2013).

Besides their potential role in evolution, viruses have facilitated the understanding of various basic concepts and phenomena in biology (Pumplin and Voinnet, 2013; Scholthof, 2014). However, they have also long been considered as disease-causing entities and are regarded as major causes of considerable losses in food crop production. Pathogenic viruses imperil food security by decimating crop harvests as well as reducing the quality of produce, thereby lowering profitability. This is particularly so in the tropics and subtropics where there are ideal conditions throughout the year for the perpetuation of the pathogens along with their vectors. Viruses account for almost half of the emerging infectious plant diseases (Anderson *et al.*, 2004). Moreover, technologies of DNA and RNA deep sequencing (Wu *et al.*, 2010; Adams *et al.*, 2012; Grimsley *et al.*, 2012; Zhuo *et al.*, 2013; Barba *et al.*, 2014; Kehoe *et al.*, 2014), as well as genomics and metagenomics (Adams *et al.*, 2009; Kristensen *et al.*, 2010; Roossinck *et al.*, 2010; Rosario *et al.*, 2012), have allowed for the discovery of new species of plant viruses – some of which have been isolated from symptomless plants (Roossinck, 2005, 2011; Kreuze *et al.*, 2009; Wylie *et al.*, 2013; Saqib *et al.*, 2014). Recent investigations suggest that some viruses actually confer a

*E-mail: fermin@ula.ve

range of ecological benefits upon their host plants (Mölken and Stuefer, 2011; Roossinck, 2011, 2012; Prendeville *et al.*, 2012; Mac-Diarmid *et al.*, 2013), for example, traits such as tolerance to drought (Xu *et al.*, 2008; Palukaitis *et al.*, 2013) and cold (Meyer, 2013; Roossinck, 2013). Studies of viruses associated with non-crop plants have only just begun, but findings so far indicate that overall very little is known about viruses infecting plants (Wren *et al.*, 2006). It is becoming increasingly evident that the view of viruses as mere pathogens is outdated. These entities possess the potential for facilitating a variety of interactions among macroscopic life. Therefore, a lot of work is needed in terms of research dealing with the diversity, evolution and ecology of viruses to truly comprehend their rich contribution to all human endeavours, including agriculture and food security. This introductory section focuses on some of the topics that are of current interest and relevance to tropical and subtropical regions where a number of plant diseases that threaten food security are caused by viruses.

1.2   Biology: Structure, Taxonomy and Diversity

Of the ca. 2000 viruses listed in the 2013 report of the International Committee for the Taxonomy of Viruses, less than 50%, or ca. 1300, are plant viruses. Viruses, which by definition contain either a RNA or DNA genome surrounded by a protective, virus-coded protein coat (CP) are viewed as mobile genetic elements, and characterized by a long co-evolution with their host. Many plant viruses have a relatively small genome; one of the smallest among plant viruses is a nanovirus with a genome of about 1 kb while the closterovirus genome can be up to 20 kb. Despite this apparent simplicity, nearly every possible method of encoding information in nucleic acid is exploited by viruses, and their biochemistry and mechanisms of replication are more varied than those found in the bacterial, plant and animal kingdoms (Mac-Naughton and Lai, 2006; Koonin, 2009).

According to Baltimore (Fig. 1.1), classification of viruses comprises seven independent classes, based on the nature of the nucleic acid making up the virus particle: double-stranded (ds) DNA, single stranded (ss) DNA, dsRNA, ss (+) RNA, ss (–) RNA, ssRNA (RT) or ssDNA (RT). The entities are further categorized by the Committee on Taxonomy of Viruses into five hierarchically arranged ranks: order, family, subfamily, genus and species. The polythetic species concept (van Regenmortel, 1989) as applied to the definition of the virus species recognizes viruses as a single species if they share a broad range of characteristics while making up a replicating lineage that occupies a specific ecological niche (Kingsbury, 1985; van Regenmortel, 2003). There are also proposals for the consideration of virus architecture in the higher-order classification scheme (Abrescia *et al.*, 2009). Arguably, the defining feature of a virus is the CP, the structure of which is restricted by stereochemical rules (almost invariably icosahedral or helical) and genetic parsimony. Hurst (2011) introduced another proposal, namely the consideration of dividing life into two domains (i.e. the cellular domain and the viral domain), and thus the adoption of a fourth domain for viruses, along with entities such as viroids and satellites. It is opined that leaving viruses out of evolutionary, ecological, physiological or conceptual studies of living entities presents an incomplete understanding of life at any level. The proposed title of this domain is Akamara, which is of Greek derivation and translates to *without chamber* or *without void*; aptly referring to the absence of a cellular structure.

1.2.1   Virus evolution and the emergence of new diseases

Viruses are recognized as the fastest evolving plant pathogens. Genetic variation allows for the emergence and selection of new, fitter virus strains, as well as shapes the dynamics surrounding plant–virus and plant–vector interactions. Genetic changes are typically accomplished by mutations, the rate of which is greatest among RNA viruses because of non-proofreading activity of their replicases (i.e. RNA-dependent RNA polymerases). Recombination, either homologous or heterologous, is another source of virus variation. Recombination in potyviruses, for instance,

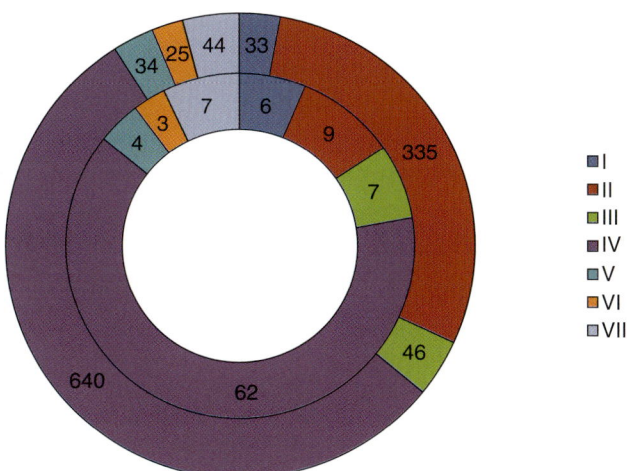

Fig. 1.1. The current plant virosphere (from the term viriosphere coined by Suttle in 2005) is comprised of (pathogenic) viruses belonging to all groups under Baltimore's classification. Group I (viruses with dsDNA genomes) include members of the family *Phycodnaviridae*; Group II (viruses with ssDNA genomes) those of the families *Geminiviridae* and *Nanoviridae*; Group III (viruses with dsRNA genomes) members of the families *Amalgaviridae*, *Endornaviridae*, *Partitiviridae* and *Reoviridae*; Group IV (viruses with (+) ssRNA genomes) that includes viruses from the families *Alphaflexiviridae*, *Betaflexiviridae*, *Benyviridae*, *Bromoviridae*, *Closteroviridae*, *Luteoviridae*, *Potyviridae*, *Secoviridae*, *Tombusviridae*, *Tymoviridae*, and *Virgaviridae*; Group V (viruses with (–) ssRNA genomes) with virus species of the families *Bunyaviridae*, *Rhabdoviridae* and *Ophioviridae*; Group VI (ssRNA-RT viruses with a DNA intermediate in their replication cycle) that consists of the plant virus families *Pseudoviridae* and *Metaviridae*; and finally, Group VII (dsRNA-RT viruses possessing an RNA intermediate in their replication cycle) with the members of the family *Caulimoviridae*. The inner circle provides the number of genera per group, while the outer circle includes the total number of species per group as of 2014. ds, double-stranded; RT, reverse transcriptase; ss, single-stranded.

has been shown to be especially frequent (Chare and Holmes, 2006). In other groups, like the family *Bunyaviridae*, reassortment of their genome segments seems to represent the underlying source of variation (Briese *et al.*, 2013).

Once variation is introduced, selection pressures that range from the action of host resistance genes to host shifts and environmental changes, or other mechanisms of genetic drift, contribute to changes in the genetic makeup of the virus population. Complementation between viruses in mixed infections can also lead to the maintenance of viruses with deleterious mutations, and hence increase the availability of variants that selection can act upon. Finally, current thinking suggests that genome organization, particularly in viruses showing 'overprinting', that is, gene overlapping, also plays a role. Gene overlapping, which allows for genome com-

pression, can increase the deleterious effect of mutations in viruses as more than one gene is affected resulting in reduced evolutionary rates and adaptive capacity (Chirico *et al.*, 2010; Sabath *et al.*, 2012).

1.2.2 Wild or non-crop plants as reservoirs and targets of 'new' causal agents of disease

Many viruses and their respective vectors are associated with non-crop reservoirs that potentially act as bridges between crop plants. Conversely, crop viruses have the capacity to infect non-crop plants with similar probability (Vincent *et al.*, 2014). In either scenario, the simplicity of plant virus genomes allows for quick adaptation of viruses to new hosts, and generalist viruses tend to exhibit greater potential to cause more damage than specialist viruses. While the scenario of increased virus invasion of native species is worrying and raises concern for the survival of endangered species, equally worrying is the effective jumping of viruses between native and crop species. Recent findings suggest that native plant communities are likely to contain potentially damaging viral pathogens (Kehoe *et al.*, 2014). Increased frequency of these reports is expected as new contact between native plants and introduced crops or weeds continues because of mans' activities and climate change.

1.2.3   Virus–virus interactions

Co-infection is another factor involved in shaping the genetic structure and diversity of plant viruses resulting in variations in symptom expression, infectivity, accumulation and/or vector transmissibility. Co-infections naturally occur due to the geographic overlap of distinct pathogenic types and appear to be the rule rather than the exception. The outcome of the mixed infection depends mainly on the plant species, virus strains, the order of infection and initial amount of inoculum. Antagonistic interactions between closely related viruses can lead to cross-protection and mutual exclusion. However, infections with different viruses in the same host can result in the appearance of more severe symptom expression than either single infection alone (viral synergism). Co-infection with *Clover yellow vein virus* (*Potyviridae*) and *White clover mosaic virus* (*Alphaflexiviridae*), for example, causes more severe disease development in pea (*Pisum sativum*), probably due to some unknown action of the potyvirus P3N-PIPO protein (Hisa *et al.*, 2014). Co-infection opens the possibility for inter-specific recombination or reassortment, and thus the generation of new viral species. Presumably virus–virus interactions are not only formidable forces that shape virus evolution, but also sources of emerging diseases in cases where viruses (including helper viruses or pseudotype viruses) do not share the same geographical distribution, but enter into contact because

of germplasm movement, the introduction of vectors, habitat disturbance, etc. or a combination thereof (Da Palma *et al.*, 2010).

1.2.4   Plant–virus interactions

As alluded to earlier, bottleneck events limit genetic variation in virus populations. Various barriers in plants impose severe bottlenecks on populations of invading viruses. One such barrier is the host genetic restriction of virus colonization *in planta* and the disruption of long-distance movement (for reviews, see Waigmann and Heinlein, 2007; Kubinak and Potts, 2013). Another barrier is achieved via the reduction in the number of initial infection events to which a plant or plant population is exposed as well as concurrent interactions with alternate host reservoirs (Acosta-Leal *et al.*, 2011). Transmission events, both horizontal and vertical, also represent events that may impose a bottleneck. Work with *Cucumber mosaic virus* illustrates the complex interplay between the mode of transmission and host-parasite co-evolution in determining virulence evolution (Pagán *et al.* 2014). *Cucumber mosaic virus* is an ss (+) RNA virus that has the broadest host range described for a plant virus. It infects more than 1200 species in more than 100 plant families and is transmitted in a non-persistent manner by more than 80 species of aphids (Hemiptera: Aphididae) and through seed. Under experimental conditions, vertical passaging led to an adaptation to vertical transmission and a concomitant decrease in virus accumulation and virulence. This was attributed to reciprocal host adaptation. On the contrary, horizontal passaging was shown to have no effect on either virus accumulation or virulence.

1.3   Plant Virus–Vector Interactions

Virus entry into plant cells is only possible through the disruption of the cuticle and plant cell wall either by mechanical processes (wind, rain, hail or human- or herbivore-induced wounds) or by vectors. The latter include a number of sap-sucking species of arthropods, for example, which deliver virus particles directly into the cell cytoplasm (and the vascular system) leading to the rapid dissemination of the virus through the whole plant. Although most viruses are naturally transmitted by vectors, only few plant–virus systems are well studied and characterized (Bragard *et al.*, 2013). The degree to which virus replication determines the rate of transmission and virulence (Froissart *et al.*, 2010), the effect of environmental impacts such as climate change on virus–vector interactions, among others, are mostly unexplored.

In general, plant viruses are hosted by many plant species, but are transmitted by very few specific vectors (Power and Flecker, 2003). Diverse members of the phyla Arthropoda (vastly represented by insects of the order Hemiptera) and Nematoda, as well as zoosporic species belonging to the kingdoms Fungi and Stramenopiles and some protists *sensu lato* (including plasmodiophorids) are known to transmit plant viruses. A puzzling case of mosquitoes harbouring tymoviruses expands the repertoire of insects serving as plant virus vectors (Wang *et al.*, 2012). Aphids are, however, among the most studied of the insect vectors (Powell *et al.*, 2006) – they easily feed on plants using their piercing–sucking mouthparts and become viruliferous after brief probing on an infected plant. Since in many, if not all cases, viruses are transmitted as intact virions, the CP represents the first and most important virus protein that interacts with the vector and determines the specificity of virus transmission. Depending on the virus group, other proteins play a role in the first steps of contact between the virus and its vector, like the helper component-proteinase (HC-Pro) of potyviruses. After making contact with the aphid's stylet, virions are retained for a period thereafter and then released by salivation. In the case of circulative viruses, it has been postulated that insect cell receptors mediate the internalization of the circulating virions. In other cases, where propagation also occurs, interactions are more complex and necessitate the intervention of host-specific proteins to guarantee virus replication. In some

insects, plant viruses can be transmitted via sexual reproduction. The whitefly, *Bemisia tabaci* B biotype, for example, transmits *Tomato yellow leaf curl virus* (*Geminiviridae*) between males and females.

Broadly speaking, non-circulative viruses only interact with the mouthparts of their vectors; acquisition occurs in minutes, inoculation periods are in the range of seconds to minutes and there are equally short retention periods. On the contrary, in the circulative and propagative modes of transmission, interaction between the virus and the vector involves the haemocoel and replication of the virus within the vector. In both cases, however, the acquisition time ranges from minutes to hours, and once viruliferous, virus transmission to other plants occurs after a few days and up to weeks. In the circulative non-propagative mode of transmission, the vector remains viruliferous for hours to weeks, while the vector remains viruliferous during its lifespan in the propagative mode of transmission. In the latter case, the virus can be inherited by the progeny of the viruliferous vector. Irrespective, the mode of transmission possibly affects the evolution of virus virulence, as well as the virus' ability to colonize and exploit vectors in order to facilitate their own transmission (Froissart *et al.*, 2010; Gray *et al.*, 2014).

Thrips-transmitted viruses belong to four genera, *Tospovirus, Ilarvirus, Carmovirus* and *Sobemovirus*. Transmission in the latter three genera is characterized by movement of infected pollen and entry of the viruses through wounds generated during feeding. Tospoviruses, on the other hand, are persistently and propagatively transmitted. A distinguishing feature is the acquisition of the viruses only by larvae of the thrips species. The virus passes from the larvae to the adult during pupation (Wijkamp *et al.*, 1995; Whitefield *et al.*, 2015). Virus replication occurs in both larval stages and adults. Much effort has been directed to understanding the intricate mechanisms that underlie the circulation of the viruses through the developing animal.

Transmission by mites is semi-persistent and in some cases circulative. Both processes of acquisition and transmission involve the virus CP. Whiteflies, however, feed on phloem cells and if virus-infected, facilitate a persistent or semi-persistent relationship with the virus. CP also plays a fundamental role in retention during transmission. In the case of circulative begomoviruses, virus particles on their way from the haemolymph to the salivary glands interact with a GroEL homologue produced by an endosymbiont in the insect (reviewed by Kliot and Ghanim, 2013). Presumably, the interaction protects against proteolysis. Hoppers can persistently transmit different species of plant viruses belonging to a wide range of families (mostly to monocots) in a circulative and propagative manner.

Some 30 species of nematodes are known to transmit at least 14 species of viruses. These viruses were initially classified in the genera *Nepovirus* and *Tobravirus*; however, reclassification to other genera was performed when transmission by aphids or mites, not nematodes, was demonstrated (Bragard *et al.*, 2013). Plant virus-transmitting nematodes feed mostly near or at the root tip using a spear-shaped structure at the anterior part of the body, and this allows the animal to puncture the plant cell and extract cell contents – including virions if the plant is infected. Virions are retained on the surface of the spear and in the area surrounding the oesophageal cavity via the CP or some other virus-encoded protein.

Finally, a limited number of soil-borne zoosporic endoparasites belonging to the plasmodiophorids (Rhizaria: Cercozoa) and chytrid fungal (Fungi: Chitridiomycota) groups are known to transmit several plant viruses belonging to the families *Potyviridae* and *Virgaviridae* and the genus *Benyvirus*, as well as the families *Ophioviridae* and *Tombusviridae* and the genera, *Potexvirus* and *Varicosavirus*, respectively (Bragard *et al.*, 2013). These viruses are acquired externally (e.g. the chytrid *Olpidium* sp.) or internally within infected plant tissue and carried by resting spores and zoospores. Glycoprotein receptors seem to play a role in attachment of the virions in a CP-dependent manner. Mechanisms of delivery to plant cells and the involvement of other virus and cell factors are not clear.

Plant resistance mechanisms against vectors by antixenosis (modification of vector behaviour in terms of feeding preferences) or antibiosis (increased mortality or reduced fitness or reproductive capacity of the vector) have been reported (Gómez *et al.*, 2009). In both cases pre-existing physical barriers, metabolites or deterrents act to prevent transmission from the vector to the plant. Additionally, resistance to aphids, nematodes or whiteflies exists at different functional and morphological levels (Montero-Astúa *et al.*, 2014; Sundaraj *et al.*, 2014). Viruses can also affect plant hosts in a manner that favours vector attraction or behaviour, and hence, transmission (Palukaitis *et al.*, 2013). As mentioned earlier, the CP plays an integral role in virus–vector interactions and transmission (see Urcuqui-Inchima *et al.*, 2001; Ni and Cheng Kao, 2013). CPs not only give structure to the virions (encapsidating and protecting the virus genome), but also facilitate interactions with receptors, chaperones and other factors of the vector and the virus itself during acquisition, movement, replication and transmission. Additionally, it has recently been shown that the structure of *Potato virus A* (*Potyviridae*) virions is characterized by the presence of a significant fraction of disordered segments in its intravirus CP subunits (Ksenofontov *et al.*, 2013). It is posited that since intrinsically disordered segments of proteins enlarge the range of their specifically recognized partners, such 'promiscuity' might explain in part the spectacular efficiency of this protein in all interactions it establishes with plant (and vector) factors. This finding gives support to prior observations that vector transmission of plant viruses requires conformational changes of virions (Kakani *et al.*, 2004). Nonetheless, CP interactions alone do not explain virus transmission in all cases. For many viruses, if not all, the presence of virus inclusions or aggregates of different sizes in infected cells have been demonstrated. These aggregates apparently participate in virus transmission by the controlled release and uptake of virions. They seem to be essential for the successful transmission of *Cauliflower mosaic virus* by its aphid vector (Moshe and Gorovits, 2012; Bak *et al.*, 2013). Moreover, the generation of more ordered, complex virus-derived structures within the vector itself facilitates, for example, the intercellular spread of *Rice dwarf virus* through leafhopper cells and transmission of the virus by this insect (Chen *et al.*, 2012). Although not a direct consequence of the interaction between a virus and its insect vector, some insects induce the production of a volatile alcohol (methanol) in plants on feeding. As a consequence, methanol sensitizes the plant and allows for virus entry and spread within the plant and between plants by insect vectors (Komarova *et al.*, 2014).

Many challenges lie ahead in terms of our understanding of virus–vector interactions and the ways we can use this knowledge to design control strategies relevant for multi-host plant viruses. For example, expansion of investigations into the role of ubiquitination-related enzymes linked to viral infection to a system wide analysis involving virus vectors could provide insights into how these mechanisms can be exploited for the development of new antiviral strategies (Alcaide-Loridan and Jupin, 2012). A complete understanding of the mechanisms and factors surrounding phloem transport of plant viruses (Hipper *et al.*, 2013) could also facilitate manipulation or avoidance of vector feeding and thus control virus transmission. Recently, it was demonstrated that the expression of viral glycoproteins in transgenic plants interfered with virus acquisition and effectively blocked virus transmission by insect vectors (Montero-Astúa *et al.*, 2014).

1.4 Diagnosis and Crop Protection Technologies

Because effective management of virus diseases requires an integrated approach aimed at preventing or delaying infection, timely and accurate diagnosis of virus infections is of paramount importance. There is the added challenge of discrimination of unrelated strains and the reliable detection and characterization of related strains. International attempts to develop and standardize diagnostic protocols for plant viruses, are coordinated by the European and Mediterranean Plant Protection Organization and by the International Plant Protection Convention.

Traditionally, the detection of virus infections has relied on biological testing or indexing. Indexing is based on the detection of the virus pathogen and associated symptoms following grafting on an appropriate indicator plant. The technique is still widely used as part of the certification programs against certain pathogens (e.g. *Citrus tristeza virus*, and *tomato spotted wilt*, *impatiens necrotic spot* and *watermelon silver mottle* tospoviruses) (EPPO, 2014). Nonetheless, it is necessary that visual inspection for symptoms is accompanied with other confirmatory tests to ensure accurate diagnosis. Among the various diagnostic techniques, immuno-based methods are routinely used for virus detection (Hull, 2002), specifically some form of antibody-based enzyme immunoassay utilizing polyclonal antibodies that have been generated against purified viral CP (van Regenmortel, 1982) or viral proteins expressed as recombinant fusion proteins in instances where the virus is intrinsically poorly immunogenic or is difficult to purify from host tissues (Raikhy *et al.*, 2007; Lee and Chang, 2008; Gulati-Sakhuja *et al.*, 2009; Rani *et al.*, 2010; Rana *et al.*, 2011; Khatabi *et al.*, 2012; Mandal *et al.*, 2012). More recently, tests employing a cocktail of polyclonal antibodies also derived from fusion constructs of viral gene sequences of two or three different viruses are being developed (Kapoor *et al.*, 2014). This approach will facilitate the detection of mixed virus infections which are usually observed in the field.

Other diagnostic tests include the PCR, RT-PCR and hybridization-based techniques (Gilbertson *et al.*, 1991). These tests have proven rapid, sensitive and reasonably inexpensive to conduct. Degenerate primers are used typically in PCR (Rojas *et al.*, 1993; Wyatt and Brown, 1996). Degenerate primers have facilitated the identification of, for example, most geminiviruses, but mixed infections and the presence of satellite DNA, which are commonly found in association with monopartite begomoviruses in South-East Asia (Dry *et al.*, 1997; Mansoor *et al.*, 2003), interfere with the identification of viruses present in samples. Often combinations of ELISA and PCR technologies are employed in an attempt to improve sensitivity and to avoid problems with inhibitors. Advances in real-time quantitative PCR technology have enabled large-scale detection of many plant RNA and DNA viruses. Recent developments in multiplex real-time PCR show promise for future identification, genotyping and quantitation of viral targets in a single, rapid reaction (Fageria *et al.*, 2013). But for now, microarray technologies provide the option of multi-pathogen detection (Hammond *et al.*, 2015). Labelled nucleic acids isolated from samples are hybridized to a large number of diagnostic probes spotted on a platform. The array is subsequently scanned to produce a file of fluorescence intensities for the probes (Nam *et al.*, 2014). An amplification step prior to hybridization is often included to increase the sensitivity for low titre viruses.

A new cadre of techniques are emerging, which unlike the traditional methods, do not require an *a priori* prediction of the viruses likely to be present in the sample. They include for example, rolling-circle amplification coupled with restriction fragment length polymorphism and next-generation sequencing of small RNAs isolated from infected plants. Although these approaches are powerful and flexible, they may not prove suitable for routine diagnostic procedures, and are more likely to facilitate the identification of novel or unknown viruses (Schubert *et al.*, 2007; Kreuze *et al.*, 2009; Hagen *et al.*, 2012). Before long, the field of nanotechnology is likely to bring on board new diagnostic tools. Electrochemical DNA biosensors, for example, provide a novel technique for the recognition of target DNA by hybridization (Malecka *et al.*, 2014). Essentially, target DNA is captured in a recognition layer. The probe–target complex then triggers a signal for electronic display and analysis. Potential advantages of these devices include rapid detection, portability and adaptability.

Efforts to identify and implement control strategies against virus diseases vary with the crop and the region. Typically, dissemination within and between regions is often addressed through quarantine controls in addition to other government interventions that restrict the movement of plant materials within the region. As regards to on farm practices, these range from the interference of vector-mediated virus transmission, the implementation of biological and cultural management practices and the development of host–plant resistance. Prevalent among farmers, however, is the policy of 'living with the disease'. There is willingness on their part to change to varieties that offer more tolerance or are resistant, and until they become available, to continue with the existing varieties and harvest as much as possible or increase the area under production to achieve the production required. But tolerant and/or resistant varieties are not always available or they are not readily combined with other desirable horticultural attributes. Alternate approaches to the development of host resistance have emerged that utilize molecular techniques either in the form of linked molecular markers to speed up and simplify the selection of resistance genes or pathogen-derived or transgenic resistance.

There are two categories of transgenic resistance in host plants that show significant promise for disease management. First is the use of plant-derived genetic resistance that was reviewed by Truniger *et al.* (2008) and Fraile and Garcia-Arenal (2010). The second exploits the post-transcriptional gene silencing machinery to generate small interfering RNAs (siRNAs) that target viral genomes or critical host factors for degradation. One favoured strategy for engineering resistance to plant viruses is the expression of hairpin (hp) RNA constructs composed of inversely repeated viral RNA sequences separated by an intron spacer. The hpRNAs are processed by Dicer into siRNAs and these can provide whole plant resistance to virus infection. This strategy has shown greater than 90% effectiveness in combating virus infection. For *Plum pox virus* resistance, several constructs consisting of overlapping portions of P1/HC-Pro, HC-Pro, and HC-Pro/P3 coding regions were generated and tested (Hily *et al.*, 2004; Di Nicola-Negri *et al.*, 2005, 2010; Kundu *et al.*, 2008; Ilardi and Nicola-Negri, 2011). The 5′ UTR/P1 fragment was found to be the most effective for broad-spectrum transgenic resistance. A related strategy was used to create resistance to cucurbit-infecting potyviruses. Here an inverted repeat construct was prepared using a large fragment of the *Zucchini yellow mosaic virus* (ZYMV) HC-Pro gene which also showed substantial similarity with that of *Watermelon mosaic virus* (WMV) and *Papaya ringspot virus* serotype W (PRSV-W). Transgenic cucumber and melon lines inoculated with ZYMV or WMV failed to accumulate viral RNAs, while plants inoculated with PRSV-W exhibited significantly lower levels of virus than non-transformed plants (Leibman *et al.*, 2011). This is an exciting example of the engineering of small RNAs for resistance to related virus strains or even related species.

Another silencing approach that is proving to be effective is the silencing of host factors that are crucial for virus susceptibility. One of the most common factors used by members of the family *Potyviridae* is an isoform of the translation initiation factor 4E (eIF(iso)4E). Mutation in eIF4E family is a common component of recessive resistance against plant viruses. The mechanism of recessive resistance to potyviruses, especially mediated by eIF4E and eIF(iso)4E is explained in detail by Truniger and Aranda (2009). This type of resistance blocks virus multiplication in inoculated leaves. Examples of recessive eIF4E-mediated resistance to species members of the *Potyvirus* supergroup include: *mo1* (*Lettuce mosaic virus*) in lettuce, *lsp1* (*Tobacco etch virus*) in *Arabidopsis*, *cum1-1* (*Clover yellow vein virus*) in cucumber, *pvr2* (*Pepper veinal mottle virus*) in pepper and *sbm-1* (*Pea seed-borne mosaic virus*) in pea. Interestingly, eIF4E and IF(iso)4E resistance is the result of failed interactions with the potyvirus VPg. There is only one reported example of recessive eIF4E-mediated resistance to members of the *Bean common mosaic virus* (BCMV) supergroup and that is *bc-3* (BCMV) in bean (Naderpour *et al.*, 2010). Silencing eIF(iso)4E can confer resistance to ZYMV and *Moroccan watermelon mosaic virus*, which are both members of the BCMV supergroup (Rodríguez-Hernández *et al.*, 2012). This exciting advance in engineered resistance demonstrates that certain recessive resistance mechanisms provide broad-spectrum resistance to potyvirus infection that can extend to other members of the BCMV supergroup. Therefore, in crops where recessive resistance genes are not available for breeding elite cultivars, siRNA or hpRNA silencing can be used to provide protection against infection either by targeting the virus itself or a critical host factor (Truniger *et al.*, 2008).

1.5 Virus Diseases Threaten Food Security in Tropical and Subtropical Regions

Although accurate figures for crop losses due to virus infections are not readily available, it is widely accepted that among the plant pathogens, viruses are second only to fungal pathogens with respect to economic losses. Human actions are extensively implicated in virus disease outbreaks and epidemics, as is the appearance of new viruses that switched host species or new variants of classic viruses that acquired new virulence factors or different epidemiological patterns. While technological advances in, for example, diagnostic and agronomic practices have reduced the risk of epidemics in developed countries more so than developing countries, virus diseases remain a threat to global food security and have the potential to be widespread with subsequent economic, social and environmental impacts.

Plant protection plays an important role in minimizing the losses incurred by virus diseases and improving food security, that is, in satisfying the demand worldwide for both the quality and quantity of agricultural goods (Savary *et al.*, 2012). There are many possible intervention points in the crop–pathogen interaction, but decisions on which are to be prioritized will depend on a combination of feasibility and likely effects. Nonetheless, interventions require initial investment in capacity and resource building accompanied with cost estimates of adoption. Other costs will be incurred from investments in evaluation research and diagnostic programs (Oerke, 2006), as well as education programmes aimed at scientists and regulators on the diseases and prophylactic approaches. Since the basic biology of some cultivated plants and their pathogens are still poorly understood, particularly in the developing world, the emergence of new diseases adds a complicating dimension to food production and availability. Thus the challenge that lies ahead in terms of food security involves increased investment in basic and applied research, particularly in the fields of plant, vector and virus gene expression and the identification of new viruses, as well as biodiversity, distribution, adaptation and ecology of the biotic protagonists (Wren *et al.*, 2006; Mehta *et al.*, 2008; Kundu *et al.*, 2013; MacDiarmid *et al.*, 2013). However, none of these objectives will be effectively attained if the use of technologies already developed are not maximized and accompanied with the generation and exploitation of new scientific discoveries (Schumann, 2003; Walthall *et al.*, 2012; UK Plant Science, 2014). Genomics along with the other '-omics' technologies facilitate the identification of genes affecting important traits and a greater understanding of how they function, which invariably will contribute to the transfer of genes to elite varieties via marker assisted breeding or transgenic approaches. The latter technology has spurred considerable public debate over recent years that is likely to continue in the broader context of other uses of biotechnology and their consequences for human societies. Issues such as cost, safety and benefit ought to be dispassionately evaluated (Thomson, 2002, 2008; Ronald, 2011). Finally, global partnerships must also be fostered if we are to honestly pursue the final goal of nutritious, cheap and widely available food for all. Prevention and remediation of the impact of plant diseases is high and a burden for countries less prepared. Nonetheless, it has been estimated that the benefits associated with prevention and protection programs for virus transmitted diseases far surpass the costs of the protection program (Cembali *et al.*, 2003, 2004). Additionally, disease control can mitigate effects of climate change in addition to contributing to sustainable crop production (Mahmuti *et al.*, 2009).

The chapters that follow provide up-to-date information on selected viruses of important crops, including their distribution, their biological and molecular characteristics, and the approaches that control the diseases they elicit and sustain productive agricultural systems. These entities were chosen based on their potential impact on food security. They differ considerably in host range, their longevity in the host and dissemination. Many of the viruses, as discussed in this book, belong to the family *Potyviridae* (Chapters 4, 7, 8, 9, 10, 11 and 16) and others of Group IV (*Bromoviridae*, Chapter 6; *Closteroviridae*, Chapters 14 and 17; and *Secoviridae*, Chapter 15). Viruses belonging to the most important family of plant viruses, at least in terms of the number of species, the family *Geminiviridae* (Group II), are covered in Chapters 3, 5 and 13. The impressively successful *Tomato spotted wilt virus* (family *Bunyaviridae*, Group V) is examined in Chapter 12, while other important viruses belonging to Groups II (*Nanoviridae*) and VII (*Caulimoviridae*) are reviewed in Chapters 2 and 15, respectively. The overall impact of the virus diseases on crop production is considered in the individual chapters. These crops (rice, wheat, maize, potato, cassava, soybean, yam, sweet potato, tomato, citrus, banana and plantain, and pineapple, among others) are regarded as important staples in tropical and subtropical areas worldwide. They are mainly consumed directly and are major contributors to human calories and proteins. They are also targets of a diverse array of viruses (Rybicki and Pietersen, 1999; Kumar *et al.*, 2013; Rybicki, 2015). Notable examples of virus pathogens that challenge food security in sub-Saharan Africa are the mosaic viruses of cassava. The tuberous roots of cassava are the major source of dietary starch in sub-Saharan Africa. The crop was presumably introduced to the western coast of Africa in about the sixteenth century by Portuguese traders as a safeguard against periods of famine that consistently plague the region (Alabi *et al.*, 2011). Today, cassava is considered the crop of the future not only because of its contribution to food security, but also because it represents a significant income earner for smallholders, and promises immense potential as a source of industrial raw materials like glucose and starch. The crop is widely used in many countries of Africa, where unfortunately the prevalence of viral disease is high; however, these viruses are not known in South America, which is the centre of origin of cassava. Finally, although not a crop essential for food security (debatable as this statement might be), papaya, and its worst enemy, *Papaya ringspot virus* (*Potyviridae*), was included because it represents a case where the use and implementation of modern strategies of disease control cannot be defined as other than successful.

References

Abrescia, N.G.A., Grimes, J.M., Fry, E.E., Ravantti, J.J., Bamford, D.H. and Stuart, D.I. (2009) What does it take to make a virus: the concept of the viral 'self'. In: Stockley, P.G. and Twarock, R. (ed.) *Emerging Topics in Physical Virology*. World Scientific, Singapore, pp. 35–38.

Acosta-Leal, R., Duffy, S., Xiong, Z., Hammond, R.W. and Elena, S.F. (2011) Advances in plant virus evolution: translating evolutionary insights into better disease management. *Phytopathology* 101, 1136–1148.

Adams, I.P., Glover, R.H., Monger, W.A., Mumford, R., Jackeviciene, E., *et al.* (2009) Next-generation sequencing and metagenomic analysis: a universal diagnostic tool in plant virology. *Molecular Plant Pathology* 10, 537–545.

Adams, I.P., Mianob, D.W., Kinyuab, Z.M., Wangaib, A., Kimanic, E., *et al.* (2012) Use of next-generation sequencing for the identification and characterization of *Maize chlorotic mottle virus* and *Sugarcane mosaic virus* causing maize lethal necrosis in Kenya. *Plant Pathology* 62, 741–749.

Alabi, O.J., Kumar, P.L. and Naidu, R.A. (2011) Cassava mosaic disease: a curse to food security in Sub-Saharan Africa. APSnet Features. Available at: http://www.apsnet.org/publications/apsnetfeatures/Pages/cassava.aspx (accessed 30 October 2014).

Alcaide-Loridan, C. and Jupin, I. (2012) Ubiquitin and plant viruses, let's play together! *Plant Physiology* 160, 72–82.

Anderson, P.K., Cunningham, A.A., Patel, N.G., Morales, F.J., Epstein, P.R. and Daszak, P. (2004) Emerging infectious diseases of plants: pathogen pollution, climate change and agrotechnology drivers. *Trends in Ecology and Evolution* 19, 535–544.

Bak, A., Gargani, D., Macia, J.L., Malouvet, E., Vernerey, M.S., *et al.* (2013) Virus factories of *Cauliflower mosaic virus* are virion reservoirs that engage actively in vector transmission. *Journal of Virology* 87, 12207–12215.

Barba, M., Czosnek, H. and Hadidi, A. (2014) Historical perspective, development and applications of next-generation sequencing in plant virology. *Viruses* 6, 106–136.

Bragard, C., Caciagli, P., Lemaire, O., Lopez-Moya, J.J., MacFarlane, S., Peters, D., Susi, P. and Torrance, L. (2013) Status and prospects of plant virus control through interference with vector transmission. *Annual Review of Phytopathology* 51, 177–201.

Briese, T., Calisher, C.H. and Higgs, S. (2013) Viruses of the family *Bunyaviridae*: are all available isolates reassortants? *Virology* 446, 207–216.

Brüssow, H. (2009) The not so universal tree of life or the place of viruses in the living world. *Philosophical Transactions of the Royal Society Biological Sciences* 364, 2263–2274.

Cembali, T., Folwella, R.J., Wandschneider, P., Eastwellb, K.C. and Howellb, W.E. (2003) Economic implications of virus prevention program in deciduous tree fruits in the US. *Crop Protection* 22, 1149–1156.

Cembali, T., Folwell, R.J. and Wandschneider, P.R. (2004) Economic evaluation of viral disease prevention programs in tree fruits. *Mediterranean Journal of Economics, Agriculture and Environment* 4, 23–29.

Chare, E.R. and Holmes, E.C. (2006) A phylogenetic survey of recombination frequency in plant RNA viruses. *Archives of Virology* 151, 933–946.

Chen, Q., Chen, H., Mao, Q., Liu, Q., Shimizu, T., *et al.* (2012) Tubular structure induced by a plant virus facilitates viral spread in its vector insect. *PLoS Pathogens* 8, e1003032.

Chirico, N., Vianelli, A. and Belshaw, R. (2010) Why genes overlap in viruses. *Proceedings of the Royal Society Biological Sciences* 277, 3809–3817.

Claverie, J.-M. and Abergel, C. (2012) *Mimiviridae*. In: King, A.M.Q., Adams, M.J., Carstens, E.B. and Leftowitz, E.J. (eds) *Virus Taxonomy – Ninth Report of the International Committee on Taxonomy of Viruses*. Academic Press, Elsevier, San Diego, California, pp. 223–228.

Da Palma, T., Doonan, B.P., Trager, N.M. and Kasman, L.M. (2010) A systematic approach to virus-virus interactions. *Virus Research* 149, 1–9.

Di Nicola-Negri, E., Brunetti, A., Tavazza, M. and Ilardi, V. (2005) Hairpin RNA-mediated silencing of *Plum pox virus* P1 and HC-Pro genes for efficient and predictable resistance to the virus. *Transgenic Research* 14, 989–994.

Di Nicola-Negri, E., Tavazza, M., Salandri, L. and Ilardi, V. (2010) Silencing of *Plum pox virus* 5'UTR/P1 sequence confers resistance to a wide range of PPV strains. *Plant Cell Reports* 29, 1435–1444.

Dry, I.B., Krake, L.R., Rigden, J.E. and Rezaian, M.A. (1997) A novel subviral agent associated with a geminivirus: the first report of a DNA satellite. *Proceedings of the National Academy of Sciences* 94, 7088–7093.

EPPO (2014) Diagnostic protocols for regulated pests. Available at http://archives.eppo.int/EPPOStandards/diagnostics.htm (accessed 8 April 2015).

Fageria, M.S., Singh, M., Nanayakkara, U., Pelletier, Y., Nie, X., *et al.* (2013) Monitoring current-season spread of *Potato virus Y* in potato fields using ELISA and real-time RT-PCR. *Plant Disease* 97, 641–644.

Fraile, A. and Garcia-Arenal, F. (2010) The coevolution of plants and viruses: resistance and pathogenicity. *Advances in Virus Research* 76, 1–32.

Froissart, R., Doumayrou, J., Vuillaume, F., Alizon, S. and Michalakis, Y. (2010) The virulence-transmission trade-off in vector-borne plant viruses: a review of (non-)existing studies. *Philosophical Transactions of the Royal Society Biological Sciences* 365, 1907–1918.

Gilbertson, R.L., Hidayat, S.H., Martinez, R.T., Leong, S.A., Faria, J.C., *et al.* (1991) Differentiation of bean-infecting geminiviruses by nucleic-acid hybridization probes and aspects of bean golden mosaic in Brazil. *Plant Disease* 75, 336–342.

Gómez, P., Rodríguez-Hernández, A.M., Moury, B. and Aranda, M.A. (2009) Genetic resistance for the sustainable control of plant virus diseases: breeding, mechanisms and durability. *European Journal of Plant Pathology* 125, 1–22.

Gray, S., Cilia, M. and Ghanim, M. (2014) Circulative, 'nonpropagative' virus transmission: an orchestra of virus-, insect-, and plant-derived instruments. *Advances in Virus Research* 89, 141–199.

Grimsley, N.H., Thomas, R., Kegel, J.U., Jacquet, S., Moreau, H., *et al.* (2012) Genomics of algal host-virus interactions. *Advances in Botanical Research* 64, 343–381.

Gulati-Sakhuja, A., Sears, J.L., Nuñez, A. and Liu, H.Y. (2009) Production of polyclonal antibodies against *Pelargonium zonate spot virus* coat protein expressed in *Escherichia coli* and application for immuno-diagnosis. *Journal of Virological Methods* 160, 29–37.

Hagen, C., Frizzi, A., Gabriels, S., Huang, M., Salati, R., *et al.* (2012) Accurate and sensitive diagnosis of geminiviruses through enrichment, high-throughput sequencing and automated sequence identification. *Archives of Virology* 157, 907–915.

Hammond, J., Henderson, D.C., Bagewadi, B., Jordan, R.L., Perry, K.L., *et al.* (2015) Progress in the development of a universal plant virus microarray for the detection and identification of viruses. *Acta Horticulturae* 1072, 149–156.

Hily, J.M., Scorza, R., Malinowski, T., Zawadzka, B. and Ravelonandro, M. (2004) Stability of gene silencing-based resistance to *Plum pox virus* in transgenic plum (*Prunus domestica* L.) under field conditions. *Transgenic Research* 13, 427–436.

Hipper, C., Brault, V., Ziegler-Graff, V. and Revers, F. (2013) Viral and cellular factors involved in phloem transport of plant viruses. *Frontiers in Plant Science* 4, 154.

Hisa, Y., Suzuki, H., Atsumi, G., Choi, S.H., Nakahara K.S., *et al.* (2014) P3N-PIPO of *Clover yellow vein virus* exacerbates symptoms in pea infected with *White clover mosaic virus* and is implicated in viral synergism. *Virology* 449, 200–206.

Hull, R. (2002) *Matthews' Plant Virology*. Academic Press, London, 1001 pp.

Hurst, C.J. (2011) An introduction to viral taxonomy with emphasis on microbial and botanical hosts and the proposal of Akamara, a potential domain for the genomic acellular agents. In Hurst C. (ed.) *Studies in Viral Ecology: Microbial and Botanical Host Systems* Vol. 1. Wiley-Blackwell, Hoboken, New Jersey, pp. 41–65.

Ilardi, V. and Nicola-Negri, E.D. (2011) Genetically engineered resistance to *Plum pox virus* infection in herbaceous and stone fruit hosts. *GM Crops* 2, 24–33.

Kakani, K., Reade, R. and Rochon, D. (2004) Evidence that vector transmission of a plant virus requires conformational change in virus particles. *Journal of Molecular Biology* 338, 507–517.

Kapoor, R., Mandala, R., Paulb, K.P., Chigurupatia, P. and Jaina, R.K. (2014) Production of cocktail of polyclonal antibodies using bacterial expressed recombinant protein for multiple virus detection. *Journal of Virological Methods* 196, 7–14.

Kehoe, M.A., Coutts, B.A., Buirchell, B.J. and Jones, R.A.C. (2014) Plant virology and next generation sequencing: experiences with a *Potyvirus*. *PLoS ONE* 9, e104580.

Khatabi, B., He, B. and Hajimorad, M.R. (2012) Diagnostic potential of polyclonal anti-bodies against bacterially expressed recombinant coat protein of *Alfalfa mosaic virus*. *Plant Disease* 96, 1352–1357.

Kingsbury, D.W. (1985) Species classification problems in virus taxonomy. *Intervirology* 24, 62–70.

Kliot, A. and Ghanim, M. (2013) The role of bacterial chaperones in the circulative transmission of plant viruses by insect vectors. *Viruses* 5, 1516–1535.

Komarova, T.V, Sheshukova, E.V. and Dorokhov, Y.L. (2014) Cell wall methanol as a signal in plant immunity. *Frontiers in Plant Science* 5, 101.

Koonin, E.V. and Dolja, V.V. (2013) A virocentric perspective on the evolution of life. *Current Opinion in Virology* 3, 546–557.

Koonin, E.V. (2009) On the origin of cells and viruses: primordial virus world scenario. *Annals of the New York Academy of Sciences* 1178, 47–64.

Kreuze, J.F., Perez, A., Untiveros, M., Quispe, D., Fuentes, S., *et al*. (2009) Complete viral genome sequence and discovery of novel viruses by deep sequencing of small RNAs: a generic method for diagnosis, discovery and sequencing of viruses. *Virology* 388, 1–7.

Kristensen, D.M., Mushegian, A.R., Dolja, V.V. and Koonin, E.V. (2010) New dimensions of the virus world discovered through metagenomics. *Trends in Microbiology* 18, 11–19.

Ksenofontov, A.L., Paalme, V., Arutyunyan, A.M., Semenyuk, P.I., Fedorova, N.V., *et al*. (2013). Partially disordered structure in intravirus coat protein of *Potyvirus Potato virus A*. *PLoS One* 8, e67830.

Kubinak, J.L. and Potts, W.K. (2013) Host resistance influences patterns of experimental viral adaptation and virulence evolution. *Virulence* 4, 410–418.

Kumar, P.L., López, K. and Njuguna, C. (2013) Evolution, ecology and control of plant viruses – Book of Abstracts (Compilators). *12th International Symposium on Plant Virus Epidemiology*. Arusha, Tanzania (January–February 2013).

Kundu, J.K., Briard, P., Hily, J.M., Ravelonandro, M. and Scorza, R. (2008) Role of the 25–26 nt siRNA in the resistance of transgenic *Prunus domestica* graft inoculated with *Plum pox virus*. *Virus Genes* 36, 215–220.

Kundu, S., Chakraborty, D., Kundu, A. and Pal, A. (2013) Proteomics approach combined with biochemical attributes to elucidate compatible and incompatible plant-virus interactions between *Vigna mungo* and *Mungbean yellow mosaic India virus*. *Proteome Science* 11, 15.

Lee, S.-C. and Chang, Y.-C. (2008) Performances and application of antisera produced by recombinant capsid proteins of *Cymbidium mosaic virus* and *Odontoglossum ringspot virus*. *European Journal of Plant Pathology* 122, 297–306.

Legendre, M., Bartoli, J., Shmakova, L., Jeudy, S., Labadie, K., *et al*. (2014) Thirty-thousand-year-old distant relative of giant icosahedral DNA viruses with a pandoravirus morphology. *Proceedings of the National Academy of Sciences, USA* 111, 4274–4279.

Leibman, D., Wolf, D., Saharan, V., Zelcer, A., Arazi, T., *et al*. (2011) A high level of transgenic viral small RNA is associated with broad potyvirus resistance in cucurbits. *Molecular Plant-Microbe Interactions* 24, 1220–1238.

MacDiarmid, R., Rodoni, B., Melcher, U., Ochoa-Corona, F. and Roossinck, M. (2013) Biosecurity implications of new technology and discovery in plant virus research. *PLoS Pathogens* 9, e1003337.

MacNaughton, T.B. and Lai, M.M. (2006) HDV RNA replication: ancient relic or primer? *Current Topics in Microbiology and Immunology* 307, 25–45.

Mahmuti, M., West, J.S., Watts, J., Gladders, P. and Fitt, B.D.L. (2009) Controlling crop disease contributes to both food security and climate change mitigation. *International Journal of Agricultural Sustainability* 7, 189–202.

Malecka, K., Michalczuk, L., Radecka, H. and Radecki, J. (2014) Ion-channel genosensor for the detection of specific DNA sequences derived from *Plum pox virus* in plant extracts. *Sensors* 14, 18611–18624.

Mandal, B., Kumar, A., Rani, P. and Jain, R.K. (2012) Complete genome sequence, phylo-genetic relationships and molecular diagnosis of an Indian isolate of *Potato virus X*. *Journal of Phytopathology* 160, 1–5.

Mansoor, S., Briddon, R.W., Zafar, Y. and Stanley, J. (2003) Geminivirus disease complexes: an emerging threat. *Trends in Plant Science* 8, 128–134.

Mehta, A., Brasileiro, A.C.M., Souza, D.S.L., Romano, E., Campos, M.A., *et al.* (2008) Plant-pathogen interactions: what is proteomics telling us? *The FEBS Journal* 275, 3731–3746.

Meyer, J.R. (2013) Sticky bacteriophage protect animal cells. *Proceedings of the National Academy of Sciences, USA* 110, 10475–10476.

Mölken, T. and Stuefer, J.F. (2011) The potential of plant viruses to promote genotypic diversity via genotype x environment interactions. *Annals of Botany* 107, 1391–1397.

Montero-Astúa, M., Rotenberg, D., Leach-Kieffaber, A., Schneweis, B.A., Park, S., *et al.* (2014) Disruption of vector transmission by a plant-expressed viral glycoprotein. *Molecular Plant–Microbe Interactions* 27, 296–304.

Moshe, A. and Gorovits, R. (2012) Virus-induced aggregates in infected cells. *Viruses* 4, 2218–2232.

Naderpour, M., Lund, O.S., Larsen, R. and Johansen, E. (2010) Potyviral resistance derived from cultivars of *Phaseolus vulgaris* carrying *bc-3* is associated with the homozygotic presence of a mutated *eIF4E* allele. *Molecular Plant Pathology* 11, 255–263.

Nam, M., Kim, J.S., Lim, S., Park, C.Y., Kim, J.G., *et al.* (2014) Development of the large-scale oligonucleotide chip for the diagnosis of plant viruses and its practical use. *Plant Pathology Journal* 30, 51–57.

Ni, P. and Cheng Kao, C. (2013) Non-encapsidation activities of the capsid proteins of positive-strand RNA viruses. *Virology* 446, 123–132.

Oerke, E.-C. (2006) Crop losses to pests. *Journal of Agricultural Science* 144, 31–43.

Pagán, I., Montes, N., Milgroom, M.G. and García-Arenal, F. (2014) Vertical transmission selects for reduced virulence in a plant virus and for increased resistance in the host. *PLoS Pathogen* 10 (7), e1004293.

Palukaitis, P., Groen, S.C. and Carr, J.P. (2013) The Rumsfeld paradox: some of the things we know that we don't know about plant virus infection. *Current Opinion in Plant Biology* 16, 513–519.

Philippe, N., Legendre, M., Doutre, G., Couté, Y., Poirot, O., *et al.* (2013) Pandoraviruses: Amoeba viruses with genomes up to 2.5 Mb reaching that of parasitic eukaryotes. *Science* 341, 281–286.

Powell, G., Tosh, C.R. and Hardie, J. (2006) Host plant selection by aphids: behavioral, evolutionary and applied perspectives. *Annual Review of Entomology* 51, 309–330.

Power, A.G. and Flecker, A.S. (2003) Virus specificity in disease systems: are species redundant? In: Kareiva, P. and Levin, S.A. (eds) *The Importance of Species: Perspectives on expendability and triage*. Princeton University Press, Princeton, New Jersey, pp. 330–351.

Prendeville, H.R., Ye, X., Morris, T.J. and Pilson, D. (2012) Virus infections in wild plant populations are both frequent and often unapparent. *American Journal of Botany* 99, 1033–1042.

Pumplin, N. and Voinnet, O. (2013) RNA silencing suppression by plant pathogens: defence, counter-defence and counter-counter-defence. *Nature Reviews Microbiology* 11, 745–760.

Raikhy, G., Hallan, V., Kulshrestha, S., Zaidi, A.A. (2007) Polyclonal antibodies to the coat protein of *Carnation etched ring virus* expressed in bacterial system: production and use in immunodiagnosis. *Journal of Phytopathology* 155, 616–622.

Rana, T., Chandel, V., Hallan, V. and Zaidi, A.A. (2011) Expression of recombinant *Apple chlorotic leaf spot virus* coat protein in heterologous system: production and use in immunodiagnosis. *Journal of Plant Biochemistry and Biotechnology* 20, 138–141.

Rani, P., Pant, R.P. and Jain, R.K. (2010) Serological detection of *Cymbidium mosaic* and *Odontoglossum ringspot* viruses in orchids with polyclonal antibodies developed against their recombinant coat proteins. *Journal of Phytopathology* 158, 542–545.

Rodríguez-Hernández, A.M., Gosalvez, B., Sempere, R.N., Burgos, L., Aranda, M.A., *et al.* (2012) Melon RNA interference (RNAi) lines silenced for Cm-eIF4E show broad virus resistance. *Molecular Plant Pathology* 13, 755–763.

Rojas, M.R., Gilbertson, R.L., Russell, D.R. and Maxwell, D.P. (1993) Use of degenerate primers in the polymerase chain reaction to detect whitefly-transmitted geminiviruses. *Plant Disease* 77, 340–347.

Ronald, P. (2011) Plant genetics, sustainable agriculture and global food security. *Genetics* 188, 11–20.

Roossinck, M.J. (2005) Symbiosis versus competition in plant virus evolution. *Nature Reviews Microbiology* 3, 917–924.

Roossinck, M.J. (2010) Lifestyles of plant viruses. *Philosophical Transactions of the Royal Society Biological Sciences* 365, 1899–1905.

Roossinck, M.J. (2011) The good viruses: viral mutualistic symbioses. *Nature Reviews Microbiology* 9, 99–108.

Roossinck, M.J. (2012) Even viruses can be beneficial microbes. *Microbiology Australia* 3, 111–112.

Roossinck, M.J. (2013) Plant virus ecology. *PLoS Pathogens* 9, e1003304.

Roossinck, M.J., Saha, P., Wiley, G.B., Quan, J., White, J.D., *et al.* (2010) Ecogenomics: using massively parallel pyrosequencing to understand virus ecology. *Molecular Ecology* 19, 81–88.

Rosario, K., Duffy, S. and Breitbart, M. (2012) A field guide to eukaryotic circular single-stranded DNA viruses: insights gained from metagenomics. *Archives of Virology* 157, 1851–1871.

Rybicki, E.P. (2015) A top ten list for economically important plant viruses. *Archives of Virology* 160, 17–20.

Rybicki, E.P. and Pietersen, G. (1999) Plant virus disease problems in the developing world. *Advances in Virus Research* 53, 127–175.

Sabath, N., Wagner, A. and Karlin, D. (2012) Evolution of viral proteins originated *de novo* by overprinting. *Molecular Biology and Evolution* 29, 3767–3780.

Saqib, M., Wylie, S.J. and Jones, M.G.K. (2014) Serendipitous identification of a new *Iflavirus*-like virus infecting tomato and its subsequent characterization. *Plant Pathology* doi: 10.1111/ppa.12293.

Savary, S., Ficke, A., Aubertot, J.-N. and Hollier, C. (2012) Crop losses due to diseases and their implications for global food production losses and food security. *Food Security* 4, 519–537.

Scholthof, K.-B.G. (2014) Making a virus visible: Francis O. Holmes and a biological assay for *Tobacco mosaic virus*. *Journal of the History of Biology* 47, 107–145.

Schubert, J., Habekuss, A., Kazmaier, K. and Jeske, H. (2007) Surveying cereal-infecting geminiviruses in Germany – diagnostics and direct sequencing using rolling circle amplification. *Virus Research* 127, 61–70.

Schumann, G.L. (2003) *Annual Review in Phytopathology* 41, 377–398.

Sundaraj, S., Srinivasan, R., Culbreath, A.K., Riley, D.G. and Pappu, H.R. (2014) Host plant resistance against *Tomato spotted wilt virus* in peanut (*Arachis hypogaea*) and its impact on susceptibility to the virus, virus population genetics, and vector feeding behavior and survival. *Phytopathology* 104, 202–210.

Suttle, C. (2005) The viriosphere: the greatest biological diversity on Earth and driver of global processes. *Environmental Microbiology* 7, 481–482.

Thiel, G., Moroni, A., Blanc, G. and Van Etten, J.L. (2013) Potassium ion channels: could they have evolved from viruses? *Plant Physiology* 162, 1215–1224.

Thomson, J.A. (2002) Research needs to improve agricultural productivity and food quality, with emphasis on biotechnology. *The Journal of Nutrition* 132, 3441S–3442S.

Thomson, J.A. (2008) The role of biotechnology for agricultural sustainability in Africa. *Philosophical Transactions of the Royal Society Biological Sciences* 363, 905–913.

Truniger, V. and Aranda, M.A. (2009) Recessive resistance to plant viruses. *Advances in Virus Research* 75, 119–159.

Truniger, V., Nieto, C., Gonzalez-Ibeas, D. and Aranda, M. (2008) Mechanism of plant eIF4E-mediated resistance against a *Carmovirus* (*Tombusviridae*): cap-independent translation of a viral RNA controlled *in cis* by an (a) virulence determinant. *Plant Journal* 56, 716–727.

UK Plant Science (2014) Current status and future challenges. A report by the UK Plant Sciences Federation. Society of Biology, London.

Urcuqui-Inchima, S., Haenni, A.L. and Bernardi, F. (2001) Potyvirus proteins: a wealth of functions. *Virus Research* 74, 157–175.

van Regenmortel, M.H.V. (1982) *Serology and Immunochemistry of Plant Viruses*. Academic Press Inc., New York, 302 pp.

van Regenmortel, M.H.V. (1989) Applying the species concept to plant viruses. *Archives of Virology* 104, 1–17.

van Regenmortel, M.H.V. (2003) Viruses are real, virus species are man-made, taxonomic constructions. *Archives of Virology* 148, 2481–2488.

Vincent, S.J., Coutts, B.A. and Jones, R.A.C. (2014) Effects of introduced and indigenous viruses on native plants: exploring their disease causing potential at the agro-ecological interface. *PLoS One* 9, e91224.

Waigmann, E. and Heinlein, M. (2007) *Viral transport in plants*. Springer-Verlag, Berlin/Heidelberg, pp.188.

Walthall, C.L., Hatfield, J., Backlund, P., Lengnick, L., Marshall, E., *et al.* (2012) Climate change and agriculture in the United States: effects and adaptation. *USDA Technical Bulletin* 1935.

Wang, L., Lv, X., Zhai, Y., Fu, S., Wang, D., *et al.* (2012) Genomic characterization of a novel virus of the family *Tymoviridae* isolated from mosquitoes. *PLoS One* 7, e39845.

Whitefield, A.E., Falk, B.W. and Rotenberg, D. (2015) Insect vector-mediated transmission of plant viruses. *Virology* (in press) doi:10.1016/j.virol.2015.03.026.

Wijkamp, I., Almarza, N., Goldbach, R. and Peters, D. (1995) Distinct levels of specificity in thrips transmission of tospoviruses. *Phytopathology* 85, 1069–1074.

Wren, J.D., Roossinck, M.J., Nelson, R.S., Scheets, K., Palmer, M.W., *et al.* (2006) Plant virus biodiversity and ecology. *PLoS Biology* 4, e80.

Wu, Q., Luo, Y., Lu, R., Lau, N., Lai, E.C., *et al.* (2010) Virus discovery by deep sequencing and assembly of virus-derived small silencing RNAs. *Proceedings of the National Academy of Sciences USA* 107, 1606–1611.

Wyatt, S.D. and Brown, J.K. (1996) Detection of subgroup III geminivirus isolates in leaf extracts by degenerate primers and polymerase chain reaction. *Phytopathology* 86, 1288–1293.

Wylie, S.J., Li, H. and Jones, M.G.K. (2013) *Donkey orchid symptomless virus*: a viral 'platypus' from Australian terrestrial orchids. *PLoS One* 8, e79587.

Xu, P., Cheng, F., Mannas, J.P., Feldman, T., Sumner, L.W., *et al.* (2008) Virus infection improves drought tolerance. *New Phytologist* 180, 911–921.

Yutin, N., Colson, P., Raoult, D. and Koonin, E.V. (2013) *Mimiviridae*: clusters of orthologous genes, reconstruction of gene repertoire evolution and proposed expansion of the giant virus family. *Virology Journal* 10, 106.

Zhuo, Y., Gao, G., Shi, J.A., Zhou, X. and Wang, X. (2013) miRNAs: biogenesis, origin and evolution, functions on virus-host interaction. *Cellular Physiology and Biochemistry* 32, 499–510.

2 Banana Bunchy Top

Niyongere Célestin,[1]* Aman Bonaventure Omondi[2] and Guy Blomme[3]

[1]ISABU, Bujumbura, Burundi; [2]BIOVERSITY International, Bujumbura, Burundi; [3]BIOVERSITY International, Addis Ababa, Ethiopia

2.1 Introduction

Banana bunchy top disease (BBTD) is caused by *Banana bunchy top virus* (BBTV), which is transmitted by the aphid vector *Pentalonia nigronervosa* Coquerel and through infected planting materials. It is one of the most economically important diseases in many banana-producing areas of Africa, Asia and the South Pacific (Furuya *et al.*, 2005; Hooks *et al.*, 2009). Between 1913 and 1920, the banana-growing industry in Australia was almost completely destroyed by the disease (Magee, 1927; Hooks *et al.*, 2009). In the 1990s, the first severe outbreak of BBTD in Africa was estimated to have reduced banana production in the Nkhatabay and Nkhotakota districts of Malawi from 3500 ha to about 800 ha (Soko *et al.*, 2009; Kumar *et al.*, 2011). In the Great Lakes countries of Africa, about 90% yield loss has been reported in severely BBTD-infected banana plantations in the Rusizi valley in Burundi (Niyongere *et al.*, 2011). Due to the highly destructive potential of the disease it causes, BBTV was listed as one of the world's 100 worst invasive species, and the International Plant Protection Convention included it as a pathogen which should be subject to rigorous quarantine measures (Kumar *et al.*, 2011).

2.2 Importance of Banana as the Main Host Plant of Banana Bunchy Top Disease

Banana (*Musa* spp.) is cultivated in more than 130 countries in the tropics and subtropics and is a staple food crop for millions of people, particularly in Africa (Frison and Sharrock, 1998). Banana is also the most important fruit crop used as dessert, and about 16 million tonnes are exported from these banana-growing countries each year (Daniells, 2009). It is a source of carbohydrates to about 70 million people in Africa. In some countries of Africa like those of the Great Lakes region of Africa, which includes Rwanda, Burundi, Uganda and the Democratic Republic of the Congo, the dependence on banana production is particularly evident by the high levels of per capita annual consumption of 382, 236, 70 and 69 kg/capita/year in these countries, respectively (Frison and Sharrock, 1998). Banana is also used to produce beer in Burundi and Rwanda, making the crop one of the main dependable cash crops in these two countries (Karamura *et al.*, 1998). Banana produces fruits all year round, thus supplying food and income to the farmer on a continuous basis. In this respect, the crop is a major contributor to food security worldwide (Olorunda, 1998).

*E-mail: niyocelestin@gmail.com

Banana cultivars are mostly triploid and polyploid hybrids of two wild banana species, *Musa acuminata* (AA) and *Musa balbisiana* (BB) (Stover and Simmonds, 1987). Five main genomic groups of cultivated banana designated as AA, AAA, AB, AAB and ABB, and four tetraploid hybrids designated as AAAA, AAAB, AABB and ABBB, have been genetically developed (Heslop-Harrison and Schwarzacher, 2007).

Unfortunately, banana yields of the different genotypes are low due to different constraints, including poor soil fertility, as well as inefficient management of fungal, bacterial and, particularly, viral diseases. About 20 virus species belonging to 5 families have been reported to infect banana and plantain worldwide. These include BBTV (genus *Babuvirus*, family *Nanoviridae*); several species of banana streak virus (genus *Badnavirus*, family *Caulimoviridae*) responsible for streak disease and *Banana bract mosaic virus* (genus *Potyvirus*, family *Potyviridae*) are the most economically important banana viruses. There are other viruses with minor impact on banana production such *Cucumber mosaic virus* (genus *Cucumovirus*, family *Bromoviridae*), *Banana mild mosaic virus* (family *Betaflexiviridae*) and *Banana virus X* (family *Betaflexiviridae*) (Pietersen and Thomas, 2000; Kumar *et al.*, 2015). Among diseases caused by viruses, BBTD caused by *Banana bunchy top virus* is considered to be the most serious disease affecting banana crops worldwide and the livelihoods of people (Dale, 1987; IITA, 2010; Islam *et al.*, 2010; Kumar *et al.*, 2011).

2.3 Current Distribution of Banana Bunchy Top Disease

The history of BBTD has shown that its spread can mainly be attributed to the exchange of planting materials. The disease, reported for the first time in the Fiji Islands in 1889 (Magee, 1927), has since been recorded in 36 countries; 15 are in Africa and 22 others are in Asia, Australia and the South Pacific Islands (Kumar *et al.*, 2011; Blomme *et al.*, 2013). Central and South America apparently remain free of the disease (Ferreira *et al.*,

1997; Amin *et al.*, 2008). In Africa, BBTD was first reported in Egypt in 1901, and subsequently in 1958 in sub-Saharan Africa at the Institut National pour l'Etude Agronomique au Congo Belge, Yangambi Agricultural Research Station in central Democratic Republic of the Congo (Wardlaw, 1961; Fouré and Manser, 1982). It was also reported in 1964 in Eritrea (Saverio, 1964). Cases of the disease were recorded in 1987 in the Rusizi valley encompassing parts of Burundi and Rwanda (Sebasigari and Stover, 1988). In 1982, BBTD outbreaks were observed in Gabon, Congo-Brazzaville and Equatorial Guinea (Fouré and Manser, 1982). The disease was described in Malawi and Angola (Kumar and Hanna, 2008) in the early 1990s, and also in Cameroon, Central African Republic, Zambia, Benin and Nigeria making a total of 15 African countries affected by banana bunchy top disease (IITA, 2010).

2.4 Description of *Banana bunchy top virus*

Although BBTD was first reported from the Fiji Islands in 1889, its causal agent was only identified some 100 years later (Magee 1927; Kumar *et al.*, 2011), and was given the name of *Banana bunchy top virus* (Karan, 1995; Vetten *et al.*, 2005). The virus, BBTV, contains at least six circular single-stranded DNA components, each about 1.1 kb encoding for a single open reading frame in the virion sense strand (Hu *et al.*, 2007; Sharman *et al.*, 2008). These six DNA components have been consistently associated with BBTV worldwide in all geographical isolates (Dale *et al.*, 2000; Horser *et al.*, 2001a).

BBTV genomic components, initially labelled as DNA-1, DNA-2, DNA-3, DNA-4, DNA-5 and DNA-6, have been renamed to better represent the function of the encoded proteins (Vetten *et al.*, 2005). DNA-R encodes the viral replicase (rolling-circle replication initiation protein), DNA-S (capsid protein), DNA-M (movement protein), DNA-C (cell cycle link protein), DNA-N (nuclear shuttle protein) and the function of the DNA-U3 is not yet known (Hafner *et al.*, 1995; Wanitchakorn *et al.*, 1997). Each of these components is approximately 1 kb in length and all are individually encapsulated in icosahedral virions of 18–20 nm in diameter and share a common genome organization. In addition, each circular single-stranded DNA encodes a single open reading frame, except BBTV DNA-R which encodes two (Burns *et al.*, 1995; Beetham *et al.*, 1997). All components contain a conserved stem-loop common region with a conserved nine nucleotide (TAT-TATTAC) loop sequence that likely marks the origin of DNA replication, a major common region, a potential TATA box and a polyadenylation signal associated with each gene (Beetham *et al.*, 1997; Su *et al.*, 2003). In addition, the DNA-R component of BBTV contains conserved iterons that are the sequences involved in sequence-specific interaction (Herrera *et al.*, 2006; Amin *et al.*, 2008).

DNA-R has been identified in all BBTV isolates where it encodes the 'master' Rep (M-Rep) that directs self-replication in addition to replication of other BBTV genome components (Karan *et al.*, 1994; Theresia, 2008). The DNA-S component encodes the coat protein (CP) of 20.5 kDa for the integral BBTV genome (Horser *et al.*, 2001b). Based on sequence analysis of the DNA-R and DNA-S (CP) components, Karan *et al.* (1994) and Wanitchakorn *et al.* (2000) demonstrated that BBTV isolates cluster into two distinct groups. The 'South Pacific group' (isolate variability of 0.7–3.8%) comprises isolates from Australia and the South Pacific region, South-Eastern Asia that includes India and Pakistan, and Africa; the 'Asian group'

(isolate variability of 0.7–8%) comprises isolates from China, Indonesia, Japan, the Philippines, Taiwan and Vietnam (Wanitchakorn *et al.*, 2000; Kumar *et al.*, 2011). The difference between the two groups is approximately 10% based on DNA-R genome (Karan *et al.*, 1994). In general, various BBTV isolates characterized so far around the world have >85% homology (Banerjee *et al.*, 2014).

2.5 Transmission of *Banana bunchy top virus* in Banana Fields

2.5.1 Disease spread

BBTD spreads with exchange of infected suckers from region to region, and via the banana aphid, *P. nigronervosa* Coquerel (Hemiptera: Aphididae), from plant to plant (Ferreira *et al.*, 1997; Robson *et al.*, 2006). In contrast to the aetiological agents of other banana diseases, BBTV is not transmitted mechanically through the use of contaminated garden tools (Wardlaw, 1961; Kumar *et al.*, 2011).

2.5.2 *P. nigronervosa*: forms and distribution

P. nigronervosa exists as either wingless or winged aphids (Fig. 2.1). It is widely distributed and associated with banana plantations throughout tropical and subtropical areas irrespective of the presence of bunchy top disease (Hu *et al.*, 2007; Foottit *et al.*, 2010).

Fig. 2.1. Typical appearance of banana aphid *P. nigronervosa*. (a) a colony comprising wingless females and nymphs of different stages; (b) wingless adult; and (c) winged adult (UH-CTAHR, Nelson, 2004).

Dispersing winged adults establish new colonies on other new host banana plants.

2.5.3   *P. nigronervosa* on *Musa* spp.

The aphid was described on banana for the first time by Coquerel in 1859 in India (Bhadra and Agarwala, 2010). Aphids were initially reported to have occurred on member species of the Zingiberaceae and Araceae families, but morphological and morphometric studies confirmed that this aphid represented a separate species, named *Pentalonia caladii* van der Goot (Bhadra and Agarwala, 2010; Foottit *et al.*, 2010). *P. caladii* van der Goot has been shown to transmit BBTV under experimental inoculation conditions, but at a lower level of efficiency compared to *P. nigronervosa* (Watanabe *et al.*, 2013). *P. nigronervosa* Coquerel was, therefore, confirmed to have high host specificity to *Musa* spp. (Kumar *et al.*, 2011). Its reproduction is almost totally asexual (Foottit *et al.*, 2010). The lifespan of an aphid ranges from 19 to 26 days, and during this period the aphid will produce up to 20 descendants in optimal conditions of 24–28°C (Yasmin *et al.*, 1999). The winged aphids, which often develop after seven to ten generations of wingless individuals, are most likely responsible for the spread of the virus (Nelson, 2004; Young and Wright, 2005). Ambient temperatures above 14°C enable aphid flights (Jones *et al.*, 2010) and transmission of the virus to a healthy banana plant by feeding for as little as 15 minutes to almost 2 hours (Dale, 1987; Hu *et al.*, 1996). *P. nigronervosa* retains BBTV throughout its adult life in persistent mode and transmits it in a circulative manner, but there is no evidence of transovarial transmission to progeny (Nelson, 2004; Anhalt and Almeida, 2008).

2.5.3   Aphid distribution and location

These aphids are not strong fliers and the mean distance between new infections and their source of inoculum in an established plantation was estimated at 17.2 m by Allen in 1987. However, it was reported that they may be carried over considerable distances by especially strong winds (Allen, 1987; Ferreira *et al.*, 1997). Aphids are more frequently observed near the base of banana plants, in between leaf sheaths, and at the base of the youngest unfurled leaf (Robson *et al.*, 2006). In addition, Young and Wright (2005) reported a spatial edge effect, with larger aphid colonies observed at the edge of plantations.

2.5.4   The disease incubation period

The time until the appearance of the diagnostic dark-green leaf streaks on the leaf lamina varies from 19 days in the summer (25–33°C) to 125 days in the winter (June to August with –2°C to 11.5°C) in Australia (Magee, 1927; Allen, 1987). Previous research showed that this incubation period, after screen-house inoculation, is positively correlated with the age of the host plant, and negatively correlated with the number of viruliferous aphids feeding on the plant (Robson *et al.*, 2006; Hooks *et al.*, 2008). The incubation period is also influenced by banana variety as some genotypes with genome B (AAB and ABB) express symptoms more slowly than those with the A genome (AA and AAA), such as the Cavendish (AAA) subgroup (Robson *et al.*, 2006; Hooks *et al.*, 2008).

2.5.5   Incidence of BBTD

Smith *et al.* (1998) report on the exponential increase of BBTD incidence, which, together with the high transmission efficiency associated with the vector *P. nigronervosa*, underscore the importance of aphid population control in the management of the disease (Hu *et al.*, 1996; Robson *et al.*, 2006). The occurrence of symptomless infections on certain genotypes (e.g. *Musa* ABB genome, such as the 'Saba' variety) further complicates disease management and increases the risk of the inadvertent spread of the pathogen, since these plants act as reservoirs for BBTV (Allen, 1987; Drew *et al.*, 1989).

2.6 Banana Bunchy Top Disease Symptoms

BBTD symptoms are easily distinguishable, under field conditions, from those induced by other virus diseases of banana (Fig. 2.2). These symptoms initially include the characteristic development of 'Morse code' streaking of variable length in the leaf veins, midribs and petioles, followed by progressive dwarfing of leaves and the development of marginal leaf chlorosis, upright and crowded leaves at the apex of the plant, hence the name bunchy top disease (Magee, 1927; Ferreira et al., 1997). In infected plants, the phloem and its associated parenchyma tissue show excessive and irregular cell divisions. Symptoms develop more quickly at higher temperatures (above 24°C) than at lower temperatures (below 10°C) both in the field and in controlled environment greenhouses (Dale et al., 2000).

Plants infected by BBTV at an early growth stage are unable to produce bunches, whereas those infected at later stages of growth produce small bunches often of poor quality (Dale, 1987; Su et al., 2003). The suckers raised from BBTD-infected mats are usually severely stunted, with leaves that do not expand normally and remain bunched at the top of the pseudostem. Suckers with these symptoms do not bear fruits (Dale, 1987; Su et al., 2003). On the other hand, secondary infection of plants occurs as a result of aphid transmission following an initial period of BBTV-free plant growth. The symptoms of secondary infections are generally milder and evident in new growth developed after infection (Magee, 1927).

Despite distinctive and easily distinguished symptoms of the disease, symptoms on their own are not enough for BBTD diagnosis. The limitation of visual diagnosis of BBTV is the incubation period of the

Fig. 2.2. Banana bunchy top disease symptoms on a (a) banana leaf, (b) petiole, (c) a young banana plant and (d) an older plant bearing fruits, compared to a healthy banana plant.

pathogen, which lasts for about one month. This complicates disease management in banana fields, since there is an increasing risk of pathogen spread through the exchange of symptomless suckers (Allen, 1987; Drew *et al.*, 1989).

2.7   Diagnosis of *Banana bunchy top virus*

Sensitive detection methods have been developed to confirm diagnosis when BBTV is suspected based on symptom expression. ELISA is commonly used for BBTV detection, although reports indicate limited sensitivity particularly where there are very low virus titres in banana tissues (Wanitchakorn *et al.*, 2000). The triple antibody sandwich ELISA is also used to detect the presence of BBTV in banana leaf samples and to determine virus concentration.

The most sensitive methods currently available are based on the PCR techniques using primers that amplify specific DNA sequences of the virus, namely DNA-R and CP (Su *et al.*, 2003). Various studies have tested different types of templates as starting materials to achieve optimal test performance and result consistency, for example, purified total DNA extracts (Furuya *et al.*, 2005) as well as the inclusion of an initial step of immunocapture (Sharman *et al.*, 2000) to allow for the immobilization of virions on the walls of PCR reaction tubes using specific antibodies. The protocol of PCR detection using banana crude extracts has, however, considerably simplified disease detection (Busogoro *et al.*, 2009).

2.8   Banana Bunchy Top Disease Control Options

Under suitable conditions, aphid vector populations progressively increase resulting in a gradual upsurge in diseased mats and corresponding inoculum levels (Magee, 1938; Hooks *et al.*, 2009). Controlling these aphids with contact insecticide sprays is very difficult because the spray must drench

the region behind the leaf sheaths and reach protected areas to kill the aphids (Ferreira *et al.*, 1997). Once the disease has been introduced into an area, Robson *et al.* (2006) reported that eradication is very difficult. In addition, resistant genotypes do not exist, though some studies have reported differences in susceptibility (Magee, 1948; Daniells, 2009). Cultivars in the AA and AAA genomic groups are highly susceptible to BBTD with the exception of 'Gros Michel', whereas cultivars containing two B genomes are regarded as less susceptible (Magee, 1948; Ariyatne and Liyanage, 2002). The deployment of tolerant cultivars may be an option for the integrated management of BBTD in regions where small-scale agriculture dominates. There are no available transgenic varieties resistant to BBTD.

The recommended strategies for controlling the disease include:

1. The identification and prompt uprooting of symptomatic plants to eliminate BBTV-infected mats that are the likely source of secondary infection and spread of the pathogen (Dale, 1987; Hooks *et al.*, 2008; Jones, 2009).
2. Replanting with virus-free tissue culture plantlets indexed using serological or molecular tools such as triple antibody sandwich ELISA and PCR, respectively (Geering, 2009; Ikram *et al.*, 2009). Indeed, the established nurseries in BBTD-affected areas should be covered with mosquito nets to minimize the access of aphids to plants, coupled with regular applications of insecticides. Planting materials derived from tissue culture should be tested and found virus free before the transfer and establishment of new banana plots at no less than 30 m distance away from existing diseased fields so as to prevent early infection and possible inoculum build up (C. Niyongere, personal observations).
3. A holistic approach taking into account the banana variety and regular field sanitation (de-leafing, de-suckering and weeding) needs to be enforced in order to maintain aphid colonies under the threshold at which more winged aphids are produced and are able to spread the virus (Kumar *et al.*, 2011).
4. Quarantine measures should be also established between BBTD-affected and non-affected areas (Thresh, 2003).

5. Creating awareness on BBTD transmission and management measures through training of different stakeholders including extension staff and farmers should be adopted for an integrated management strategy especially under small-scale farmer conditions.

2.9 Concluding Remarks

BBTD remains a serious threat to banana production in most affected countries and continues to spread in the absence of any meaningful control measures. The development of integrated management strategies should consider various factors, including agroecological practices such as the use of tolerant varieties and climatic factors, as well as socioeconomic aspects based on collective/community approaches to the eradication of affected mats to reduce virus inoculum, together with the availability and accessible cost of indexed virus-free planting materials (Mgenzi *et al.*, 2003; Blomme *et al.*,

2013). Therefore, raising farmers' awareness on BBTD management practices that target the reduction of aphid colonies under the threshold at which winged aphids are produced and spread the virus (Kumar *et al.*, 2011) and the use of virus-free planting material is best carried out through campaigns at the local community level. Additionally, as the spread of the virus pathogen has been attributed to the exchange of planting materials across regions and countries, quarantine measures ought to be established by policy makers so as to prevent virus spread into new areas not yet affected by BBTD.

Acknowledgements

The authors thank BIOVERSITY International (CGIAR member) and ISABU (Institut des Sciences Agronomiques du Burundi) for funding and facilitating, respectively, the Banana Bunchy Top disease research activities presented in this chapter.

References

Allen, R.N. (1987) Further studies on epidemiological factors influencing control of banana bunchy top disease and evaluation of control measures by computer simulation. *Australian Journal of Agriculture Research* 38, 373–382.

Amin, I., Qazi, J., Mansoor, S., Ilyas, M. and Briddon, R.W. (2008) Molecular characterisation of *Banana bunchy top virus* from Pakistan. *Virus Genes* 36, 191–198.

Anhalt, M.D. and Almeida, R.P.P. (2008) Effect of temperature, vector life stage, and plant access period on transmission of banana bunchy top virus to banana. *Phytopathology* 98, 743–748.

Ariyatne, I. and Liyanage, T. (2002) Survey on incidences and severity of virus diseases of banana in Sri Lanka. *Annals of the Sri Lanka Department of Agriculture* 4, 245–254.

Banerjee, A., Roy, S., Beherea, G.T., Roy, S.S., Dutta, S.K. *et al.* (2014) Identification and characterization of a distinct *Banana bunchy top virus* isolate of Pacific-Indian Oceans group from North-East India. *Virus Research* 183, 41–49.

Beetham, P.R., Hafner, G.J., Harding, R.M. and Dale, J.L. (1997) Two mRNAs are transcribed from *Banana bunchy top virus* DNA-1. *Journal of General Virology* 78, 229–236.

Bhadra, P. and Agarwala, B.K. (2010) A comparison of fitness characters of two host plant-based congeneric species of the banana aphid, *P. nigronervosa* and *P. caladii. Journal of Insect Science* 10, 140.

Blomme, G., Ploetz, R., Jones, D., De Langhe, E., Price, N., *et al.* (2013) A historical overview of the appearance and spread of *Musa* pests and pathogens on the African continent: highlighting the importance of clean *Musa* planting materials and quarantine measures. *Annals of Applied Biology* 162, 4–26.

Burns, T.M., Harding, R.M. and Dale, J.L. (1995) The genome organisation of *Banana bunchy top virus*: analysis of six ssDNA components. *Journal of General Virology* 76, 1471–1482.

Busogoro, J.P., Dutrecq, O. and Jijakli, H. (2009) Development of simple molecular protocols to detect *Banana bunchy top virus* with the PhytoPass system. *Acta Horticulturae ISHS* 828, 213–218.

Dale, J.L. (1987) Banana bunchy top: an economically important tropical plant virus disease. *Advances in Virus Research* 33, 301–325.

Dale, J.L., Horser, C.L., Karan, M. and Harding, R.M. (2000) Additional rep-encoding DNAs associated with *Banana bunchy top virus*. *Archives of Virology* 146, 71–86.

Daniells, J.W. (2009) Global banana disease management – getting serious with sustainability and food security. *Acta Horticulturae* 828, 411–416.

Drew, R.A., Moisander, J.A. and Smith, M.K. (1989) The transmission of *Banana bunchy top virus* in micro-propagated bananas. *Plant Cell, Tissue and Organ Culture* 16, 187–193.

Ferreira, S.A., Trujillo, E.E. and Ogata, D.Y. (1997) *Banana Bunchy Top Virus*. College of Tropical Agriculture and Human Resources, University of Hawaii (CTAHR). Available at: www.issg.org/database/species/reference_files/BBTV/CoExSe.pdf (accessed 18 July 2012).

Foottit, R.G., Maw, H.E.L, Pike, K.S. and Miller, R.H. (2010) The identity of *Pentalonia nigronervosa* Coquerel and *P. caladii* van der Goot (Hemiptera: Aphididae) based on molecular and morphometric analysis. *Zootaxa* 2358, 25–38.

Fouré, E. and Manser, P.D. (1982) Note sur l'apparition au Gabon d'une grave maladie virale des bananiers et plantains: Le Bunchy Top. *Fruits* 37, 409–414.

Frison, E. and Sharrock, S. (1998) The economic, social and nutritional importance of banana in the world. In: Picq, C., Fouré, E. and Frison, E.A. (eds) *Bananas and Food Security*. International Symposium, Douala, Cameroon, 10–14 November 1998. INIBAP, Montpellier, France, pp. 21–35.

Furuya, N., Kawano, S. and Natsuaki, K.T. (2005) Characterization and genetic status of *Banana bunchy top virus* isolated from Okinawa, Japan. *Journal of General Plant Pathology* 71, 68–73.

Geering, A.D.W. (2009) Viral pathogens of banana: outstanding questions and options for control. *Acta Horticulturae* 828, 39–50.

Hafner, G.J., Harding, R.M. and Dale, J.L. (1995) Movement and transmission of *Banana bunchy top virus* component one in banana. *Journal of General Virology* 76, 2279–2285.

Herrera, V.V.A., Dugdale, B., Harding, R.M. and Dale, J.L. (2006) An iterated sequence in the genome of *Banana bunchy top virus* is essential for efficient replication. *Journal of General Virology* 87, 3409–3412.

Heslop-Harrison, J.S. and Schwarzacher, T. (2007) Domestication, genomics and the future for banana. *Annals of Botany* 100, 1073–1084.

Hooks, C.R.R., Wright, M.G., Kabasawa, D.S., Manandhar, R. and Almeida, R.P.P. (2008) Effect of *Banana bunchy top virus* infection on morphology and growth characteristics of banana. *Annals of Applied Biology* 153, 1–9.

Hooks, C.R.R., Manandhar, R., Perez, E.P., Wang, K.H. and Almeida, R.P.P. (2009) Comparative susceptibility of two banana cultivars to *Banana bunchy top virus* under laboratory and field environments. *Journal of Economic Entomology* 102, 897–904.

Horser, C.L., Harding, R.M. and Dale, J.L. (2001a) Banana bunchy top nanovirus DNA-1 encodes the master replication initiation protein. *Journal of General Virology* 82, 459–464.

Horser, C.L., Karan, M., Harding, R.M. and Dale, J.L. (2001b) Additional Rep-encoding DNAs associated with *Banana bunchy top virus*. *Archives of Virology* 146, 71–86.

Hu, J.S., Wang, M., Sether, D., Xie, W. and Leonhardt, K.W. (1996) Use of Polymerase Chain Reaction (PCR) to study transmission of banana bunchy top by the banana aphid (*Pentalonia nigronervosa*). *Annals of Applied Biology* 128, 55–64.

Hu, J.M., Fu, H.C., Lin, C.H., Su, H.J. and Yeh, H.H. (2007) Reassortment and concerted evolution in *Banana bunchy top virus* genomes. *The Journal of Virology* 81, 1746–1761.

IITA (International Institute of Tropical Agriculture) (2010) African *Musa* under attack! Two deadly banana diseases in Eastern and Central Africa. Available at: http://old.iita.org/cms/details/news_details.aspx?articleid=3478&zoneid=81 (accessed 10 July 2012).

Ikram, H., Dahot, M.U., Khan, S. and Kousar, N. (2009) Screening of banana bunchy top diseased plants: a way to control its spreading. *Plant Omics Journal* 2, 175–180.

Islam, M.N., Naqvi, A.R., Jan, A.T. and Haq, Q.M.R. (2010) Genetic diversity and possible evidence of recombination among *Banana bunchy top virus* (BBTV) isolates. *International Research Journal of Microbiology* 1, 1–12.

Jones, D.R. (2009) Disease and pest constraints to banana production. *Acta Horticulturae* 828, 21–36.

Jones, R.A.C., Salm, M.U., Maing, T.J., Diggle, A.J. and Thackray, D.J. (2010) Principles of predicting plant virus disease epidemics. *The Annual Review of Phytopathology* 48, 179–203.

Karamura, E., Frison, E., Karamura, D.A. and Sharrock, S. (1998) Banana production systems in Eastern and Southern Africa. In: Picq, C., Fouré, E. and Frison, E.A. (eds) *International Symposium on Bananas and Food Security*, Douala, Cameroon, 10–14 November 1998. INIBAP, Montpellier, France, pp. 401–412.

Karan, M. (1995) Sequence diversity of DNA components associated with *Banana bunchy top virus*. PhD thesis, Queensland University of Technology, Australia.

Karan, M., Harding, R.M. and Dale, J.L. (1994) Evidence for two groups of *Banana bunchy top virus* isolates. *Journal of General Virology* 75, 3541–3546.

Kumar, P.L. and Hanna, R. (2008) *Banana bunchy top virus* in sub-Saharan Africa: established or emerging problem? Poster presentation 'Banana and Plantain in Africa: Harnessing international partnerships to increase research impact', workshop held October 5–9, Mombasa Leisure Lodge Resort, Kenya.

Kumar, P.L., Hanna, R., Alabi, O.J., Soko, M.M., Oben, T.T., *et al.* (2011) *Banana bunchy top virus* in sub-Saharan Africa: investigations on virus distribution and diversity. *Virus Research* 159, 171–182.

Kumar, P.L., Selvarajan, R., Iskra-Caruana, M.L., Chabanne, M. and Hanna, R. (2015) Biology, etiology, and control of virus diseases of banana and plantain. *Advances in Virus Research* 91, 229–269.

Magee, C.J. (1927) Investigation on the bunchy top disease of the banana. *Council for Scientific and Industrial Research Bulletin* 30, 1–64.

Magee, C.J. (1938) *Bunchy top of bananas*. Plant disease leaflet Number 54. Department of Agriculture, New South Wales, Australia.

Magee, C.J. (1948) Transmission of bunchy top to banana varieties. *The Journal of the Australian Institute of Agricultural Science* 14, 18–24.

Mgenzi, S.R.B., Mkulila, S.I., Blomme, G., Gold, C.S., Karamura, E.B., *et al.* (2003) The effect of pest management practices on banana pests in the Kagera of Tanzania. In: Blomme, G., Gold, C. and Karamura, E. (eds) *Proceedings of the Workshop on Farmer-Participatory testing of IPM Options for Sustainable Banana Production in Eastern Africa*. Seeta, Uganda, 8–9 December 2003, pp. 43–52.

Nelson, S.C. (2004) Banana bunchy top: detailed signs and symptoms. CTAHR (College of Tropical Agriculture and Human Resources), University of Hawaii. Available at: http://www.ctahr.hawaii.edu/bbtd/downloads/bbtv-details.pdf (accessed 12 June 2012).

Niyongere, C., Ateka, E., Losenge, T., Blomme, G. and Lepoint, P. (2011) Screening *Musa* genotypes for banana bunchy top disease resistance in Burundi. *Acta Horticulturae* 897, 439–447.

Olorunda, A.O. (1998) Bananas and food security in sub-Saharan Africa: the role of postharvest food technologies. In: Picq, C., Fouré, E. and Frison, E.A. (eds), *International Symposium on Bananas and Food Security*, Douala, Cameroon, 10–14 November 1998. INIBAP, Montpellier, pp. 375–382.

Pietersen, G. and Thomas, J.E. (2001) Overview of *Musa* virus diseases. In: Hughes, J.d'A. and Odu, B.O. (eds) *Plant Virology in Sub-Saharan Africa*. Proceedings of a Conference organized by IITA, 4–8 June 2001, Ibadan, Nigeria, pp. 50–60. Available at: http://www.iita.org/plant-virology (accessed 15 January 2015).

Robson, J.D., Wright, M.G. and Almeida, R.P.P. (2006) Within-plant spatial distribution and binomial sampling of *Pentalonia nigronervosa* (Hemiptera: Aphididae) on banana. *Journal of Economic Entomology* 99, 2185–2190.

Saverio, B. (1964) Banana cultivation in Eritrea and its problems. *Edagricole* 8, 5–56.

Sebasigari, K. and Stover, R.H. (1988) *Banana diseases and pests in East Africa. Report of a Survey in 1987*. INIBAP, Montpellier, France, pp. 3–6.

Sharman, M., Thomas, J.E. and Dietzen, R.G. (2000) Development of a multiplex immunocapture PCR with colourimetric detection for viruses of banana. *Journal of Virological Methods* 89, 75–88.

Sharman, M., Thomas, J.E., Skabo, S. and Holton, T.A. (2008) *Abaca bunchy top virus*, a new member of the genus *Babuvirus* (family *Nanoviridae*). *Archives of Virology* 153, 135–147.

Smith, M.C., Holt, J., Kenyon, L. and Foot, H. (1998) Quantitative epidemiology of *Banana bunchy top virus* disease and its control. *Plant Pathology* 47, 177–187.

Soko, M.M., Dale, J., Kumar, P.L., James, A., Izquierdo, L., *et al.* (2009) Banana bunchy top disease survey reports, previous control efforts and the way forward. Bvumbwe Agricultural Research Station, Limbe, Malawi, pp. 200.

Stover R.H. and Simmonds N.W. (1987) *Bananas*. 3rd edn. Longman Scientific & Technical, New York and Wiley, Harlow, UK, pp. 314–315.

Su, H.J., Tsao, L.Y., Wu, M.L. and Hung, T.H. (2003) Biological and molecular categorisation of strains of *Banana bunchy top virus*. *Journal of Phytopathology* 151, 290–296.

Theresia, T.H.T. (2008) Towards the development of transgenic *Banana bunchy top virus* (BBTV)-resistant banana plants: interference with replication. PhD thesis, Queensland University of Technology, Australia.

Thresh, J.M. (2003) Control of plant virus diseases in sub-Saharan Africa: the possibility and feasibility of an integrated approach. *African Crop Science Journal* 11, 199–223.

Vetten, H.J., Chu, P.W.G., Dale, J.L., Harding, R., Hu, J., *et al.* (2005) Family *Nanoviridae*. In: Fauquet, C.M., Mayo, M.A., Maniloff, J., Desselberger, U. and Ball, L.A. (eds) *Virus Taxonomy, VIIIth Report of the International Committee on Taxonomy of Viruses*. Elsevier Academic Press, San Diego, California, and London, pp. 343–352.

Wanitchakorn, R., Harding, R.M. and Dale, J.L. (1997) *Banana bunchy top virus* DNA-3 encodes the viral coat protein. *Archives of Virology* 142, 1673–1680.

Wanitchakorn, R., Harding, R.M. and Dale, J.L. (2000) Sequence variability in the coat protein gene of two groups of banana bunchy top isolates. *Archives of Virology* 145, 593–602.

Wardlaw, C.W. (1961) The virus diseases: bunchy top. In: *Banana Diseases, Including Plantains and Abaca*. Longmans, Green and Co, London, pp. 68–115.

Watanabe, S., Greenwell, A.M. and Bressan, A. (2013). Localization, concentration, and transmission efficiency of *Banana bunchy top virus* in four asexual lineages of Pentalonia aphids. *Viruses* 5, 758–775.

Yasmin, T.H.E., Khalid, S. and Mailik, S.A. (1999) Some studies on biology of *Pentalonia nigronervosa* Conquarrel, the vector of *Banana bunchy top virus*. *Pakistan Journal of Biological Sciences* 2, 1398–1400.

Young, C.L. and Wright, M.G. (2005) Seasonal and spatial distribution of banana aphid, *Pentalonia nigronervosa* (Hemiptera: Aphididae), in banana plantations on Oahu. *Proceedings of the Hawaiian Entomological Society* 37, 73–80.

3

Wheat Dwarf

Isabelle Abt[1,2] and Emmanuel Jacquot[1]*

[1]INRA-Cirad-SupAgro Montpellier, Montpellier, France;
[2]Bayer S.A.S./Bayer CropScience, Lyon, France

3.1 Introduction

Cereals can be infected by a vast range of pathogens of which *Wheat dwarf virus* (WDV, family *Geminiviridae*, genus *Mastrevirus*), the aetiological agent of wheat dwarf disease (WDD), is one of the most damaging of all. WDV is exclusively transmitted from plant to plant by leafhoppers. Very few options are available for the farmers to control this pathogen in the field. Indeed, the lack of genetic resistance against WDV in cereal germplasm, the rare sources of genetic tolerance and the absence of anti-viral molecules lead to the use of indirect management methods such as the modification of cultivation practices and/or use of insecticides to protect cereal crops from WDV infections.

Even though WDD can be associated with important economical and agronomical impacts, scientific knowledge on this pathosystem and the epidemiology of the disease are still limited. Only a few studies on WDV have been published after three decades between the first report of wheat dwarf-like symptoms back in the 1960s and the description of WDD outbreaks in the 1990s (e.g. Vacke, 1972; MacDowell *et al.*, 1985; MacDonald *et al.*, 1988; Schalk *et al.*, 1989). However, the recent increase in the prevalence of WDD in European, African and Asian cereal-growing areas has positively impacted the number of research projects carried out on WDD. The

data generated by these projects have helped to improve the knowledge of the scientific community studying the wheat dwarf pathosystem, and have facilitated the development of innovative management strategies. This chapter presents an overview of WDD, its causal agent, its leafhopper vectors and the available management strategies.

3.2 The Disease

WDD is reported to occur mainly in barley (*Hordeum vulgare* L.), wheat (*Triticum aestivum* L.) and oat (*Avena sativa* L.) fields. It is associated with the development of a range of symptoms such as dwarfing, yellowing, mottling, streaking of leaves, suppressed heading and severe stunting (Vacke, 1972), and infected winter cereals can succumb to the disease and die during cold winter periods. Symptomatic plants often occur in patches in fields but sometimes, when favourable conditions occur, the disease spreads from reservoirs (i.e. alternative cereal hosts, ground keepers, volunteer plants) or from infected to healthy plants within the field, and in some cases the entire field can display disease symptoms. Mean yield losses due to WDD have been estimated to reach up to 35% of the grain production expected from healthy plants; however, local epidemics can be associated with higher yield losses of

*E-mail: emmanuel.jacquot@rennes.inra.fr

up to 90% (Fohrer *et al.*, 1992; Lindsten and Lindsten, 1999; Lindblad and Waern, 2002; Sirlova *et al.*, 2005) resulting in devastating crop failures. Because the incidence of the disease in winter cereals differs greatly from year to year, between field locations, and between cereal species and cultivars, WDD has long been considered as a minor sanitary problem by growers (Fig. 3.1).

Early records indicate the presence of dwarf-like symptoms in wheat in Sweden prior to the 1960s. However, the association between a virus, a vector and the wheat dwarf symptoms was reported for the first time in the 1960s using materials sampled from a wheat field located in the western part of the former Czechoslovak Socialist Republic (Vacke, 1961). The prevalence of WDD in cereal fields was relatively low in the 20th century. Indeed, before 1980s there are very few descriptions in the scientific literature of dwarf-like symptoms, sometimes associated with severe outbreaks on wheat, suggesting that WDD was uncommon in cereal-growing areas at that time. However, after three decades the disease has re-emerged in a number of European countries (e.g. Gaborjanyi *et al.*, 1988; Lapierre *et al.*, 1991; Conti, 1993; Jilaveanu and Vacke, 1995;

Najar *et al.*, 2000). Following this re-emergence in the late 1990s, the distribution of WDD has increased to currently include numerous European countries (Hungary (Bisztray *et al.*, 1989), France (Bendahmane *et al.*, 1995), Sweden (Lindsten and Lindsten, 1999), Germany (Huth, 2000), Poland (Jezewska, 2001), Finland (Lemmetty and Huusela-Veistola, 2005), Spain (Achon and Serrano, 2006), Bulgaria (Tobias *et al.*, 2009), Ukraine (Tobias *et al.*, 2011), UK (Schubert *et al.*, 2014) and Austria (Schubert *et al.*, 2014)), as well as regions in the Middle East (Turkey (Koklu *et al.*, 2007) and Iran (Behjatnia *et al.*, 2011)), Africa (Tunisia (Najar *et al.*, 2000) and Zambia (Kapooria and Ndunguru, 2004)), Western Asia (Syria (Ekzayez *et al.*, 2011)) and Asia (China (Xie *et al.*, 2007; Wang *et al.*, 2008)). It is important to note that for some of these reports of WDD, the virus associated with the wheat dwarf-like symptoms has not been fully characterized. According to the current worldwide distribution of WDD, the disease is now considered a serious problem for grain production.

The reason for the increase in the distribution of WDD has, to date, not been clearly determined. Currently, the reporting of wheat dwarf in a WDD-free cereal-growing area is

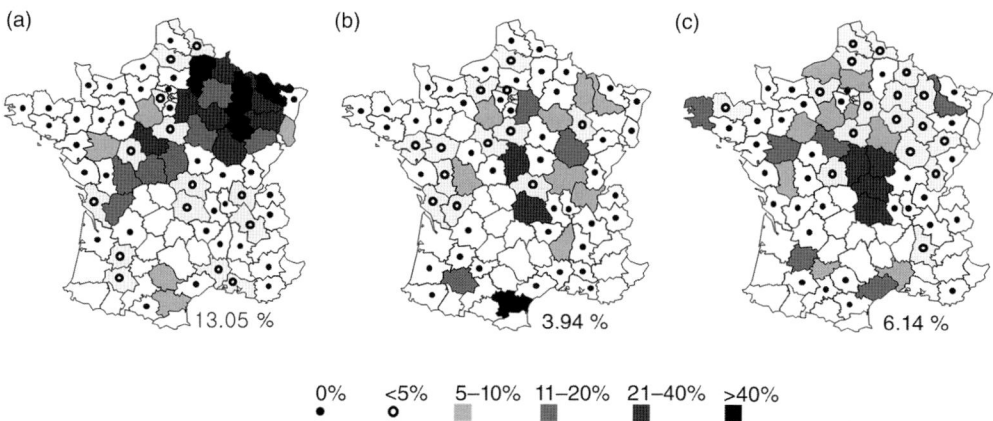

Fig. 3.1. Prevalence of *Wheat dwarf virus* in France. Field surveys were performed in wheat and barley production areas in France during the spring of (a) 2008, (b) 2009 and (c) 2010. A total of about 6000 cereal leaves per year were sampled and analysed using serological (ELISA) procedures. For each department, the percentage of WDV-infected leaves is represented using the code inserted in the figure. Departments that were not included in the surveys are presented in white. The mean prevalence of WDV in cereal fields, estimated using our data-set, is presented under each map.

a result of the findings of surveys conducted for the first time using sensitive diagnostic tools (see Diagnostic methods for wheat dwarf disease). However, the occurrence and detection of wheat dwarf in a cereal field depends on numerous parameters:

1. The characteristics (size and sanitary status) of local reservoirs.
2. The behaviour (feeding, short- and long-distance movements and size of populations) of vectors present in the area. Primary spread of WDD results from the migration of viruliferous vectors (see Transmission of the virus) from wild and/or cultivated reservoirs to newly sown cereal fields (Felix *et al.*, 1992; Lindsten and Lindsten, 1999).
3. The susceptibility of the cultivated hosts.
4. Certain cultivation practices.

In addition to these parameters, global climate change may also be involved; contributing to changes in the distribution of this vector-borne disease (Habekuβ *et al.*, 2009). Insect vectors are primarily affected as increases in temperatures foster the colonization of new environments and/or hosts.

As recent publications suggest, the distribution of WDD is still increasing. Field surveys monitoring WDD must be carried out regularly, especially in cereal-growing regions where wheat dwarf has yet to be reported. This would help to identify region(s) where the disease is emerging.

An analysis of wild reservoir hosts (i.e. wild Poaceae, see Host range of *Wheat dwarf virus* isolates) has revealed low frequencies of WDV-infected plants, suggesting their minor role as a source of the virus in cereal fields and in WDV epidemics. Self-sown cereals, host to numerous insects and pathogens (Manurung *et al.*, 2005), appear to provide a stable reservoir for the virus and play some role in the epidemiology of WDD (Manurung *et al.*, 2004). The secondary spread of the virus by vector populations produced within the field is also important from an epidemiological point of view. Surveys performed in Sweden show that the primary infection of a field at a rate of 5% can lead to high incidence of up to 50% of the disease

at the end of the growing season (Lindblad and Sigvald, 2004). Additionally, some of the recent innovations in cultivation practices expose the crops to higher risks of primary WDV infections. For example, early sowing leads to the production of young plants during periods of favourable conditions (e.g. high temperatures) and the movement of insects within and between fields, which in turn increases the probability for young plants to be visited by viruliferous vectors. Simultaneously, early sowing is associated with longer periods of exposure of emerging cereal plants to insects. This leads to the production of vector progenies, that is, eggs that are laid on the visited young plants overwinter on leaves and hatch the following growing season. These progenies invariably contribute to an increase in the rate of infection in the field in the following spring. These data show that cultivation practices strongly impact both primary (from reservoirs to the newly sown cereal fields) and secondary (between host plants within a field) spread of WDD and must be considered as key parameters in the epidemiology of WDD. In addition to cultivation practices, other parameters such as the susceptibility of recently deployed cereal genotypes can play a role in both the emergence and the spread of wheat dwarf in cereal-growing regions.

3.3 Main Characteristics of the Causal Agent of Wheat Dwarf Disease: The *Wheat dwarf virus* (es)

The pathogen responsible for WDD belongs to a complex of viruses originally described as *Wheat dwarf virus* (family *Geminiviridae*, genus *Mastrevirus*; Fig. 3.2). Members of the *Geminiviridae* family are classified into seven genera (i.e. *Becurtovirus*, *Begomovirus*, *Curtovirus*, *Eragrovirus*, *Mastrevirus*, *Topocuvirus* and *Turncurtovirus*) according to their genome organization, host range, insect vector and genomic sequence (Fauquet *et al.*, 2003; Bernardo *et al.*, 2013; Varsani *et al.*, 2014). Viral species belonging to the *Mastrevirus* genus: (i) have a monopartite single-stranded (ss)

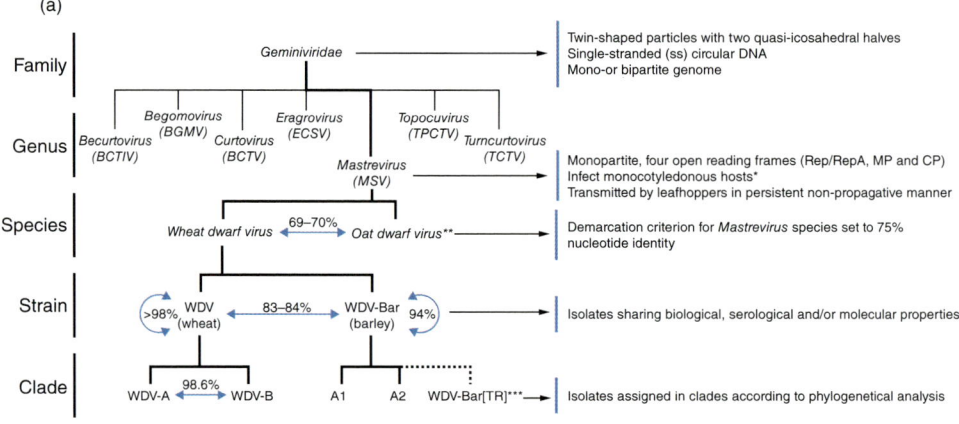

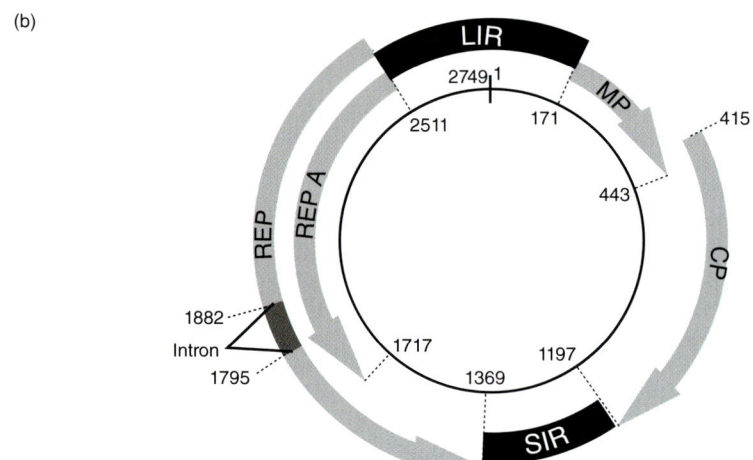

Fig. 3.2. Classification and genomic organization of *Wheat dwarf virus* (WDV). (a) Viruses belonging to the *Geminiviridae* family are classified into seven genera (i.e. *Becurtovirus, Begomovirus, Curtovirus, Eragrovirus, Mastrevirus, Topocuvirus* and *Turncurtovirus*) according to their molecular and biological characteristics (genome organization, host range and insect vector). The WDV species: (i) is a *Mastrevirus* member; (ii) consists of two main strains (wheat and barley); and (iii) contains clades of isolates. The genomic organization of a WDV wheat strain isolate is presented (b). The black circle illustrates the circular ssDNA genome (2749 bases). The four ORFs are presented by arrows with the names of the corresponding encoded proteins (MP, movement protein; CP, coat protein; Rep/RepA, replicase). The putative ORF Cx proposed by Kvarnheden *et al.* (2002) is not presented. The long and short intergenic sequences (LIR and SIR, respectively) involved in replication and expression of the viral genome are shown. Nucleotide positions of ORFs, LIR, SIR, and of the intron located within Rep sequence are according to Enkoping 1 sequence (GenBank accession number AJ311031). Type members of the seven genera of the *Geminiviridae* family are: *Beet curly top Iran virus, Bean golden yellow mosaic virus, Beet curly top virus, Eragrostis curvula streak virus, Maize streak virus, Tomato pseudo-curly top virus* and *Turnip curly top virus*. The percentage of nucleotide similarity between species, strains and clades are presented. *, except *Bean yellow dwarf virus, Tobacco yellow dwarf virus, Chickpea chlorosis virus* and *Chickpea redleaf virus*; **, *Oat dwarf virus* is a *Mastrevirus* species causing wheat dwarf disease, at least on oat plants; ***, WDV-Bar[TR] is a recombinant isolate between a barley strain isolate and a still unknown *Mastrevirus* member.

circular DNA genome of 2.6–2.8 kb in length packaged in twin-shaped particles with two quasi-icosahedral halves (Zhang *et al.*, 2001; Boulton, 2002); (ii) mostly infect monocotyledonous hosts, but there are several exceptions such as the recognized species *Tobacco yellow dwarf virus* (Morris *et al.*, 1992) and *Chickpea redleaf virus* (Thomas *et al.*, 2010) and the proposed species *Chickpea chlorosis virus* (Thomas *et al.*, 2010) that can infect susceptible dicot hosts; and (iii) are transmitted by leafhoppers in a persistent non-propagative manner. The WDV genome is about 2.75 kb (Macdowell *et al.*, 1985), is encapsidated in a particle that is 30 nm in length and 18 nm in width. The WDV genome encodes four proteins. The two replication-associated proteins (Rep and RepA) are both expressed from the complementary sense transcript. The presence of an intron in the Rep gene leads to the production of the two different forms of the replication protein (Accotto *et al.*, 1989; Schalk *et al.*, 1989; Mullineaux *et al.*, 1990; Wright *et al.*, 1997). RepA is translated directly from the native RNA transcript, whereas a splicing step of the mRNA molecule is required to allow for the production of the Rep protein. Consequently, Rep and RepA proteins have identical N-terminal sequences. The movement protein (MP) and the coat protein (CP) are translated from the virion-sense transcript. The MP is involved in systemic infection of hosts and the CP is involved in different functions in the viral infection cycle including the virus/vector interactions involved in transmission (Wang *et al.*, 2014). A further putative open reading frame was described for some WDV isolates (Kvarnheden *et al.*, 2002). However, it is still unclear whether this putative open reading frame encodes a functional protein. The WDV genome also contains two non-coding sequences corresponding to the long and short intergenic regions (LIR and SIR, respectively). These regions contain sequences that are important for viral replication and for the regulation of gene expression (Hofer *et al.*, 1992). The LIR sequence, located between the 5'-ends of Rep/RepA and MP genes, is involved in the initiation of replication of the genomic (+) DNA strand (Fenoll *et al.*, 1988). Indeed, LIR contains a highly

conserved nonanucleotide motif, partly included in the head of a stem loop structure, which is cleaved by Rep at the 'TAATATT/ AC'-specific sequence to initiate the production of the (+) DNA strand through a rolling-circle replication process (Kammann *et al.*, 1991; Heyraud *et al.*, 1993; Laufs *et al.*, 1995). The synthesis of the complementary minus (–) DNA strand (second strand) starts at the 3'-end of a short complementary primer, packaged in viral particles, capable of hybridizing to a sequence located in the SIR region (Kammann *et al.*, 1991). The double-stranded DNA molecule obtained after the synthesis of the minus strand acts as a template for the production of complementary- and virus-sense transcripts. This bidirectional transcription is controlled by promoter (transcription initiation step) and terminator (transcription termination step) sequences present in the LIR and SIR regions, respectively (Morris-Krsinich *et al.*, 1985; Dekker *et al.*, 1991).

Phylogenetic analyses carried out with WDV sequences (currently 228 at the GenBank public database as 13 December 2014) from isolates sampled on different host species showed that WDV: (i) forms a clade distinctly separate from the other *Mastrevirus* species; and (ii) consists of several strains including the originally described wheat (WDV) and the barley (WDV-Bar) strains (Lindsten and Vacke, 1991; Commandeur and Huth, 1999). Findings of the field surveys carried out in Sweden have allowed for the description of wheat strains of WDV (Schubert *et al.*, 2014). It was suggested that different WDV strains are not homogeneously present in the areas known to be affected by WDV. The WDV barley and wheat strains share 83–84% nucleotide similarity (Koklu *et al.*, 2007) with LIR and SIR being the most variable portions of the WDV genomes (Schubert *et al.*, 2007). As the demarcation criterion for *Mastrevirus* is set at 78% and 94% nucleotide similarity for species and strains, respectively (Muhire *et al.*, 2013), wheat and barley strains are currently considered as members of the same WDV species. The biological significance of the 16–17% nucleotide difference observed between strains has yet to be determined.

Barley WDV isolates present a minimum of 94% nucleotide similarity between them, while wheat WDV isolates show at least 98% of sequence similarity (Kvarnheden *et al.*, 2002; Schubert *et al.*, 2007; Ramsell *et al.*, 2008; Kundu *et al.*, 2009; Liu *et al.*, 2012). The high degree of similarity between the genomes of isolates originating in wheat was also observed in comparative analyses performed using sequences from isolates sampled over a 30-year period, suggesting a high degree of conservation of genomic sequences of WDV wheat isolates over time (Schubert *et al.*, 2014). Molecular characteristics of WDV isolates sampled from wild Poaceae plants were described as being similar to those present in cultivated crops. However, it appears that this population of WDV can be divided into two clades (A and B) of the WDV wheat strain with a mean nucleotide difference of 1.4% (Ramsell *et al.*, 2008). No clear correlation was observed between these clades and the geographical origin of the studied WDV isolates (Liu *et al.*, 2012). A similar approach performed with WDV barley isolates showed that they can be divided into three subtypes (central Europe, Spain and Turkey) based on genetic relationships and geographical origin (Koklu *et al.*, 2007). Finally, based on analyses performed with sequences corresponding to the SIR region, it was demonstrated that WDV barley isolates can also be grouped into the clusters A1 and A2 (Schubert *et al.*, 2014). In more recent phylogenetic analyses carried out to refine the *Mastrevirus* species classification, Muhire and co-workers (2013) proposed a revision of the categories of WDV isolates into five strain groups (A to E), assigning WDV strains that preferentially infect barley and wheat, respectively, to groups A and C.

It has been shown that diverse populations of WDV can be found within a single host (Schubert *et al.*, 2014). The absence of antagonism between WDV isolates in the same host might favour recombination between viral sequences during host infection. The WDV-Bar[TR] isolate, a variant of the WDV barley strain, has been described from infected barley in Turkey (Koklu *et al.*, 2007). The genome organization of WDV-Bar[TR] is the same as a typical *Mastrevirus*

member, and has 83–84% nucleotide similarity to wheat strains of WDV. Complete analysis of the WDV-Bar[TR] genomic sequences highlighted its recombinant status. Indeed, WDV-Bar[TR] appears to be a barley WDV isolate with part of its genome corresponding to a new WDV-like *Mastrevirus* species (Ramsell *et al.*, 2009). From the numerous complete WDV sequences deposited in public databases, the WDV-Bar[TR] isolate might represent the first case of recombination suspected for this viral species. This contrasts with results associated with other members of the genus *Mastrevirus*, such as *Maize streak virus* species where numerous putative recombinant isolates have been reported (Owor *et al.*, 2007). However, analyses of sequence alignments performed by Schubert and co-workers (2014) suggest that, in addition to the recombination identified in the WDV-Bar[TR] genome, other recombinant WDV isolates from barley exist. Visual inspection of sequence alignments of genomic data from field isolates highlighted regions of the viral genome with short (few nucleotides) recombination patterns between wheat/barley strains. Analysis of these recombination patterns suggests the replacement of barley strain sequences by homologous sequences from wheat strains (Schubert *et al.*, 2014). In addition to wheat/barley chimeric isolates, intra-specific recombinant genomes were detected with two WDV wheat strains sampled in China (Wu *et al.*, 2008). It is important to note that defective forms of wheat and barley strains, containing at least part of the SIR and LIR sequences, have also been identified in WDV-infected plants (MacDonald *et al.*, 1988; Schubert *et al.*, 2014). Based on sequences of isolates collected during surveys of cereal fields, a new *Mastrevirus* species, the *Oat dwarf virus* (ODV) closely related to the WDV species but distinct from wheat and barley strains, has been described in *Avena fatua* in Germany. This virus seems to be one of the causal agents of WDD on oat (Schubert *et al.*, 2007). The genome of *Oat dwarf virus* isolate shares only 69–70% nucleotide sequence similarity with the wheat and barley strains of WDV. This characteristic, below the demarcation criterion established for *Mastrevirus* species (Fauquet *et al.*, 2008), allows for the assignment of *Oat dwarf virus* as a new viral species in the genus *Mastrevirus*.

3.4 Host Range of *Wheat dwarf virus* Isolates

As with most of the described members of the genus *Mastrevirus*, WDV infects monocotyledonous plants (ICTV Report, 2012; Muhire *et al.*, 2013). The viruses of the WDV species are able to infect a wide range of hosts belonging to the family Poaceae, such as the economically important cereals like wheat, barley, oat and rye, as well as maize and many wild grasses, including *Apera spicaventi*, *A. fatua*, *Lolium multiflorum* and *Poa pratensis* (Vacke, 1972; Lindsten and Vacke, 1991; Vacke and Cibulka, 1999; Fuchs *et al.*, 2001). The host range of isolates from WDV wheat and barley strains overlap, but they seem to use wheat and barley as their preferred cereal host, respectively. The WDV wheat strain induces symptoms in wheat similar to those induced by WDV barley strain in barley (Ramsell *et al.*, 2009). However, the ability of each strain to infect the other's preferred host is a matter of debate. Some epidemiological data associated with WDD indicate that WDV strains of the barley do not infect wheat plants (Tobias *et al.*, 2011), whereas field surveys reported WDV barley strains in wheat and WDV wheat strains in barley (Commandeur and Huth, 1999; Schubert *et al.*, 2007; Koklu *et al.*, 2007; Kundu *et al.*, 2009). In order to test the ability of WDV strains to infect the main cereal species, the host range of each strain has been tested using leafhopper-mediated transmission experiments. Results showed that under experimental conditions, the WDV barley strain was not able to infect wheat plants, while barley plants were infected by the WDV wheat strain (Lindsten and Vacke, 1991).

These data are in accordance with molecular analyses carried out on samples from field surveys. They show that barley strains are preferentially found on barley, whereas wheat strains can be described on both wheat and barley hosts. However, these results were not confirmed by a recent host range study carried out using an agroinfectious clone of a barley strain. Indeed, in this work, Ramsell and co-workers (2009) showed that, in addition to barley, rye and oat, the barley strain used in the study successfully infected wheat plants. Moreover, it has been shown that sequences of barley and wheat strains sampled from wheat and barley plants, respectively, are similar to those of barley and wheat strains sampled in barley and wheat, respectively (Wang *et al.*, 2008). These different reports seem to suggest that barley strains are associated with low infection rates on wheat. However, once the infection occurs, extreme dwarfing (Wang *et al.*, 2008) and a high mortality of infected wheat plants can occur (Ramsell *et al.*, 2009). Altogether, these data suggest that both strains can infect, at least occasionally, the other's preferred host.

3.5 Transmission of the Virus

Wheat dwarf virus is transmitted from plant to plant by the leafhopper *Psammotettix alienus* (Dahlbom) (Hemiptera: Cicadellidae: Deltocephalinae), a holarctic species commonly found in cereal fields and grasslands (Raatikainen and Vasarainen, 1973; Lindblad and Areno, 2002). The leafhopper-mediated transmission of WDV occurs in a circulative, non-propagative manner (Brault *et al.*, 2010). This means that the viruses ingested by leafhoppers during feeding on infected plants: (i) pass through different organs and tissues (e.g. midgut) of the insect vector; (ii) are released into the haemolymph; and (iii) eventually the salivary glands. The viruses can then be inoculated into healthy plant hosts when the insect salivates during feeding and initiates new infections. The time required for the virus to complete this cycle is termed the latency period. It has been recently demonstrated that WDV has a very short latency period in leafhopper vectors, because during the first minutes of acquisition of viral particles, the latter can pass directly through the sheath of the filter chamber to reach the salivary glands and be readily transmissible after 5 minutes following a 5-minute acquisition access period (Wang *et al.*, 2014). This transient direct transfer of particles to salivary glands lasts for a few minutes and is then followed by the standard circulative, non-propagative pathway (Brault *et al.*, 2010) recruiting the anterior and middle midgut organs of the leafhopper.

Once the virus is acquired by the vector, the viruliferous leafhopper can transmit WDV particles to new hosts during each feeding episode on a plant. As viral particles are not lost at moulting, they persist in the vector for its entire lifespan. However, even though the virus directly interacts with different organs of the insect, it is generally accepted that WDV does not multiply in the vector and is not transmitted to the progeny produced by viruliferous female leafhoppers (i.e. there is no vertical transmission of viral particles). However, this has yet to be unequivocally established for the *Wheat dwarf virus–Psammotettix* pathosystem. When an egg hatches, non-viruliferous leafhopper larva starts to feed on plants and consequently there is the opportunity to acquire WDV from infected hosts. Leafhoppers are known for their high capacity to spatially spread. Even at low density, there is the possibility of viruliferous insects inoculating a large number of host plants. The host preference of *P. alienus* is not known, but it is able to transmit the virus to many commonly occurring wild grasses and cultivated cereals (see Host range of *Wheat dwarf virus* isolates). *P. alienus* is able to acquire both wheat and barley strains of WDV and to transmit them to cereals (Manurung *et al.*, 2004).

3.5.1 *Psammotettix* vectors

Different criteria linked to the morphological characteristics of the head, the abdomen and the wings of the insects, can be easily used for the identification of leafhoppers belonging to the genus *Psammotettix* (Vilbaste, 1982; Della Giustina, 1989). However, accurate identification of *Psammotettix* species requires the morphological description of the male genitalia (i.e. characteristics of

aedeagus) that does not allow for the assignment of females to *Psammotettix* species. Numerous *Psammotettix* species have been described in the literature. However, contradictory information is published on the description of some species such as *P. alienus*, *P. striatus* and *P. provincialis*. While most published data indicate that *P. alienus* (Dahlb.) is the WDV vector (e.g. Zhang *et al.*, 2010), some authors have documented transmission with *P. provincialis* (Ribaut) (Ekzayez *et al.*, 2011). Due to the complex taxonomy of the species belonging to this genus, leafhopper species used in WDV studies are often poorly described. This could lead to some conflicting results, especially for those linked to the efficiency of WDV transmission. Descriptions of *Psammotettix* species using molecular tools (e.g. COI; Le Roux and Rubinoff, 2009) or vibrational communications data (Derlink *et al.*, 2014) have to be considered in order to improve our knowledge on the role of each *Psammotettix* species in the epidemiology of WDV. Indeed, if different leafhopper species can transmit the WDV, these data would be very important for the future analysis of the WDD epidemiology.

The life cycle of *P. alienus* has been well studied by Manurung *et al.* (2005) and Guglielmino and Virla (1997). The cycle begins in autumn with eggs laid by adult gravid females in the mesophyll of cereal leaves (Fig. 3.3). The number of eggs produced by a leafhopper female is influenced by different parameters, including temperature and plant host species. Leafhoppers overwinter in the egg stage. The embryonic development of eggs occurs when the temperature and day length increase. Egg development lasts for 18 days at 20°C, 70–95% relative humidity, 18 hours of light/day, and ends with the hatching of a larva (L1 stage). This larva evolves through five successive stages (from L1 to L5). Under optimal conditions, the complete development of the larva needs 32 days. The last step for leafhopper development (i.e. the production of male and female adults), is reached in early summer. Asexual reproduction, widely used by aphids, does not occur in leafhoppers. Fecundation and first egg laying can occur

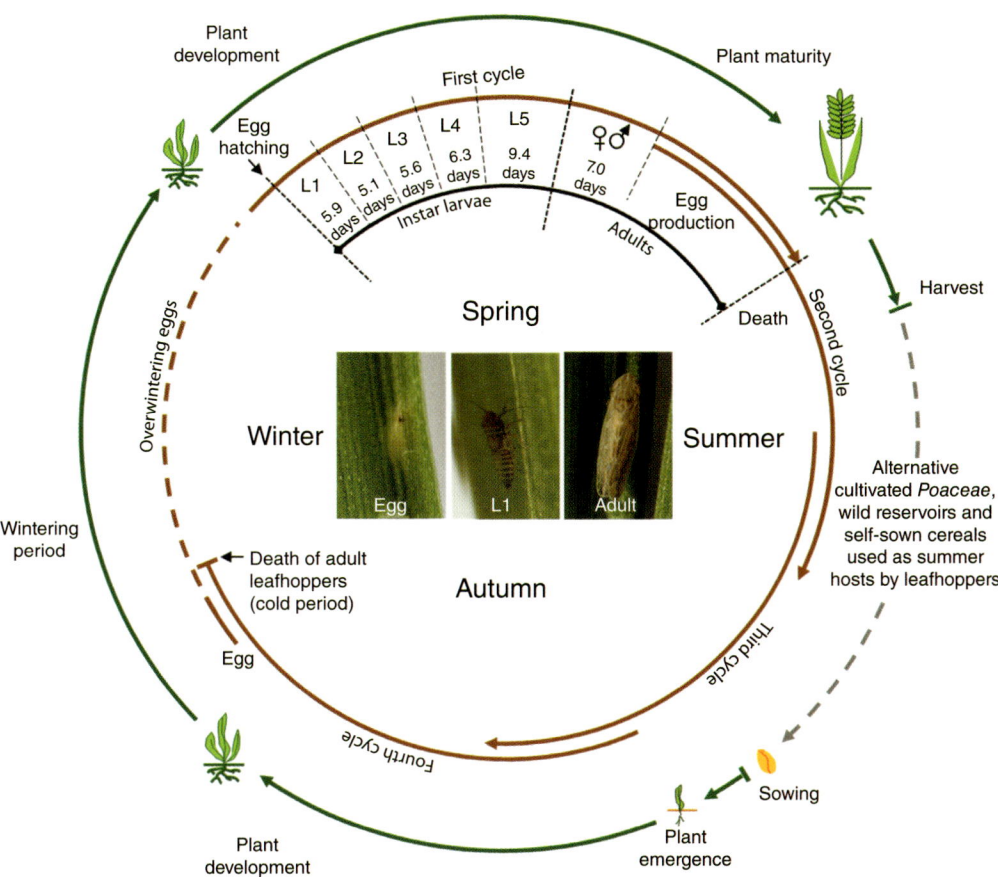

Fig. 3.3. Schematic representation of winter cereal and *Psammotettix alienus* life cycles. The main development stages (from sowing to harvest) of a winter cereal are presented by the outer circle. The four successive and the overlapping biological cycles of *P. alienus* are presented by overlapping arrows of the inner circle. Under optimal conditions (20°C, 70–95% relative humidity, 18/6 light/dark hours), the length of *Psammotettix* life cycles (from egg to the death of the adult) is 71 days (Guglielmino and Virla, 1997). Eggs produced in autumn overwinter on cereals and hatch the following growing season (next spring). The five larval stages (L1 to L5) last, according to Manurung *et al.* (2005), 5.9, 5.1, 5.6, 6.3 and 9.4 days, respectively. The 7-day-old adults can copulate to produce the next generation of insects. Leafhoppers are mainly present in cultivated cereal hosts during the favourable periods (spring, early summer and autumn), but they also use, during the summer period, alternative cultivated or wild hosts as denoted by the dashed line present in the outer circle of the diagram.

after the seventh day of the adult stage. Consequently, the duration of a complete *P. alienus* life cycle from egg to egg is about 58 days. Several studies connected to the dynamics of *Psammotettix* populations have indicated that the density of individuals can reach up to 43 adults/m² (Manurung *et al.*, 2005), and that the sex ratio of adult populations is close to 1.

Environmental conditions of cereal-growing regions in France allow leafhoppers to achieve four complete life cycles during the spring-to-autumn period, whereas in Northern Europe and in north-western China only two leafhopper generations are produced per year (Schiemenz, 1969). Larvae and adults can be found on young actively growing crops in spring and in summer on maturing crops. Adults are also found in autumn on newly sown crops. The biological activity of the adults declines with decreasing temperatures (Lindblad and Areno, 2002). Finally, all individuals die during the cold winter season.

Adults and the different larval stages are able to efficiently transmit WDV (Lindsten and Vacke, 1991; Abt et al., 2015). However, it is important to note that plant-to-plant movement of larvae in the fields is more important than that of winged leafhopper adults. This characteristic impacts the dynamics of disease spread. Indeed, even at a low density, these insects, and especially the larvae, can cause important yield losses by transmitting the virus to the numerous host plants they visit. An analysis carried out on the interaction between leafhoppers and other insects revealed that the presence of aphids in cereal fields modifies the behaviour of leafhoppers leading to an increase in the number of plants visited by each individual (Alla et al., 2001). This antagonistic interaction between leafhoppers and aphids, two insects commonly found simultaneously in cereal fields, indirectly optimizes the spread of WDV.

3.6   Diagnostic Methods for Wheat Dwarf Disease

The symptoms associated to WDD cannot be easily used to accurately identify the presence of WDV infections. Moreover, as the host range of WDV barley and wheat strains overlap (see Host range of *Wheat dwarf virus* isolates), the use of biological approaches (i.e. pathotyping) is not recommended for accurate assignment of WDV isolates in their appropriate strain designations. Different serological and molecular techniques are available for the detection of strains in host plant or vector samples. Characteristics of viral compounds (i.e. capsid proteins and nucleic acids) generally allow for the differentiation between WDV strains. In serological tests, the high level of sequence similarity between the CP of WDV isolates do not allow for distinction between viral strains using polyclonal antisera (Vacke and Cibulka, 2000),

but this specific characterization can be achieved with monoclonal antibodies (Rabenstein et al., 2005). Several well-established molecular methods including standard PCR (Commandeur and Huth, 1999; Schubert et al., 2002), PCR-restriction fragment length polymorphism (RFLP) (Kundu et al., 2009) and rolling-circle amplification-restriction fragment length polymorphism (Schubert et al., 2007) have been utilized in the identification of WDV strain-specific sequences. Moreover, a recent description of a direct sequencing procedure applied to rolling-circle amplification products provides an alternative for rapid detection and characterization of WDV isolates (Schubert et al., 2007). The list of detection tools for WDV diagnosis has been recently completed by the isothermal recombinase polymerase amplification procedure (Glais and Jacquot, 2015) and by molecular-based quantification assays. Real-time PCR assay using a TaqMan probe and targeting a conserved region of the CP gene sequence has been published (Zhang et al., 2010). This tool allows for the detection of 30 copies of the WDV genome in wheat tissues and in leafhoppers. Real-time PCR tools for the specific detection, characterization and quantification of WDV wheat and barley strains have also been published (Gadiou et al., 2012). The genomic region targeted by these assays encompasses the 12 nucleotide sequence (located in the Rep gene) known to be deleted in the genome of WDV barley strains. Depending on the primers and probe used in the assay, quantification protocols make it possible to target specific WDV strains in samples.

3.7   Management Strategies

Very few options are available for farmers to control WDV in fields. Due to the lack of WDV-resistant and WDV-tolerant sources in cultivated cereals, WDV protection relies on agrotechnical measures, and on occasion the use of chemical treatments (i.e. insecticides to control vectors).

Grasslands are mostly composed of closely related species within the family Poaceae. These WDV host plants increase the risk for WDD spread between infected weeds and cereal crops. Self-sown cereals are important for maintaining leafhopper populations and they act as virus reservoirs. Ploughing and the removal of self-sown cereals is an important means of reducing leafhopper populations, and consequently, the risk associated with WDD. Vacke (1972) reported that young plants are more susceptible to WDD than mature plants. When inoculated at different stages of development, winter wheat plants were observed to develop mature plant resistance at the first node stage (stage DC31) (Lindblad and Sigvald, 2004). As most plants are susceptible to virus infection only during certain periods of their development, the sowing time of a crop and the vector movement between fields are key components of integrated management strategies. This suggests that late sowing and/or the use of cultivars with rapid development (i.e. reaching maturity stage that is associated with virus resistance more quickly) would help to protect against WDD infections (Lindblad and Waern, 2002).

Insecticides can be directly applied on seeds before sowing or sprayed on leaves during the growing season. While insecticide treatments benefit the crops by reducing the number of insects in the protected areas and prevents the spread of insect-transmitted viruses, their use causes severe impacts on the environment (pollution, detrimental effects on non-target beneficial insect species, and selection for resistance against the chemicals) and represents a financial burden. Due to these direct and indirect costs, several ecological, agronomical and economical parameters must be taken into account before using any insecticide-based control strategy. However, to prevent excessive use of chemicals, it is necessary to accurately identify when and how cultivated areas are exposed to the targeted virus. Thus, it is important to acquire knowledge on biotic (variability of viruses, diversity of vector species, susceptibility of host plants, characteristics of virus/vector/host interactions) and abiotic (environment characteristics) parameters involved in the prevalence and the spread of the disease.

Efforts have been made to identify WDD-resistant germplasm among the available diverse genetic wheat and barley accessions. In fact, cultivation of resistant cultivars is one of the cheaper and more environmentally friendly methods for managing viral diseases. Numerous varieties of winter wheat have been tested for their ability to resist WDV infections. However, only minor quantitative differences between the tested hosts and susceptible reference genotypes have been reported so far (Habekuß et al., 2009). Screening for resistance based on yield reductions has been applied to winter wheat genotypes, which has allowed for the definition of groups of susceptible genotypes (Sirlova et al., 2005). The least susceptible group was, however, associated with 78% yield reduction when infected by WDV. Most of the tested materials were susceptible to WDV infection (Vacke and Cibulka, 2000). Only rare genotypes have been classified as moderately resistant. One of the more recently conducted screening procedures in winter wheat for the identification of WDV-resistant genotypes revealed that infected plants of the cultivars 'MvDalma' and 'MvVekni' remained asymptomatic. Moreover, the viral titre of the infected plants, as determined by real-time PCR technique, was lower in these two wheat cultivars than in the susceptible reference hosts. These two cultivars may represent the first sources of resistance described in wheat for WDV (Benkovics et al., 2010).

3.8 Concluding Remarks

Research carried out on wheat dwarf during the last three decades has allowed for the description of the main characteristics of the pathogens and vectors involved in WDD. So far, the wheat dwarf pathosystem appears to be simpler than the better described barley yellow dwarf pathosystem, another important vector-borne cereal disease that is induced by many different virus species (barley yellow dwarf viruses, *Wheat yellow dwarf virus* and cereal yellow dwarf viruses) and is transmitted by several aphid vectors (e.g. *Rhopalosiphum padi, Sitobion avenae, Metopolophium dirhodum*) (Krueger *et al.*, 2013). However, the epidemiology of WDD is more complex than initially proposed with: (i) the presence of strains, recombinants and distinct viral species, all capable of inducing WDD; (ii) the complex taxonomy of the vector(s), *Psammotettix* spp.; and (iii) the contradictory reports on WDV host range. Moreover, new wheat dwarf-like virus variants, including emerging recombinants, presumably exist in reservoirs and/or cultivated crops. The routine use of the latest developed WDV-specific tools in epidemiological studies carried out in the field and cereal-growing areas will improve knowledge on the sanitary status of cultivated hosts and leafhopper vectors. The generation of such data is highly important for forecasting the spread of this viral disease, and consequently, for the development of effective management methods against wheat dwarf.

Acknowledgments

The authors thank Anders Kvarnheden, Christophe Lacomme, Annie De Keyzer and Marc Letroublon for critical reading of the manuscript. The plant material used to produce data presented in Fig. 3.1 was collected from cereal fields in France by Bayer Crop Science technicians.

References

Abt, I., Julian, C., Souquet, M., Derlink, M., Mabon, R., *et al.* (2015) Characterization of *Psammotettix* species, vectors of *Wheat dwarf virus*. XVth Plant Virology Meeting, 18–22 January, Aussois, France.

Accotto, G.P., Donson, J. and Mullineaux, P.M. (1989) Mapping of *Digitaria streak virus* transcripts reveals different RNA species from the same transcription unit. *The EMBO Journal* 8, 1033–1039.

Achon, M.A. and Serrano, L. (2006) First detection of *Wheat dwarf virus* in barley in Spain associated with an outbreak of Barley yellow dwarf. *Plant Disease* 90, 970.

Alla, S., Moreau, J.P. and Frerot, B. (2001) Effects of the aphid *Rhopalosiphum padi* on the leafhopper *Psammotettix alienus* under laboratory conditions. *Entomologia Experimentalis et Applicata* 98, 203–209.

Behjatnia, S.A.A., Afsharifar, A.R., Tahan, V., Motlagh, M.H.A., Gandomani, O.E., *et al.* (2011) Widespread occurrence and molecular characterization of *Wheat dwarf virus* in Iran. *Australasian Plant Pathology* 40, 12–19.

Bendahmane, M., Schalk, H.J. and Gronenborn, B. (1995) Identification and characterization of *Wheat dwarf virus* from France using a rapid method for geminivirus DNA preparation. *Phytopathology* 85, 1449–1455.

Benkovics, A.H., Vida, G., Nelson, D., Veisz, O., Bedford, I., *et al.* (2010) Partial resistance to *Wheat dwarf virus* in winter wheat cultivars. *Plant Pathology* 59, 1144–1151.

Bernardo, P., Golden, M., Akram, M., Naimuddin, Nadarajan, N., *et al.* (2013) Identification and characterisation of a highly divergent geminivirus: evolutionary and taxonomic implications. *Virus Research* 177, 35–45.

Bisztray, G., Gaborjanyi, R. and Vacke, J. (1989) Isolation and characterization of *Wheat dwarf virus* found for the first time in Hungary. *Journal of Plant Diseases and Protection* 96, 449–454.

Boulton, M.I. (2002) Functions and interactions of mastrevirus gene products. *Physiological and Molecular Plant Pathology* 60, 243–255.

Brault, V., Uzest, M., Monsion, B., Jacquot, E. and Blanc, S. (2010) Aphids as transport devices for plant viruses. *Compte Rendus Biologies* 333, 524–538.

Commandeur, U. and Huth, W. (1999) Differentiation of strains of *Wheat dwarf virus* in infected wheat and barley plants by means of polymerase chain reaction. *Journal of Plant Diseases and Protection* 106, 550–552.

Conti, M. (1993) Leafhopper-borne plant viruses in Italy. *Memorie della Societá Entomologica Italiana* 72, 541–547.

Dekker, E.L., Woolston, C.J., Xue, Y.B., Cox, B. and Mullineaux, P.M. (1991) Transcript mapping reveals different expression strategies for the bicistronic RNAs of the *Geminivirus Wheat dwarf virus*. *Nucleic Acids Research* 19, 4075–4081.

Della Giustina, W. (1989) *Homoptères Cicadellidae 3: compléments Faune de France 73*. Fédération Française des Sociétés de Sciences Naturelles, INRA éditions, Paris, France.

Derlink, M., Pavlovcic, P., Stewart, A.J.A. and Virant-Doberlet, M. (2014) Mate recognition in duetting species: the role of male and female vibrational signal. *Animal Behaviour* 90, 181–193.

Ekzayez, A.M., Kumari, S.G. and Ismail, I. (2011) First report of *Wheat dwarf virus* and its vector (*Psammotettix provincialis*) affecting wheat and barley crops in Syria. *Plant Disease* 95, 76–76.

Fauquet, C.M., Bisaro, D.M., Briddon, R.W., Brown, J.K., Harrison, B.D., *et al.* (2003) Revision of taxonomic criteria for species demarcation in the family *Geminiviridae*, and an updated list of begomovirus species. *Archives of Virology* 148, 405–421.

Fauquet, C.M., Briddon, R.W., Brown, J.K., Moriones, E., Stanley, J., *et al.* (2008) Geminivirus strain demarcation and nomenclature. *Archives of Virology* 153, 783–821.

Felix, I., Larcher, J.M., Maraby, J., Philippeau, G. and Vinatier, K. (1992) Risques d'attaques de cicadelles et conditions d'efficacité des insecticides. *Perspectives Agricoles* 173, 98–106.

Fenoll, C., Black, D.M. and Howell, S.H. (1988) The intergenic region of *Maize streak virus* contains promoter elements involved in rightward transcription of the viral genome. *The EMBO Journal* 7, 1589–1596.

Fohrer, F., Lebrun, I. and Lapierre, H. (1992) Acquisitions récente sur le virus du nanisme du blé. *Phytoma* 443, 18–20.

Fuchs, E., Mehner, S., Manurung, B. and Grüntzig, M. (2001) Strain-spectrum of *Wheat dwarf virus* (WDV) in Saxony-Anhalt. In: *Proceedings of the Fifth Congress of the European Foundation for Plant Pathology*. Taormina–Giardini Naxos, Italy, 18–22 September 2000 (CD Rom), pp. 213–215.

Gaborjanyi, R., Vacke, J. and Bisztray, G. (1988) *Wheat dwarf virus*: a new cereal pathogen in Hungary. *Növénytermelés* 37, 495–500.

Gadiou, S., Ripl, J., Janourova, B., Jarosova, J. and Kundu, J.K. (2012) Real-time PCR assay for the discrimination and quantification of wheat and barley strains of *Wheat dwarf virus*. *Virus Genes* 44, 349–355.

Glais, L. and Jacquot, E. (2015) Detection and characterization of viral species/sub-species using isothermal Recombinase Polymerase Amplification (RPA) assays. In: Lacomme, C. (ed.) *Methods in Molecular Biology, Plant Pathology Techniques and Protocols*. Humana Press Inc., in press.

Guglielmino, A. and Virla, E.G. (1997) Postembryonic development and biology of *Psammotettix alienus* (Dahlbom) (Homoptera, Cicadellidae) under laboratory conditions. *Bollettino di Zoologia Agraria e di Bachicoltura* 29, 65–80.

Habekuß, A., Riedel, C., Schliephake, E. and Ordon, F. (2009) Breeding for resistance to insect-transmitted viruses in barley- an emerging challenge due to global warming. *Journal fur Kulturpflanzen* 61, 53–61.

Heyraud, F., Matzeit, V., Schaefer, S., Schell, J. and Gronenborn, B. (1993) The conserved nonanucleotide motif of the geminivirus stem-loop sequence promotes replicational release of virus molecules from redundant copies. *Biochimie* 75, 605–615.

Hofer, J.M.I., Dekker, E.L., Reynolds, H.V., Woolston, C.J., Cox, B.S., *et al.* (1992) Coordinate regulation of replication and virion sense gene expression in *Wheat dwarf virus*. *The Plant Cell* 4, 213–223.

Huth, W. (2000) Viruses of gramineae in Germany – a short overview. *Journal of Plant Diseases and Protection* 107, 406–414.

ICTV Report (2012) *Virus taxonomy: classification and nomenclature of viruses: Ninth Report of the International Committee on Taxonomy of Viruses*. King, A.M.Q., Adams, M.J., Carstens, E.B. and Lefkowitz, E.J (eds). Elsevier Academic Press, San Diego, California.

Jezewska, J. (2001) First report of *Wheat dwarf virus* occurring in Poland. *Phytopathologia Polonica* 21, 93–100.

Jilaveanu, A. and Vacke, J. (1995) Isolation and identification of *Wheat dwarf virus* (WDV) in Romania. *Protectia Plantelor* 23, 51–62.

Kammann, M., Schalk, H.J., Matzeit, V., Schaefer, S., Schell, J., *et al.* (1991) DNA replication of *Wheat dwarf virus*, a geminivirus, requires 2 cis-acting signals. *Virology* 184, 786–790.

Kapooria, R.G. and Ndunguru, J. (2004) Occurrence of viruses in irrigated wheat in Zambia. *Bulletin OEPP* 34, 413–419.

Koklu, G., Ramsell, J.N.E. and Kvarnheden, A. (2007) The complete genome sequence for a Turkish isolate of *Wheat dwarf virus* (WDV) from barley confirms the presence of two distinct WDV strains. *Virus Genes* 34, 359–366.

Krueger, E.N., Beckett, R.J., Gray, S.M. and Miller, W.A. (2013) The complete nucleotide sequence of the genome of Barley yellow dwarf virus-RMV reveals it to be a new Polerovirus distantly related to other yellow dwarf viruses. *Frontiers in Microbiology* 4, 1–11.

Kundu, J., Gadiou, S. and Cervena, G. (2009) Discrimination and genetic diversity of *Wheat dwarf virus* in the Czech Republic. *Virus Genes* 38, 468–474.

Kvarnheden, A., Lindblad, M., Lindsten, K. and Valkonen, J.P.T. (2002) Genetic diversity of *Wheat dwarf virus*. *Archives of Virology* 147, 205–216.

Lapierre, H., Cousin, M.T., Della Giustina, W., Moreau, J.P., Khogali, M., *et al.* (1991) Nanisme blé: agent pathogéne et vecteur. Description, biologie, interaction. *Phytoma* 432, 26–28.

Laufs, J., Jupin, I., David, C., Schumacher, S., Heyraud-Nitschke, F., *et al.* (1995) Geminivirus replication: genetic and biochemical characterization of Rep protein function, a review. *Biochimie* 77, 765–773.

Le Roux, J.J. and Rubinoff, D. (2009) Molecular data reveals California as the potential source of an invasive leafhopper species, *Macrosteles* sp. nr. *severeni*, transmitting the aster yellows phytoplasma in Hawaii. *Annals of Applied Biology* 154, 429–439.

Lemmetty, A. and Huusela-Veistola, E. (2005) First report of *Wheat dwarf virus* in winter wheat in Finland. *Plant Disease* 89, 912–912.

Lindblad, M. and Areno, P. (2002) Temporal and spatial population dynamics of *Psammotettix alienus*, a vector of *Wheat dwarf virus*. *International Journal of Pest Management* 48, 233–238.

Lindblad, M. and Sigvald, R. (2004) Temporal spread of *Wheat dwarf virus* and mature plant resistance in winter wheat. *Crop Protection* 23, 229–234.

Lindblad, M. and Waern, P. (2002) Correlation of wheat dwarf incidence to winter wheat cultivation practices. *Agriculture Ecosystems and Environment* 92, 115–122.

Lindsten, K. and Lindsten, B. (1999) Wheat dwarf- an old disease with new outbreaks in Sweden. *Journal of Plant Diseases and Protection* 106, 325–332.

Lindsten, K. and Vacke, J. (1991) A possible barley adapted strain of *Wheat dwarf virus* (WDV). *Acta Phytopathologica et Entomologica Hungarica* 26, 175–180.

Liu, Y., Wang, B., Vida, G., Cseplo-Karolyi, M., Wu, B.L., *et al.* (2012) Genomic analysis of the natural population of *Wheat dwarf virus* in wheat from China and Hungary. *Journal of Integrative Agriculture* 11, 2020–2027.

Macdonald, H., Coutts, R.H.A. and Buck, K.W. (1988) Characterization of a subgenomic DNA isolated from *Triticum aestivum* plants infected with *Wheat dwarf virus*. *Journal of General Virology* 69, 1339–1344.

Macdowell, S.W., Macdonald, H., Hamilton, W.D.O., Coutts, R.H.A. and Buck, K.W. (1985) The nucleotide sequence of cloned *Wheat dwarf virus* DNA. *The EMBO Journal* 4, 2173–2180.

Manurung, B., Witsack, W., Mehner, S., Gruntzig, M. and Fuchs, E. (2004) The epidemiology of *Wheat dwarf virus* in relation to occurrence of the leafhopper *Psammotettix alienus* in Middle-Germany. *Virus Research* 100, 109–113.

Manurung, B., Witsack, W., Mehner, S., Gruntzig, M. and Fuchs, E. (2005) Studies on biology and population dynamics of the leafhopper *Psammotettix alienus* Dahlb. (Homoptera: Auchenorrhyncha) as vector of *Wheat dwarf virus* (WDV) in Saxony-Anhalt, Germany. *Journal of Plant Diseases and Protection* 112, 497–507.

Morris, B.A.M., Richardson, K.A., Haley, A., Zhan, X.C. and Thomas, J.E. (1992) The nucleotide sequence of the infectious cloned DNA component of *Tobacco yellow dwarf virus* reveals features of geminiviruses infecting monocotyledonous plants. *Virology* 187, 633–642.

Morris-Krsinich, B.A.M., Mullineaux, P.M., Donson, J., Boulton, M.I., Markham, P.G., *et al.* (1985) Bidirectional transcription of *Maize streak virus* DNA and identification of the coat protein gene. *Nucleic Acids Research* 13, 7237–7256.

Muhire, B., Martin, D.P., Brown, J.K., Navas-Castillo, J., Moriones, E., *et al.* (2013) A genome-wide pairwise-identity-based proposal for the classification of viruses in the genus *Mastrevirus* (family *Geminiviridae*). *Archives of Virology* 158, 1411–1424.

Mullineaux, P.M., Guerineau, F. and Accotto, G.P. (1990) Processing of complementary sense RNAs of *Digitaria streak virus* in its host and in transgenic tobacco. *Nucleic Acids Research* 18, 7259–7265.

Najar, A., Makkouk, K.M., Boudhir, H., Kumari, S.G., Zarouk, R., *et al.* (2000) Viral diseases of cultivated legume and cereal crops in Tunisia. *Phytopathologia Mediterranea* 39, 423–32.

Owor, B.E., Shepherd, D.N., Taylor, N.J., Edema, R., Monjane, A.L., *et al.* (2007) Successful application of FTA® Classic Card technology and use of bacteriophage phi 29 DNA polymerase for large-scale field sampling and cloning of complete *Maize streak virus* genomes. *Journal of Virological Methods* 140, 100–105.

Raatikainen, M. and Vasarainen, A. (1973) Early- and high-summer flight periods of leafhoppers. *Annales Agriculturae Fenniae* 12, 77–94.

Rabenstein, F., Sukhacheva, E., Habekuß, A. and Schubert, J. (2005) Differentiation of *Wheat dwarf virus* isolates from wheat, triticale, and barley by means of a monoclonal antibody. In: *Proceedings of the X Conference on Viral Diseases of Gramineae in Europe*, Louvain-la-Neuve, Belgium, p. 60.

Ramsell, J.N.E., Lemmetty, A., Jonasson, J., Andersson, A., Sigvald, R., *et al.* (2008) Sequence analyses of *Wheat dwarf virus* isolates from different hosts reveal low genetic diversity within the wheat strain. *Plant Pathology* 57, 834–841.

Ramsell, J.N.E., Boulton, M.I., Martin, D.P., Valkonen, J.P.T. and Kvarheden, A. (2009) Studies on the host range of the barley strain of *Wheat dwarf virus* using an agroinfectious viral clone. *Plant Pathology* 58, 1161–1169.

Schalk, H.J., Matzeit, V., Schiller, B., Schell, J. and Gronenborn, B. (1989) *Wheat dwarf virus*, a geminivirus of graminaceous plants needs splicing for replication. *The EMBO Journal* 8, 359–364.

Schiemenz, H. (1969) Die zikadenfauna mitteleuropa aischer Trockenrasen (Homoptera, Auchenorrhyncha). *Entomologische Abhandlungen Museum für Tierkunde in Dresden* 36, 201–280.

Schubert, J., Habekuß, A. and Rabenstein, F. (2002) Investigation of differences between wheat and barley forms of *Wheat dwarf virus* and their distribution in host plants. *Plant Protection Science* 38, 43–48.

Schubert, J., Habekuss, A., Kazmaier, K. and Jeske, H. (2007) Surveying cereal-infecting geminiviruses in Germany – diagnostics and direct sequencing using rolling circle amplification. *Virus Research* 127, 61–70.

Schubert, J., Habekuss, A., Wu, B.L., Thieme, T. and Wang, X.F. (2014) Analysis of complete genomes of isolates of the *Wheat dwarf virus* from new geographical locations and descriptions of their defective forms. *Virus Genes* 48, 133–139.

Sirlova, L., Vacke, J. and Chaloupkova, M. (2005) Reaction of selected winter wheat varieties to autumnal infection with *Wheat dwarf virus*. *Plant Protection Science* 41, 1–7.

Thomas, J.E., Parry, J.N., Schwinghamer, M.W. and Dann, E.K. (2010) Two novel *Mastreviruses* from chickpea, *Cicer arietinum* in Australia. *Archives of Virology* 155, 1777–1788.

Tobias, I., Kiss, B., Bakardjieva, N. and Palkovics, L. (2009) The nucleotide sequence of barley strain of *Wheat dwarf virus* isolated in Bulgaria. *Cereal Research Communications* 37, 237–242.

Tobias, I., Shevchenko, O., Kiss, B., Bysov, A., Snihur, H., *et al.* (2011) Comparison of the nucleotide sequences of *Wheat dwarf virus* (WDV) isolates from Hungary and Ukraine. *Polish Journal of Microbiology* 60, 125–131.

Vacke, J. (1961) Wheat dwarf virus disease. *Biologia Plantarum* 3, 228–233.

Vacke, J. (1972) Host plants range and symptoms of *Wheat dwarf virus*. *Vyzkumnych Ustavu Rostlinné Vyroby Praha–Ruzyne* 17, 151–162.

Vacke, J. and Cibulka, R. (1999) Silky bent grass (*Apera spica–venti* (L.) Beauv.) – a new host and reservoir of *Wheat dwarf virus*. *Plant Protection Science* 35, 47–50.

Vacke, J. and Cibulka, R. (2000) Response of selected winter wheat varieties to *Wheat dwarf virus* infection at an early growth stage. *Czech Journal of Genetics and Plant Breeding* 36, 1–4.

Varsani, A., Navas-Castillo, J., Moriones, E., Hernández-Zepeda, C., Idris, A., *et al.* (2014) Establishment of three new genera in the family *Geminiviridae*: *Becurtovirus*, *Eragrovirus* and *Turncurtovirus*. *Archives of Virology* 159, 2193–2203.

Vilbaste, J. (1982) Preliminary key for the identification of the nymphs of North European Homoptera, Cicadinea. II. Cicadelloidea. *Annales Zoologici Fennici* 19, 1–20.

Wang, X.F., Wu, B. and Wang, J.F. (2008) First report of *Wheat dwarf virus* infecting barley in Yunnan, China. *Journal of Plant Pathology* 90, 400.

Wang, Y.J., Mao, Q.Z., Liu, W.W., Mar, T., Wei, T.Y., *et al.* (2014) Localization and distribution of *Wheat dwarf virus* in its vector leafhopper, *Psammotettix alienus*. *Phytopathology* 104, 897–904.

Wright, E.A., Heckel, T., Groenendijk, J., Davies, J.W. and Boulton, M.I. (1997) Splicing features in *Maize streak virus* virion- and complementary-sense gene expression. *Plant Journal* 12, 1285–1297.

Wu, B., Melcher, U., Guo, X., Wang, X., Fan, L., *et al.* (2008) Assessment of codivergence of mastreviruses with their host plants. *BMC Evolutionary Biology* 8, 335.

Xie, J., Wang, X., Liu, Y., Peng, Y. and Zhou, G. (2007) First report of the occurrence of *Wheat dwarf virus* in wheat in China. *Plant Disease* 91, 111.

Zhang, W., Olson, N.H., Baker, T.S., Faulkner, L., Agbandje-McKenna, M., *et al.* (2001) Structure of the *Maize streak virus* geminate particle. *Virology* 279, 471–477.

Zhang, X., Zhou, G.H. and Wang, X.F. (2010) Detection of *Wheat dwarf virus* (WDV) in wheat and vector leafhopper (*Psammotettix alienus* Dahlb.) by real-time PCR. *Journal of Virological Methods* 169, 416–419.

4 Cassava Brown Streak

James P. Legg,[1*] P. Lava Kumar[2] and Edward E. Kanju[1]

[1]International Institute of Tropical Agriculture, Dar es Salaam, Tanzania; [2]International Institute of Tropical Agriculture, Ibadan, Nigeria

4.1 Introduction: Disease and Symptoms

4.1.1 First reports and disease aetiology

The earliest studies investigating virus diseases of cassava were initiated in the northwestern part of what is now Tanzania during the 1930s. It was during this period that a 'mosaic'-like disease was observed with characteristics that were distinct from cassava mosaic disease (CMD), which had been described several decades previously (Warburg, 1894). The cassava 'brown streak' disease (CBSD), like CMD, appeared to be a graft-transmissible systemic condition, but unlike CBSD, it produced distinctive foliar symptoms that were most prominent on lower mature leaves and was associated with an unusual root rot phenomenon (Storey, 1936). Although Storey (1936) considered the disease to have a viral aetiology, it was not until several decades later that molecular studies identified the causal viruses (Monger et al., 2001a) and Koch's postulates were fulfilled (Winter et al., 2010). For much of its history, CBSD has been a little studied disease. It is only since it began to spread more widely in 2004 that it has attracted more attention from researchers and the broader agricultural community.

4.1.2 Symptoms

Nichols (1950) provided a comprehensive description of CBSD symptoms, which may be expressed on all the major parts of the cassava plant: leaves, stems, roots and fruits (Fig. 4.1). Symptoms are most frequent in leaves, common but less frequent in roots, and occasionally seen on stems and fruits. Symptom expression is highly variable and depends on the susceptibility of the cultivar, the environmental conditions and the age of the plant (Nichols, 1950). All symptom types are more apparent during periods of cooler weather and in more mature plants. Uncertainty about the link between foliar and root symptoms was clarified when a strong association between the two was confirmed in southern Tanzania (Hillocks et al., 1996).

Nichols (1950) recognized two types of leaf symptoms. In the first, a feathery chlorosis is observed in the margins of secondary veins, spreading to tertiary veins as infection progresses and coalescing to produce a more general blotchy chlorosis. The second symptom type is characterized by a more general chlorotic mottling and lacks a clear association with veins. In both cases, there is no distortion of the shape of the leaf lamina – in contrast to symptoms of CMD – and symptoms are typically confined to lower leaves. Seasonality of CBSD symptom expression,

*E-mail: j.legg@cgiar.org

Fig. 4.1. Symptoms of cassava brown streak disease (CBSD). (a) Dark brown lesions on the green portion of the stem of a severely CBSD-affected cassava plant. (b) Drying and death of axillary buds on woody stem tissue. (c) Dieback of shoot tips. (d) Feathery chlorosis. Irregular chlorosis associated with secondary and tertiary veins. (e) Dry brown necrotic rot in tuberous roots characteristic of CBSD in mature cassava plants. (f) Healthy foliage of a CBSD-resistant cultivar.

combined with the fact that symptoms are much less obvious than those of CMD, means that considerably more skill and care are required from agricultural workers or researchers when making field-based assessments of the disease. In order to facilitate this, several symptom guides have been published (CABI, 2014; CRS, 2014). There is no clear difference in CBSD symptoms in the various parts of East and Central Africa in which the disease occurs. However, there is some evidence suggesting that different virus isolates may give rise to contrasting severities of symptom expression (Mohammed *et al.*, 2012).

Root symptoms take the form of dry, corky, necrotic lesions within the cortex of tuberous roots, which enlarge and change colour from yellow to sepia to dark brown as the plant matures (Nichols, 1950). The disease may occasionally also cause radial constrictions or longitudinal fissures, but very often the dry root rot occurs in the absence of any obvious external signs. An important consequence of this symptom is that

farmers are usually not aware that their crop is affected until they cut open the tuberous roots. There is great variability between cultivars in terms of the time of onset of root symptoms and their overall severity, as well as the relative importance of foliar versus root symptoms. Severely affected roots are susceptible to secondary infections by soil-inhabiting fungal or bacterial pathogens, which may ultimately lead to the death of the plant.

The name 'brown streak' is derived from the elongated sepia brown to black lesions which may be present on the green portions of the stems of infected plants. This is the least frequent of the three main symptom types, but is often associated with the more severe infections. As cassava plants mature and stems become increasingly woody, brown streak stem lesions may become less apparent, but damage to this part of the plant may result in the death of axillary buds and necrosis of the stem from the shoot tip downwards, producing a 'dieback' effect. This can have a deleterious effect on the viability of cuttings taken from these stems for use as planting material.

CBSD symptoms are difficult to diagnose through visual assessments, although image-based scoring schemes have been developed to aid the process (Hillocks and Thresh, 2000). Field assessments of CBSD are increasingly being supported with the use of virus diagnostics to improve their accuracy and to enable the detection of infections that are not accompanied by foliar symptom expression (Adams et al., 2013).

4.2 Distribution

4.2.1 Early years (1930s–2004)

Following the earliest reports of CBSD from parts of what is now north-eastern Tanzania, it was soon realized that this disease was endemic throughout coastal areas of East Africa (Nichols, 1950). Furthermore, it was noted that CBSD was widely distributed in the southern low-altitude regions of Malawi, and that occurrences in Uganda were most likely the consequence of inadvertent introduction via germplasm from coastal Tanzania. Although a major effort was launched to eradicate CBSD from cassava in north-eastern Uganda between 1948 and 1949, it was considered at the time that this was unlikely to be wholly effective and that the disease could be considered to be endemic in Uganda. Importantly, however, there was no evidence for 'secondary' vector-mediated spread. It was initially suggested that environmental conditions at elevations greater than 1000 m above sea level (masl) were not suitable for the survival of CBSD (Nichols, 1950). Subsequent observations of cassava at elevations of up to 1700 m above sea level, however, made it clear that while CBSD-infected plants could readily survive at such elevations, there was no plant-to-plant transmission (Jennings, 1960). This provided a strong indication that the factor limiting CBSD spread at higher elevations was the absence of vector transmission.

No published information is available on how CBSD epidemics first spread to affect large parts of East Africa, mainly since there was little research attention devoted to cassava during the early part of the 20th century. A comparison of the earliest countrywide survey in Tanzania (Legg and Raya, 1998), with more recent data from East Africa (IITA, 2012; Jeremiah et al., 2015), however, does suggest that the incidence of CBSD in the 'original' East African endemic zone has increased.

4.2.2 New epidemics (2004 to the present)

CBSD was only reported once in Uganda between the completion of the eradication campaigns of the 1940s and the new millennium. This single observation was made from one location in southern Uganda in 1994 (Hillocks and Jennings, 2003). In 2004, however, many plants expressing CBSD-like symptoms were noted in south-central districts of Uganda, and diagnostic tests confirmed these to be caused by cassava brown streak viruses (CBSVs) (Alicai et al., 2007). Although these first observations were confined to two districts (Mukono and Wakiso), and only 2 out of 120 fields surveyed were found to be infected, the situation rapidly deteriorated as the incidence increased in the two initially affected districts and new occurrences were reported elsewhere. By 2007, CBSD was present in 8.1% of 493 fields assessed and infection was present in 10 of the 26 surveyed districts (Ntawuruhunga and Legg, 2007). Four years later, in 2011, an assessment of 15 of the most important cassava-growing districts in Uganda revealed an average incidence of >50% (IITA, 2012). As this growing epidemic developed in Uganda, new occurrences of CBSD were reported from neighbouring regions and countries, including western Kenya (Akhwale et al., 2010), north-western Tanzania (Jeremiah and Legg, 2008), Rwanda (Anon, 2014), Burundi (Bigirimana et al., 2011) and eastern Democratic Republic of the Congo (DRC) (Mulimbi et al., 2012), all of which had never previously been affected by CBSD. This constituted the first documented epidemic-like spread of CBSD, and in view of the wide regional impact, led to it being referred to as the CBSD pandemic (Legg et al., 2011). There are currently confirmed reports of the presence of CBSD from nine African countries (Fig. 4.2). Most recently, reports have been made of CBSD caused by *Ugandan cassava brown streak virus* (UCBSV), from Mayotte, off the coast of East Africa (Roux-Cuvelier *et al.*, 2014). Several other countries and locations have reported CBSD-like root symptoms, including Angola, Madagascar, Gabon, Cameroon and western DRC. It is possible that these symptoms could be caused by variants of CBSVs that cannot be detected using current diagnostic methods or due to other non-viral factors. Next-generation sequencing methods are currently being deployed to determine whether there are CBSV-like sequences in affected plants.

4.3 Economic Impact

4.3.1 Plant and crop-level estimates of economic impact

Cassava brown streak disease causes economic losses in several ways as a consequence of the different types of symptoms that it produces: (i) reduction of root yield through impairment of the growth of the plant; (ii) spoilage of roots through the effects of severe root rot; (iii) premature harvesting by farmers in order to avoid spoilage of roots through rot; (iv) loss of planting material arising from premature harvesting; and (v) increased labour costs associated with peeling partially rotten roots.

Only the first two loss categories have been quantified. Hillocks *et al.* (2001), working in southern Tanzania, reported losses of up to 70% in the most sensitive cultivars due to the impairment of growth, although a more typical level of loss was considered to be 30% (Hillocks and Jennings, 2003). Data from the same study suggested that root spoilage was relatively less important, accounting for losses of up to 24% in sensitive cultivars, and *circa* 17% on average. It has been recognized for many years that reduced yields arising from early harvesting are an important indirect consequence of CBSD infection (Childs, 1957). However, its economic cost, together with that resulting from the associated loss of planting material, has yet to be estimated. Cassava is commonly planted on a 12-month cycle; therefore, stems from early harvested plants (7–10 months after planting) may be unusable as planting material by the time the

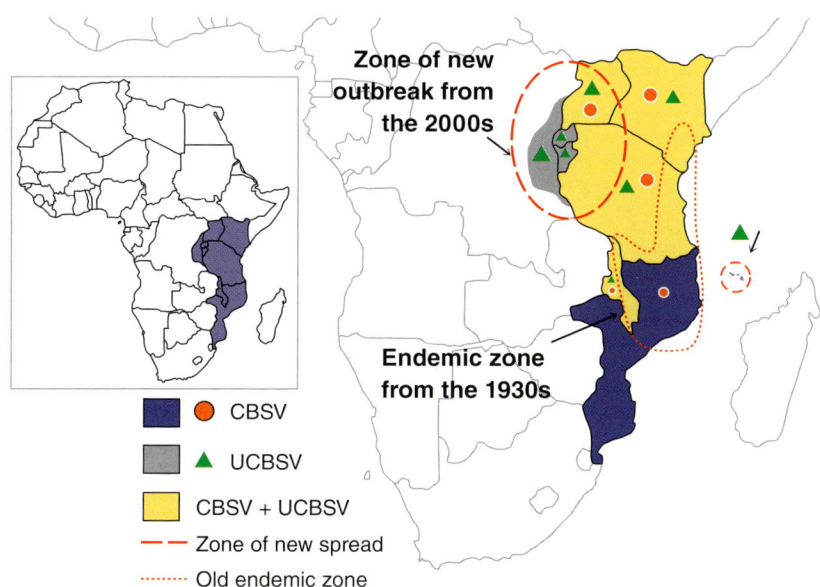

Fig. 4.2. African countries where CBSD has been reported (1930–2014). CBSV, *Cassava brown streak virus*; UCBSV, *Ugandan cassava brown streak virus*.

normal planting season comes around. A framework for the calculation of all CBSD-associated costs has been developed (Manyong *et al.*, 2012), but additional data on the aforementioned components of loss (points iii–v) will be required before this can be applied.

4.3.2 Global-level estimates of economic impact

There are very few large-scale assessments of the impact of CBSD, partly due to the lack of good survey data and partly because of the complexities of assessing CBSD yield impacts. Malawian farmers' own estimates of losses attributable to CBSD in 2001–2002 varied from nothing to 60%, with an average of 24.3% (Gondwe *et al.*, 2002). This study then used the mean CBSD incidence value of 40% for surveyed districts, and losses of 20–25% for infected plants, to make an overall production loss estimate of 137,000–172,000 t, with an approximate financial value in 2002 of US$6–7 million.

The only regional study of the economic impact of CBSD used surveillance data collected from East and Central Africa in 2009 (Burundi, eastern DRC, Kenya, Rwanda, Tanzania and Uganda) to estimate losses of 1,600,000 t, with an annual value of *circa* US$75 million (Manyong *et al.*, 2012). Although CBSD has attracted an increasing amount of interest since the development of new outbreaks in the Great Lakes region, these loss figures, when compared to those of CMD (more than US$1.9 billion annually; Legg and Fauquet, 2004), highlight the fact that CBSD currently causes less than 5% of the losses attributable to CMD.

4.4 Viral Aetiology

Based on symptoms and transmission properties, CBSD was suspected from its earliest recognition to be a virus disease (Storey, 1936; Lister, 1959); however, its exact identity remained uncertain for over 60 years. Early studies detected flexuous rod-shaped particles of about 650 nm, typical of whitefly-transmitted carlaviruses, and also pin wheel inclusion bodies, typical of potyvirus infection (Lennon *et al.*, 1985). Based on these observations, the involvement of a carlavirus, a potyvirus or a mixed infection of both was suspected in the aetiology of CBSD (Lennon *et al.*, 1985). However, tests using antiserum to the *Cowpea mild mottle virus* (family *Betaflexiviridae*, genus *Carlavirus*) isolated from cassava demonstrated an inconsistent association between *Cowpea mild mottle virus* and CBSD in cassava plants, ruling it out as the causal virus of the disease (Lennon *et al.*, 1985). Monger *et al.* (2001a) partially characterized the coat protein of a virus that was consistently associated with CBSD and demonstrated that this virus had affinities with members of the genus *Ipomovirus* (family *Potyviridae*). Based on the variation in the coat protein nucleotide sequences between different isolates collected from Mozambique and Tanzania (Monger *et al.*, 2001b), and on variable symptoms observed in different cassava cultivars (Hillocks *et al.*, 1996; Monger *et al.*, 2001b) and experimental hosts such as *Nicotiana benthamiana* (Monger *et al.*, 2001b), the occurrence of different strains of CBSD-causing viruses was suspected. Complete characterization of the virus was not achieved until a decade later (Mbanzibwa *et al.*, 2009a; Winter *et al.*, 2010), when two ipomovirus species were identified: *Cassava brown streak virus* (CBSV) and *Ugandan cassava brown streak virus*. Subsequently, several variants of these species have also been detected in the CBSD-affected countries of Southern and Eastern Africa (Mbanzibwa *et al.*, 2009b; Rwegasira *et al.*, 2011; Kamowa-Mbewe *et al.*, 2014).

4.5 Virus Taxonomy, Genome Organization and Diversity

Both CBSV and UCBSV (referred to hereafter as CBSVs) are members of the genus *Ipomovirus* (family *Potyviridae*). *Sweet potato mild mottle virus* and *Tomato mild mottle virus* are the other ipomoviruses reported from sub-Saharan Africa. The positive sense single-stranded RNA genome of CBSV consists of 8995–9008 nucleotides (GenBank Acc. No. FN434437 and GQ329864); whereas the UCBSV genome is slightly longer, comprising 9069–9070 nucleotides (GenBank Acc. No. NC_012698, FJ185044). The genomes of CBSV and UCBSV express a polyprotein of about 2902 amino acid residues encoding 11 proteins. These are (from 5′ to 3′): P1 and P3; PIPO (embedded in the open reading frame (ORF) of P3); 6K1 and 6K2 (6-kDa proteins flanking the CI, or cylindrical inclusion protein); VPg (viral genome-linked protein); NIa-Pro (the main viral proteinase); NIb (a replicase); Maf/HAM1 (a putative pyrophosphatase); and the coat protein. The protein-encoding region is flanked at the 5′ and 3′ ends each by an untranslated region, and the 3′ end finalizes with a *circa* 30 nucleotide poly-A tail (Mbanzibwa *et al.*, 2009a; Winter *et al.*, 2010).

The arrangement of ORFs on the genome of CBSVs is distinctive compared to other ipomoviruses (Patil *et al.*, 2014). The ORFs encoding for proteins 6K1, CI, 6K2, VPg, NIa-Pro and NIb are similar to members of the family *Potyviridae*. However, at the 5′ end, the P1 ORF is functionally equivalent to the P1b protein of two other ipomoviruses: *Squash vein yellowing virus* and *Cucumber vein yellowing virus*. In addition, the Maf/HAM1 protein in the CBSV and UCBSV genomes is unique and previously identified only in *Euphorbia ringspot virus*, a member of the genus *Potyvirus*. The precise function of this protein is not known. It is speculated, however, to have a role in preventing excessive RNA mutation (Mbanzibwa *et al.*, 2009a). Detailed features of these various proteins and their functions have been reviewed by Patil *et al.* (2014).

A high degree of sequence diversity between different geographical isolates of CBSVs in CBSD-affected countries has been reported (Mbanzibwa *et al.*, 2009b; Rwegasira *et al.*, 2011; Kamowa-Mbewe *et al.*, 2014; Patil *et al.*, 2014). The nucleotide sequence divergence between the full-length genomes of the two species was about 29–30%, whereas intraspecies diversity ranged from <1% to 20% for CBSV and from <1% to 17% for UCBSV (Mbanzibwa *et al.*, 2011a). Sequence-based evidence has suggested the involvement of divergent mechanisms in the evolution of UCBSV and CBSV (Mbanzibwa *et al.*, 2011a), which is also influenced by human-assisted spread of infected cassava planting material and whitefly transmission (Legg *et al.*, 2011). CBSV was found to be the predominant species associated with CBSD along the coastal zones of Kenya, Mozambique and Tanzania, whereas UCBSV was found to be the predominant species in the mid-elevation areas of Uganda, north-western Tanzania, western Kenya and the lakeshore areas of Malawi. Only UCBSV has been reported from the Great Lakes regions of Burundi, Rwanda, eastern DRC (Bigirimana *et al.*, 2011; Legg *et al.*, 2011, Mulimbi *et al.*, 2012; Patil *et al.*, 2014; Legg *et al.*, 2015) and Mayotte Island in the Indian Ocean (Roux-Cuvelier *et al.*, 2014). Co-occurrence of UCBSV and CBSV within the same fields, and even within plants, has also been demonstrated particularly around the Great Lakes (Mbanzibwa *et al.*, 2011b; Kamowa-Mbewe *et al.*, 2014). Although variable symptom reactions in experimental hosts such as *N. benthamiana* and cassava have been demonstrated (Monger *et al.* 2001b; Winter *et al.*, 2010), the biological implications of sequence diversity are not well understood. However, graft inoculations of three isolates of CBSV and three of UCBSV on to 2-month-old healthy cassava plants of five cultivars revealed significant differences in the severity of symptoms produced (Mohammed *et al.*, 2012). Although the isolates of each virus species produced variable symptoms, CBSV isolates generally elicited more severe symptoms than isolates of UCBSV. This study underscores the need for improving the understanding of the biological properties of CBSVs in different cassava genotypes, including symptomatology, interactions between UBCV and CBSV in the case of mixed infections, and the overall impacts on cultivar performance. This knowledge would also be useful in breeding for durable host resistance and in the development of strain-specific diagnostic tools.

4.6 Host Range

4.6.1 Natural hosts

The CBSVs occur naturally in cultivated cassava (*Manihot esculenta* Crantz), Ceará rubber (*M. glaziovii* Müll. Arg.) and crosses between these two that are commonly referred to as tree cassava and widely used in Africa as a leaf source for producing a green vegetable dish (Hillocks and Jennings, 2003). Although CBSD symptoms have been recognized from *M. glaziovii* and its hybrids with cultivated cassava for many years, it is only more recently that modern diagnostic methods have confirmed CBSV infection in *M. glaziovii* from coastal areas of Tanzania (Mbanzibwa *et al.*, 2011b). While it has not yet been proven that UCBSV infects this alternative host, the biological similarities between the two CBSVs suggest that it is highly improbable that UCBSV does not also infect *M. glaziovii*. Although *M. glaziovii* does represent an alternative host and potential reservoir for CBSVs, the relatively low frequency with which it occurs, in comparison to its cultivated cousin, indicates that it is unlikely to be of any significant importance in the epidemiology of CBSD.

4.6.2 Experimental hosts

Several herbaceous experimental hosts can be infected with CBSVs (Lister, 1959; Bock, 1994; Mohammed *et al.*, 2012). A diverse range of local and systemic symptoms are elicited, including necrotic lesions, chlorotic spots, vein banding and chlorosis, chlorotic stippling, leaf malformation, wilting and stunting. Experimental hosts represent members of two families: Solanaceae (*Petunia hybrida*, *N. debneyi*, *N. clevelandii*, *N. benthamiana*, *N. glutinosa*, *N. tabacum*, *N. rustica*, *Datura stramonium*, *D. ferox* and *Solanum nigrum*) and Amaranthaceae (*Chenopodium quinoa*). Both CBSV and UCBSV appear to be able to infect the same herbaceous hosts; however, isolates of the two viruses infect these plants with differing degrees of severity (Mohammed *et al.*, 2012). Of the three UCBSV and three CBSV isolates tested, CBSV isolates from Nampula (Mozambique) and Naliendele (Tanzania) gave rise to the most severe symptoms, causing the death of *N. benthamiana* and *N. clevelandii* plants. Differences of symptom severity in herbaceous hosts caused by the various isolates of CBSVs were comparable with patterns of severity caused by the same isolates in cassava plants.

4.7 Transmission and Epidemiology

4.7.1 Virus propagation and vector transmission

CBSVs infect cassava plants systemically and are consequently propagated through stem cuttings (Storey, 1936). These viruses are graft transmissible and vectored by the whitefly, *Bemisia tabaci* (Gennadius) (Maruthi *et al.*, 2005), but there is no evidence for their propagation through true seed or by mechanical means, such as through knives used during the preparation of cuttings or by hand when picking cassava leaves. The efficiency with which infection is carried from parent stem to stem cuttings has not been determined, although because of reversion it is possible that this may be less than 100%.

CBSVs, like other ipomoviruses, are transmitted by *B. tabaci* in a semi-persistent manner (Jeremiah, 2014). Transmission efficiencies of up to 45% were achieved with acquisition access periods of more than 1 hour, but the minimum acquisition access period was 5 minutes. The minimum inoculation access period was 30 minutes. Virus retention was poor, as only 7% infection in test plants was achieved by *B. tabaci* after being held on an uninfected cassava intermediary host for 24 hours, while no infections at all were recorded where this period was increased to 48 hours. This contrasts with the persistently transmitted cassava mosaic geminiviruses (CMGs), which are thought to be retained by *B. tabaci* adults for the duration of their lives (Dubern, 1994).

4.7.2   Epidemiology

Despite the relatively long history of CBSD in East Africa, there is surprisingly little published information on the epidemiology of the disease. There are several reasons for this:

1. CBSD has until recently been confined to a limited geographical range from coastal East Africa to the shores of Lake Malawi.
2. Symptoms of the disease are inconspicuous, making it relatively difficult to accurately identify infected plants.
3. Newly infected plants cannot be distinguished from those infected through cutting (in contrast to CMD where this distinction can readily be made).
4. The identity of the vector was only confirmed in the early part of the 21st century.
5. Difficulties in producing virus-free stocks of planting material have hindered the progress in setting up and running epidemiological experiments.

The first efforts to elucidate the epidemiology of CBSD were based on the observation of symptoms and the use of planting material presumed to be free from CBSVs, although these were constrained by the lack of diagnostic tests to confirm virus infection status and uncertainty about the identity of the vector. Correlations were demonstrated between the abundance of whitefly vector populations and increases in the incidence of CBSD symptoms, and plant-to-plant spread was demonstrated (Hillocks and Jennings, 2003). More recently, tissue culture techniques combined with virus diagnostics have been used to establish virus-free blocks of cassava in the field and to demonstrate steep gradients of disease spread from infected spreader plots to neighbouring test plots (Jeremiah, 2014). Within low elevation areas of occurrence, there have been contrasting reports of rates of spread: in coastal Kenya, rates of spread were considered to be low (Bock, 1994), whereas in central and southern coastal regions of Tanzania, rapid infection was reported where

conditions were favourable (Hillocks and Jennings, 2003). Strong seasonal variation in rates of CBSD spread has also been suggested for initially CBSD-free plantings of cassava in coastal Tanzania; little infection occurred in plots planted immediately before the cool dry season, whereas there was rapid spread into similar plots planted six months later – just at the start of the hottest period of the year (R. Shirima, personal communication).

At the regional level, there seems to have been little change in the epidemiological characteristics of CBSD in the 'original' endemic zone of coastal East Africa and the lakeshore areas of Malawi. This contrasts with the situation in the Great Lakes region of East and Central Africa, where there has been a rapid expansion of the CBSD pandemic from 2004 to present. Although it is clear that the cryptic symptoms of CBSD mean that it is easy for it to be inadvertently spread through infected planting material, the spread of the pandemic has been largely driven by unusually high populations of the *B. tabaci* vector (Legg *et al.*, 2011), which were also responsible for the similar region-wide spread of the severe CMD pandemic (Otim-Nape *et al.*, 2000; Legg *et al.*, 2006). The CBSD pandemic continues to spread in the eastern part of the Great Lakes region (Mulimbi *et al.*, 2012; Anon, 2014), and poses a significant threat to the important cassava-growing countries further west (Legg *et al.*, 2014a). This situation highlights the importance of the development and implementation of effective management strategies, both at local and regional levels.

4.8   Diagnostic Methods

The cryptic and seasonal nature of CBSD symptoms are such that accurate, sensitive and robust methods for diagnostic testing are vitally important. Although diagnostic tools were not available prior to the first molecular characterization of CBSV (Monger *et al.*, 2001a), rapid progress has been made in their development and application. The first diagnostic tool for CBSV was based on the reverse transcription (RT)-PCR technique (Monger *et al.*, 2001b). Further characterization of the CBSV and UCBSV genomes aided the design of several oligonucleotide primers specific to one or both virus species to enable species-specific or broadly active detection in uniplex or multiplex formats (Abarshi *et al.*, 2010, 2012; Mbanzibwa *et al.*, 2011b). A real-time RT-PCR technique has been subsequently developed for high throughput and sensitive detection and quantification of UCBSV and CBSV (Adams *et al.*, 2013). The recent development of a reverse transcription loop-mediated isothermal amplification method for the detection of CBSV and UCBSV (Tomlinson *et al.*, 2013) has raised the possibility of developing on-site diagnostic testing. Although an ELISA has been developed (DSMZ, Germany), using a combination of polyclonal and monoclonal antibodies, its usage in routine diagnosis of CBSVs is limited due to concerns over sensitivity. RT-PCR and real-time RT-PCR have become the methods of choice for routine testing.

4.9   Management Strategies

Cassava is normally propagated using stem cuttings collected from previous plantings. Therefore, diseases and pests can build up over several generations of propagation. CBSD symptoms are typically mild and mainly confined to lower leaves, making it more difficult to distinguish between healthy and infected plants than is the case for CMD. Consequently, CBSVs are more readily propagated through infected planting material (Legg *et al.*, 2011). Additionally, CBSVs are semi-persistently transmitted meaning that they are retained for relatively short periods of time, limiting the distance over which they can be carried by their whitefly vector (Jeremiah, 2014). While CBSVs may therefore be spread by vectors over relatively short distances, they can be readily carried over long distances through transport of planting material. This contrasts with the CMGs. Whiteflies can retain CMGs for longer periods (probably for several weeks) and may move long distances through a combination of active and passive wind-assisted flight.

In general, there are three possible approaches to managing losses caused by a virus disease (Thresh and Cooter, 2005): (i) decrease the proportion of plants that become infected; (ii) delay infection to such a late stage of crop growth that losses become unimportant; and (iii) decrease the severity of damage sustained after infection has occurred. The diverse ways of achieving these objectives have been discussed in detail by Thresh (2003) and Hillocks and Jennings (2003). Thresh and Cooter (2005) suggested that control measures should be simple, inexpensive and within the limited capacity of farmers. Furthermore, they advised that the measures should also be sustainable and involve little or no use of pesticides so as to avoid damage to human health, insects that provide natural control, or to the environment. The following control strategies can therefore be applied to manage CBSD.

4.9.1   Phytosanitation

This term refers to the various means of improving the health status of cassava planting material and eliminating sources of inoculum from which further spread of disease can occur through vector activity (Thresh and Cooter, 2005). There are three main components of this strategy: (i) use of disease-free planting material; (ii) planting the crop in a way that minimizes the risk of infection, such as by isolating the new field from neighbouring infected plantings; and (iii) removal (roguing) of diseased plants from within the crop (Hillocks and Jennings, 2003; Thresh and Cooter, 2005). In view of the biological characteristics of CBSD mentioned above, phytosanitation can play a major role in limiting the impact and spread of CBSD (Hillocks and Jennings, 2003). The following components are therefore recommended for CBSD control programmes:

1. The production of 'clean' stocks of planting material, including virus indexing of parent material in tissue culture, systematic

virus testing in isolated pre-basic germplasm multiplication sites and regular roguing of symptomatic plants in propagation fields.

2. The avoidance of local and regional spread through the application of quarantine measures and certification standards for large-scale distribution and sale of planting material.

3. Collective action at the community level to encourage groups of farmers growing cassava in close proximity to cooperate in implementing phytosanitary measures, including the sourcing of 'clean' planting material and its maintenance through roguing and the selection of healthy stems for replanting.

4.9.2   Use of resistant varieties

Conventional approaches to enhancing host plant resistance

Cassava breeding for host plant resistance to both CMD and CBSD started in 1935 at Amani, in north-eastern Tanzania. The most resistant hybrid developed from this programme was 46106/27, which was a third backcross derivative from a *M. esculenta* × *M. glaziovii* interspecific cross (Jennings, 1960). It is probably the most successful product of the Amani Research Programme that is presently available to farmers and whose resistance to CBSD has persisted for many years in farmers' fields in Kenya, where it is locally known as 'Kaleso', and in Tanzania where it is known as 'Namikonga' (Hillocks and Jennings, 2003). DNA fingerprinting techniques have been used to prove that 'Kaleso' is genetically identical to 'Namikonga'. This cultivar is still considered to be the best parent to use as a source of CBSD resistance in the breeding programmes of Kenya and Tanzania (Pariyo *et al.*, 2013).

Several CBSD-resistant local cultivars have been identified and recommended in Kenya ('Kaleso', 'Guzo', 'Gushe', 'Kibiriti Mweusi' and 'Ambari'), Mozambique ('Nikwaha', 'Chigoma Mafia', 'Nanchinyaya', 'Xino Nn'goe', 'Likonde', 'Mulaleia' and 'Badge') and Tanzania ('Namikonga', 'Kiroba', 'Nanchinyaya', 'Kigoma Mafia', 'Kitumbua', 'Kalulu', 'Mfaransa',

'Muzege', 'Gezaulole' and 'Kibangameno'). Some of these cultivars are the former Amani hybrids that are no longer recognized as such since they have been given local names. Most of them are better described as 'tolerant' since they readily develop foliar symptoms, but root necrosis is delayed or absent (Hillocks and Jennings, 2003). In describing plant virus resistance, breeders emphasize the effect on yield and quality, in contrast to pathologists who consider the fate of the virus in the plant (Lapidot and Friedmann, 2002). Thresh *et al.* (1998) state that:

> ...truly resistant varieties are not readily infected, even when exposed to large amounts of vector-borne inoculum; when infected they develop inconspicuous symptoms that are not associated with obvious deleterious effects on growth and yield and support low virus content and are thus likely to be a poor source of inoculum from which further spread can occur.

Efforts are being made to develop truly resistant varieties. In addition to making use of elite parents such as 'Kaleso'/'Namikonga' in conventional breeding programmes, new initiatives are also exploiting wild relatives as potential sources of novel resistance genes. Furthermore, some of the worst-affected countries in East and Southern Africa are collaborating to share their most resistant breeding lines through a region-wide germplasm exchange programme. This will provide new opportunities for enhancing host plant resistance because inter-crossing among them will concentrate resistance genes and allow recessive genes to be expressed (Hillocks and Jennings, 2003).

Molecular approaches to enhancing the selection of host plant resistance

Rapid advances have been made in sequencing technology through the early years of the 21st century, and next-generation sequencing is increasingly being used by breeders to improve the precision and speed of their variety development work. Earlier work to develop molecular markers for resistance genes is now being greatly augmented through the application of genomic selection procedures for cassava (Ly *et al.*, 2013), which are now being used specifically to target resistance to CBSVs. Transcriptomic methods have been applied to generating comparisons between infected and CBSD-free plants of susceptible (Albert) and resistant (Kaleso) cultivars. In addition to harbouring greatly lower concentrations of CBSVs, more than 700 genes were overexpressed in Kaleso compared to Albert, and some of these genes play a role in hormone signalling pathways or secondary metabolite production – both functions that are commonly associated with pathogen resistance (Maruthi *et al.*, 2014).

Although these methods offer better opportunities for improving the identification, utilization and deployment of existing resistance genes, transgenic approaches offer greater potential for introducing novel sources of resistance. Pathogen-derived resistance has been shown to be effective in transgenic plants for many years (Powell-Abel *et al.*, 1986). These techniques have been adapted for use against cassava viruses, and RNAi using near full-length CP constructs has been shown to be effective against UCBSV in both laboratory (Yadav *et al.*, 2011) and confined field trial situations (Ogwok *et al.*, 2012). More recently, constructs targeting both CBSV and UCBSV have been shown to confer high levels of resistance to both viruses as well as being successfully propagated through the vegetative cropping cycle (Odipio *et al.*, 2014). Although the regulatory environment in Africa remains weakly developed, there is a strong recognition of the potential benefits to be realized through the judicious application of transgenic technologies. CBSD-resistant cassava varieties look likely to be near the top of the list of candidate transformed plants for introduction to farming communities in CBSD-affected parts of East, Central and Southern Africa.

4.9.3 Vector control

Although the control of whiteflies is most critical to prevent the dissemination of cassava viruses, the role of vectors, as mentioned above, is slightly less important for CBSVs than it is for CMGs, since CBSVs are semi-persistently transmitted. However, developing varieties resistant to the vector has now become more important than ever in view of the recent occurrence of unusually large populations of *B. tabaci* in the Great Lakes region and their associated direct feeding damage (Thresh and Cooter, 2005). Little is currently being done to manage *B. tabaci* populations on cassava in Africa, although neonicotinoid insecticides are being effectively used in experimental fields and at planting material propagation sites, and there is some evidence for host plant resistance (Legg *et al.*, 2014b). Since *B. tabaci* transmits the viruses causing the two most economically important diseases of cassava in Africa, substantially greater attention to the challenge of developing and applying integrated whitefly control strategies is warranted.

4.10 Concluding Remarks

CBSD is an economically important constraint in important cassava-growing areas of East, Central and Southern Africa. In addition to causing a dry necrotic rot in the tuberous roots of infected plants, which can render them inedible, CBSD depresses yields through reducing plant growth. The impact of this disease has increased greatly since 2004, as a new rapidly developing outbreak has spread to previously unaffected parts of Africa, most notably in the Great Lakes region of East and Central Africa. This pandemic of CBSD is driven by the high abundance of the whitefly vector, *B. tabaci*, coupled with uncontrolled movements of infected cuttings used as planting material. The continued westwards spread of CBSD means that it is now considered to be one of the greatest plant disease threats to agriculture in Africa. Both host plant resistance and cultural methods are being used to manage the effects of the disease. Recent progress in both field- and laboratory-based experimentation to develop effective control strategies for CBSD means that there are improved prospects for tackling this important constraint to cassava production in future years.

References

Abarshi, M.M., Mohammed, I.U., Wasswa, P., Hillocks, R.J., Holt, J., *et al.* (2010) Optimization of diagnostic RT-PCR protocols and sampling procedures for the reliable and cost-effective detection of *Cassava brown streak virus. Journal of Virological Methods* 163, 353–359.

Abarshi, M.M., Mohammed, I.U., Jeremiah, S.C., Legg, J.P., Kumar, P.L., *et al.* (2012) Multiplex RT-PCR assays for the simultaneous detection of both RNA and DNA viruses infecting cassava and the common occurrence of mixed infections by two cassava brown streak viruses in East Africa. *Journal of Virological Methods* 179, 176–184.

Adams, I.P., Abidrabo, P., Miano, D.W., Alicai, T., Kinyua, Z.M., *et al.* (2013) High throughput real-time RT-PCR assays for specific detection of cassava brown streak disease causal viruses, and their application to testing of planting material. *Plant Pathology* 62, 233–242.

Akhwale, M.S., Obiero, H.M., Njarro, O.K., Mpapale, J.S. and Otunga, B.M. (2010) Participatory cassava variety selection in Western Kenya. Available at: http://www.kari.org/biennialconference/conference12/docs/PARTICIPATORY%20CASSAVA%20VARIETY%20SELECTION%20IN%20WESTERN%20KENYA.pdf (accessed 24 November 2014).

Alicai, T., Omongo, C.A., Maruthi, M.N., Hillocks, R.J., Baguma, Y., *et al.* (2007) Re-emergence of cassava brown streak disease in Uganda. *Plant Disease* 91, 24–29.

Anon (2014) Rwanda: It is a race against time as government bids to salvage fortunes of cassava farmers. Available at: http://allafrica.com/stories/201409230583.html (accessed 24 November 2014).

Bigirimana, S., Barumbanze, P., Ndayihanzamaso, P., Shirima, R. and Legg, J.P. (2011) First report of cassava brown streak disease and associated *Ugandan cassava brown streak virus* in Burundi. *New Disease Reports* 24, 26.

Bock, K.R. (1994) Studies on cassava brown streak virus disease in Kenya. *Tropical Science* 34, 134–145.

CABI (2014) Cassava brown streak viruses datasheet. Available at: http://www.cabi.org/isc/datasheet/17107 (accessed 25 November 2014).

Childs, A.H.B. (1957) Trials with virus resistant cassava in Tanga Province, Tanganyika. *East African Agricultural Journal* 23, 135–137.

CRS (2014) Cassava brown streak disease (CBSD). Available at: http://www.crsprogramquality.org/storage/pubs/agenv/cbsd-brochure-english.pdf (accessed 25 November 2014).

Dubern, J. (1994) Transmission of African cassava mosaic geminivirus by the whitefly (*Bemisia tabaci*). *Tropical Science* 34, 82–91.

Gondwe, F.M.T., Mahungu, N.M., Hillocks, R.J., Raya, M.D., Moyo, C.C., *et al.* (2002) Economic losses experienced by small-scale farmers in Malawi due to cassava brown streak virus disease. In: Legg, J.P. and Hillocks, R.J. (eds) *Cassava brown streak virus disease: past, present, and future. Proceedings of an international workshop, Mombasa, Kenya, 27–30 October*. Natural Resources International Limited, Aylesford, UK, pp. 28–36.

Hillocks, R.J. and Jennings, D.L. (2003) Cassava brown streak disease: a review of present knowledge and research needs. *International Journal of Pest Management* 49, 225–234.

Hillocks, R.J. and Thresh, J.M. (2000) Cassava mosaic and cassava brown streak virus diseases: a comparative guide to symptoms and aetiologies. *Roots* 7, 1–8.

Hillocks, R.J., Raya, M. and Thresh, J.M. (1996) The association between root necrosis and above ground symptoms of brown streak virus infection of cassava in Southern Tanzania. *International Journal of Pest Management* 42, 285–289.

Hillocks, R.J., Raya, M., Mtunda, K. and Kiozia, H. (2001) Effects of brown streak virus disease on yield and quality of cassava in Tanzania. *Journal of Phytopathology* 149, 1–6.

IITA (2012) Map database for GLCI project cassava pest and disease surveys conducted in six countries of East and Central Africa. Available at: https://www.flickr.com/photos/iita-media-library/sets/72157632798115038 (accessed 24 November 2014).

Jennings, D.L. (1960) Observations on virus disease of cassava in resistant and susceptible varieties. II Brown streak disease. *Empire Journal of Experimental Agriculture* 28, 261–269.

Jeremiah, S. (2014) The role of whitefly (*Bemisia tabaci*) in the spread and transmission of cassava brown streak disease in the field. PhD thesis. University of Dar es Salaam, Tanzania, 229 pp.

Jeremiah, S.C. and Legg, J.P. (2008) *Cassava brown streak virus*. Available at: http://www.youtube.com/watch?v=nCJdws9CnUw (Uploaded, 25 October 2008).

Jeremiah, S.C., Ndyetabula, I.L., Mkamilo, G.S., Haji, S., Muhanna, M.M., *et al.* (2015) The dynamics and environmental influence on interactions between cassava brown streak virus disease and the whitefly, *Bemisia tabaci. Phytopathology*. In press.

Kamowa-Mbewe, W., Kumar, P.L., Changadeya, W., Ntawuruhunga, P. and Legg, J.P. (2014) Diversity, distribution and effects on cassava cultivars of cassava brown streak viruses in Malawi. *Journal of Phytopathology* doi: 10.1111/jph.12339.

Lapidot, M. and Friedmann, M. (2002) Breeding for resistance to whitefly transmitted geminiviruses. *Annals of Applied Biology* 140, 109–127.

Legg, J.P. and Fauquet, C.M. (2004) Cassava mosaic geminiviruses in Africa. *Plant Molecular Biology* 56, 585–599.

Legg, J.P. and Raya, M. (1998) Survey of cassava virus diseases in Tanzania. *International Journal of Pest Management* 44, 17–23.

Legg, J.P., Owor, B., Sseruwagi, P. and Ndunguru, J. (2006) Cassava mosaic virus disease in East and Central Africa: epidemiology and management of a regional pandemic. *Advances in Virus Research* 67, 355–418.

Legg, J.P., Jeremiah, S.C., Obiero, H.M., Maruthi, M.N., Ndyetabula, I., *et al.* (2011) Comparing the regional epidemiology of the cassava mosaic and cassava brown streak virus pandemics in Africa. *Virus Research* 159, 161–170.

Legg, J.P., Shirima, R., Tajebe, L.S., Guastella, D., Simon, B., *et al.* (2014a) Biology and management of *Bemisia* whitefly vectors of cassava virus pandemics in Africa. *Pest Management Science* 70, 1446–1453.

Legg, J.P., Somado, E.A., Barker, I., Beach, L., Ceballos, H., *et al.* (2014b) A global alliance declaring war on cassava viruses in Africa. *Food Security* 6, 231–248.

Legg, J.P., Kumar, P.L., Kumar, M., Tripathi, L., Ferguson, M., *et al.* (2015) Cassava virus diseases: biology, epidemiology and management. *Advances in Virus Research* (in press).

Lennon, A.M., Aiton, M.M. and Harrison, B.D. (1985) *Cassava Viruses from Africa*. Report of the Scottish Crop Research Institute, 168. SCRI, Dundee, Scotland.

Lister, R.M. (1959) Mechanical transmission of cassava brown streak virus. *Nature* 183, 1588–1589.

Ly, D., Hamblin, M., Rabbi, I., Gedil, M., Bakare, M., *et al.* (2013) Relatedness and genotype x environment interaction affect prediction accuracies in genomic selection: A study in cassava. *Crop Science* 53, 1312–1325.

Manyong, V.M., Maeda, C., Kanju, E. and Legg, J.P. (2012) Economic damage of cassava brown streak disease in sub-Saharan Africa. In: Okechukwu, R.U. and Ntawuruhunga, P. (eds) *Tropical Root and Tuber Crops and the Challenges of Globalization and Climate Change. 11th ISTRC-AB Symposium, 4–8 2010, Kinshasa, Democratic Republic of Congo*. IITA, Ibadan, Nigeria, pp. 61–68.

Maruthi, M.N., Hillocks, R.J., Mtunda, K., Raya, M.D., Muhanna, M., *et al.* (2005) Transmission of *Cassava brown streak virus* by *Bemisia tabaci* (Gennadius). *Journal of Phytopathology* 153, 307–312.

Maruthi, M.N., Bouvaine, S., Tufan, H.A., Mohammed, I.U. and Hillocks, R.J. (2014) Transcriptional response of virus-infected cassava and identification of putative sources of resistance for cassava brown streak disease. *PLoS ONE* 9, e96642.

Mbanzibwa, D.R., Tian, Y., Mukasa, S.B. and Valkonen, J.P.T. (2009a) *Cassava brown streak virus* (*Potyviridae*) encodes a putative Maf/HAM1 pyrophosphatase implicated in reduction of mutations and a P1 proteinase that suppresses RNA silencing but contains no HC-Pro. *Journal of Virology* 83, 6934–6940.

Mbanzibwa, D.R., Tian, Y., Tugume, A.K., Mukasa, S.B., Tairo, F., *et al.* (2009b) Genetically distinct strains of *Cassava brown streak virus* in the Lake Victoria basin and the Indian Ocean coastal area of East Africa. *Archives of Virology* 154, 353–359.

Mbanzibwa, D.R., Tian, Y., Tugume, A.K., Patil, B.L., Yadav, J.S., *et al.* (2011a) Evolution of cassava brown streak disease-associated viruses. *Journal of General Virology* 92, 974–987.

Mbanzibwa, D.R., Tian, Y.P., Tugume, A.K., Mukasa, S.B., Tairo, F., *et al.* (2011b) Simultaneous virus-specific detection of the two cassava brown streak-associated viruses by RT-PCR reveals wide distribution in East Africa, mixed infections and infections in *Manihot glaziovii*. *Journal of Virological Methods* 171, 394–400.

Mohammed, I.U., Abarshi, M.M., Muli, B., Hillocks, R.J. and Maruthi, M.N. (2012) The symptoms and genetic diversity of cassava brown streak viruses infecting cassava in East Africa. *Advances in Virology* doi:10.1155/2012/795697.

Monger, W.A., Seal, S., Isaac, A.M. and Foster, G.D. (2001a) Molecular characterization of *Cassava brown streak virus* coat protein. *Plant Pathology* 50, 527–534.

Monger, W.A., Seal, S., Cotton, S. and Foster, G.D. (2001b) Identification of different isolates of *Cassava brown streak virus* and development of a diagnostic test. *Plant Pathology* 50, 768–775.

Mulimbi, W., Phemba, X., Assumani, B., Kasereka, P., Muyisa, S., *et al.* (2012) First report of *Ugandan cassava brown streak virus* on cassava in Democratic Republic of Congo. *New Disease Reports* 26, 11.

Nichols, R.F.J. (1950) The brown streak disease of cassava: distribution, climatic effects and diagnostic symptoms. *East African Agricultural Journal* 15, 154–160.

Ntawuruhunga, P. and Legg, J.P. (2007) New spread of cassava brown streak virus disease and its implications for the movement of cassava germplasm in the East and Central African region. Brief 3. Crop Crisis Control Project (C3P). Available at: http://www.crsprogramquality.org/storage/pubs/agenv/3%20cassava%20CSBD.pdf (accessed 24 July 2015).

Odipio, J., Ogwok, E., Taylor, N.J., Halsey, M., Bua, A., *et al.* (2014) RNAi-derived field resistance to cassava brown streak disease persists across the vegetative cropping cycle. *GM Crops and Food: Biotechnology in Agriculture and the Food Chain* 5, 16–19.

Ogwok, E., Odipio, J., Halsey, M., Gaitan-Solis, E., Bua, A., *et al.* (2012) Transgenic RNA interference (RNAi)-derived field resistance to cassava brown streak disease. *Molecular Plant Pathology* 13, 1019–1031.

Otim-Nape, G.W., Bua, A., Thresh, J.M., Baguma, Y., Ogwal, S., *et al.* (2000) *The Current Pandemic of Cassava Mosaic Virus Disease in East Africa and its Control*. Natural Resources Institute, Chatham, UK, 104 pp.

Pariyo, A., Tukamuhabwa, P., Baguma, Y., Kawuki, R.S., Alicai, T., *et al.* (2013) Simple sequence repeats (SSR) diversity of cassava in South, East and Central Africa in relation to resistance to cassava brown streak disease. *African Journal of Biotechnology* 12, 4453–4464.

Patil, L.B., Legg, J.P., Kanju, E. and Fauquet, C.M. (2014) Cassava brown streak disease: a threat to food security in Africa. *Journal of General Virology* doi: 10.1099/jgv.0.000014.

Powell-Abel, P., Nelson, R.S., Hoffmann, B. De N., Rodger, S.G., Fraley, R.T., *et al.* (1986) Delay of disease development in transgenic plants that express the Tobacco mosaic virus coat protein gene. *Science*, 232, 738–743.

Roux-Cuvelier, M., Teyssedre, D., Masse, D., Jade, K., Abdoul-Karime, A.L., *et al.* (2014) First report of cassava brown streak disease and associated *Ugandan cassava brown streak virus* in Mayotte. *New Disease Reports* 30, 28. Available at: http://dx.doi.org/10.5197/j.2044-0588.2014.030.028 (accessed 24 July 2015).

Rwegasira, G.M., Momanyi, G., Rey, M.E.C., Kahwa, G. and Legg, J.P. (2011) Widespread occurrence and diversity of *Cassava brown streak virus* (*Potyviridae: Ipomovirus*) in Tanzania. *Phytopathology* 101, 1159–1167.

Storey, H.H. (1936) Virus diseases on East African plants. VI. A progress report on the studies of the diseases of cassava. *East African Agricultural Journal* 2, 34–39.

Thresh, J.M. (2003) Control of plant virus diseases in sub-Saharan Africa: the possibility and feasibility of an integrated approach. *African Crop Science Journal* 11, 199–233.

Thresh, J.M. and Cooter, R.J. (2005) Strategies for controlling cassava mosaic virus disease in Africa. *Plant Pathology* 54, 587–614.

Thresh, J.M., Otim-Nape, G.W. and Fargette, D. (1998) The components and deployment of resistance to cassava mosaic virus disease. *Integrated Pest Management Reviews* 3, 209–224.

Tomlinson, J.A., Ostoja-Starzewska, S., Adams, I.P., Miano, D.W., Abidrabo, P., *et al.* (2013) Loop-mediated isothermal amplification for rapid detection of the causal agents of cassava brown streak disease. *Journal of Virological Methods* 191, 148–154.

Warburg, O. (1894) Die kulturpflanzen usambaras. *Mitteilungen aus den Deutschen Schutzgebieten* 7, 131.

Winter, S., Koerbler, M., Stein, B., Pietruszka, A., Paape, M., *et al.* (2010) Analysis of cassava brown streak viruses reveals the presence of distinct virus species causing cassava brown streak disease in East Africa. *Journal of General Virology* 91, 1365–1372.

Yadav, J.S., Ogwok, E., Wagaba, H., Patil, B.L., Bagewadi, B., *et al.* (2011) RNAi-mediated resistance to *Cassava brown streak Uganda virus* in transgenic cassava. *Molecular Plant Pathology* 12, 677–687.

5 Cassava Mosaic

Olufemi J. Alabi,[1]* Rabson M. Mulenga[2] and James P. Legg[3]

[1]Department of Plant Pathology & Microbiology, Texas A&M
AgriLife Research and Extension Center, Weslaco,
Texas, USA; [2]Zambia Agriculture Research Institute,
Mount Makulu Central Research Station, Lusaka, Zambia;
[3]International Institute of Tropical Agriculture,
Dar es Salaam, Tanzania

5.1 General Introduction

The global cassava development strategy launched by the Food and Agriculture Organization of the United Nations in Rome in 2000 concluded that:

> ... cassava could become the raw material base for an array of processed products that will effectively increase demand for the crop and contribute to agricultural transformation and economic growth in developing countries (http://www.fao.org/ag/agp/agpc/gcds/).

Although cassava is currently consumed by over 800 million people in Africa and is the third most important source of calories in the tropics, the vision of the Food and Agriculture Organization would nevertheless represent a major increase in the global significance of a crop that is still largely cultivated by resource-poor farmers utilizing traditional farming tools and practices. A native to South America, cassava (*Manihot esculenta* Crantz, family Euphorbiaceae) – also known as yuca, manioc or mandioca – is believed to have been introduced into sub-Saharan Africa by Portuguese traders during the 16th century (Carter *et al.*, 1995). Data from the Food and Agriculture Organization indicate that about 34 countries in sub-Saharan Africa currently account for ~55% of the global cassava

production of ~278 million metric tonnes (FAOSTAT, 2013). In these countries, cassava leaves and the tuberous roots are largely consumed in processed forms as a staple food. More recently, the tuberous roots of cassava have become an important raw material in the manufacture of various industrial products such as starch, flour and ethanol as biofuel.

Although cassava shows resilience against a myriad of pests and diseases, including several of viral aetiology (Thottappilly *et al.*, 2003; Alabi *et al.*, 2011), two virus disease complexes, cassava brown streak disease (CBSD) and cassava mosaic disease (CMD), are the major yield-limiting biotic constraints to its production in sub-Sahara Africa. Whereas CBSD is to date limited to countries in Eastern and Southern Africa, CMD is widespread across sub-Saharan Africa causing significant economic losses especially when plants are infected through the cutting or during early growth stages. The symptoms associated with CMD were first recorded in north-eastern Tanzania 120 years ago (Warburg, 1894). Subsequent studies have recorded the occurrence of many virus species, all of which belong to the genus *Begomovirus* (family *Geminiviridae*) (Brown *et al.*, 2012), in CMD-affected plants (Stanley and Gay, 1983; Hong *et al.*, 1993; Berrie *et al.*, 1998; Zhou *et al.*, 1998; Fondong *et al.*, 2000; Saunders *et al.*, 2002;

*E-mail: alabi@tamu.edu

Maruthi *et al.*, 2004; Bull *et al.*, 2006; Harimalala *et al.*, 2012; Tiendrébéogo *et al.*, 2012).

5.2 Symptoms

A distinctive feature of CMD is the variety of foliar symptoms produced by affected cassava, including yellow or green mosaic, mottling, and misshapen and twisted leaflets (Fig. 5.1) (Thottappilly *et al.*, 2003; Alabi *et al.*, 2011). Such symptoms may vary from plant to plant due to differences in associated virus species (and their recombinant strain or satellite molecules), the presence of single or mixed infections, the age of the plant at the time of infection, the genetic composition of the cultivar (ranging from susceptibility to tolerance to resistance), and environmental factors that may influence the host, the virus and the whitefly vector (*Bemisia tabaci* (Gennadius); Hemiptera: Aleyrodidae) (Legg and Thresh, 2000; Maruthi *et al.*, 2002; Ogbe *et al.*, 2003). Studies have shown that CMD

Fig. 5.1. Foliar symptoms apparent on cassava plant affected by cassava mosaic disease (bottom) relative to a healthy looking plant of the same age (top).

symptoms are often exacerbated in cassava harbouring mixtures of multiple viruses, recombinant variants and/or their satellites, acting synergistically in infected hosts (Fondong *et al.*, 2000; Pita *et al.*, 2001a; Owor *et al.*, 2004; Patil and Fauquet, 2010; Zinga *et al.*, 2013). Earlier studies also showed that CMD-associated viruses are capable of inducing morphological and cytological modifications in cassava and the experimental host, *Nicotiana benthamiana* Domin (Fondong *et al.*, 2000; Atiri *et al.*, 2004). As the disease develops during the cropping season, CMD-affected plants may exhibit an overall reduction in the size of leaves and plants (see Fig. 5.1) and produce few or no tuberous roots (Thottappilly *et al.*, 2003; Alabi *et al.*, 2011).

5.3 Causative Viruses

At least nine distinct established viruses, and two tentative species, have been described from cassava plants affected by CMD worldwide (Fig. 5.2). These viruses are often collectively called cassava mosaic begomoviruses or cassava mosaic geminiviruses (CMGs). They are: *African cassava mosaic virus* (ACMV; Stanley and Gay, 1983), *East African cassava mosaic virus* (EACMV; Hong *et al.*, 1993), *East African cassava mosaic Cameroon virus* (EACMCV; Fondong *et al.*, 2000), *East African cassava mosaic Kenya virus* (EACMKV; Bull *et al.*, 2006), *East African cassava mosaic Malawi virus* (EACMMV; Zhou *et al.*, 1998), *East African cassava mosaic Zanzibar virus* (EACMZV; Maruthi *et al.*, 2004), *South African cassava mosaic virus* (SACMV; Berrie *et al.*, 1998), *Indian cassava mosaic virus* (ICMV; Matthew and Muniyappa, 1992; Saunders *et al.*, 2002) and *Sri Lankan cassava mosaic virus* (SLCMV; Saunders *et al.*, 2002). However, the status of EACMCV was revised to that of an isolate of EACMV in the most recent revision of *Begomovirus* taxonomy by the *Geminiviridae* subgroup of the ICTV (Brown *et al.*, 2015). The two recently characterized tentative virus species awaiting confirmation by the International Committee on the Taxonomy of Viruses are Cassava mosaic Madagascar virus (Harimalala *et al.*, 2012)

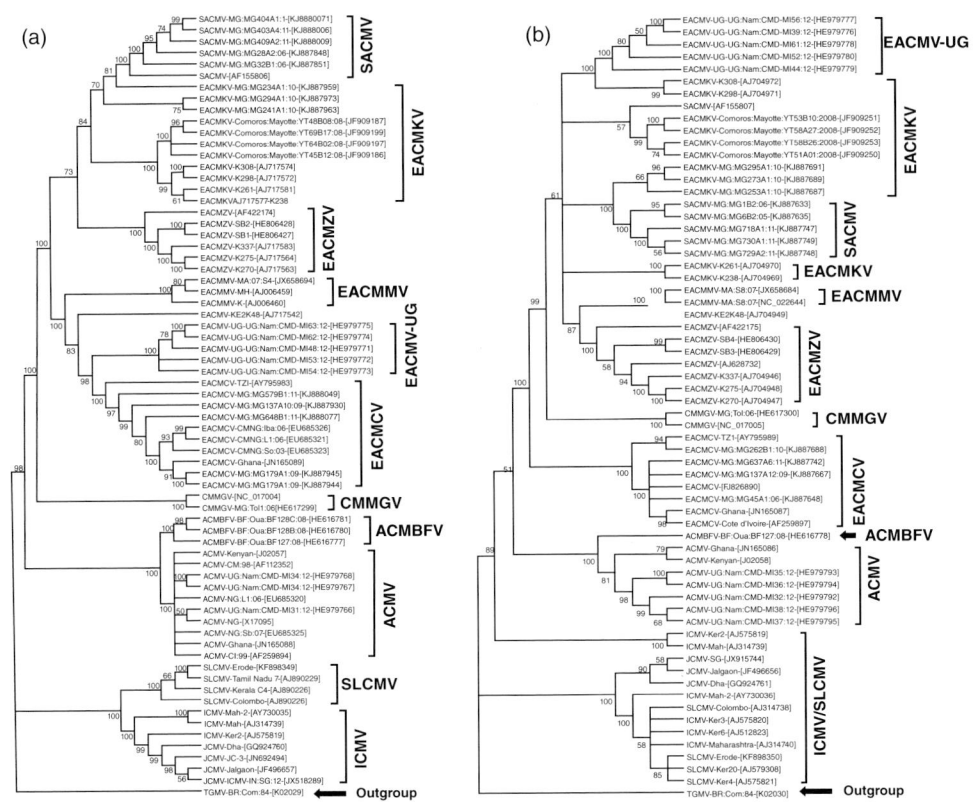

Fig. 5.2. A Neighbour-Joining phylogenetic tree depicting the diversity of (a) DNA A and (b) DNA B complete genome segments of global isolates of cassava mosaic geminiviruses (CMGs). The DNA A genome segment yielded a clear species-specific segregation of global CMG isolates relative to the DNA B-based tree. Numbers at the internodes depict bootstrap replicates supporting each internode. Each tree was rooted with corresponding genome segment sequences of an isolate of *Tomato golden mosaic virus* (TGMV) as an outgroup sequence. ACMV, *African cassava mosaic virus*; ACMBFV, African cassava mosaic Burkina Faso virus; EACMV, *East African cassava mosaic virus*; EACMCV, *East African cassava mosaic Cameroon virus*; EACMKV, *East African cassava mosaic Kenya virus*; EACMMV, *East African cassava mosaic Malawi virus*; EACMZV, *East African cassava mosaic Zanzibar virus*; ICMV, *Indian cassava mosaic virus*; SACMV, *South African cassava mosaic virus*; SLCMV, *Sri Lankan cassava mosaic virus*.

and African cassava mosaic Burkina Faso virus (Tiendrébéogo *et al.*, 2012). Seven of the approved and both tentative virus species are of sub-Saharan Africa origin, whereas ICMV and SLCMV are from the Indian sub-continent where they appear to be largely confined to date. ACMV is the most widespread of the CMGs, occurring in virtually all cassava-growing countries of sub-Saharan Africa. Isolates of ACMV appear to be almost genetically identical across their distribution range, in contrast to the EACMV-like viruses that show greater genetic variability and are more

prone to recombination leading to the evolution of new genetic variants (Thottappilly *et al.*, 2003; Patil and Fauquet, 2009; Alabi *et al.*, 2011). A recombinant virus resulting from shared genomic sequences between two parental viruses (ACMV and EACMV) and known as the Ugandan variant of *East African cassava mosaic virus* (EACMV-UG) is also widespread across many countries from sub-Saharan Africa with reports from Uganda (Otim-Nape *et al.*, 1997), Kenya and Tanzania (Karakacha *et al.*, 2001), Sudan (Harrison *et al.*, 1997), Rwanda (Legg *et al.*, 2001), the

Democratic Republic of Congo (Neuenschwander *et al.*, 2002), Burundi (Bigirimana *et al.*, 2004), Gabon (Legg *et al.*, 2004), Republic of Congo (Ntawuruhunga *et al.*, 2007), Burkina Faso (Tiendrébéogo *et al.*, 2009), Angola (Kumar *et al.*, 2008) and Cameroon (Akinbade *et al.*, 2010). More detailed historical accounts of CMGs have been the subject of several reviews (Thottappilly *et al.*, 2003; Atiri *et al.*, 2004; Legg and Fauquet, 2004; Patil and Fauquet, 2009; Alabi *et al.*, 2011).

5.4 Genome Structure, Function and Diversity

CMGs are characterized by distinct quasi-isometric geminate (twin) particles measuring about 30 × 20 nm. The virions contain circular, bipartite, single-stranded DNA (ssDNA) genomes encapsidated in protein coats of about 30 kDa (Stanley *et al.*, 2005). The two genomic components of CMGs, referred to as DNA A and DNA B (Stanley and Gay, 1983) and each approximately 2.7–2.8 kb in size, are distinct in terms of the number and function of genes each encode but both share a stretch of *circa* 200-nucleotide long sequences referred to as the common region. The common region encompasses a conserved stem-loop structure and contains several regulatory elements including the nonanucleotide TAATATTA↓C sequence (arrow denotes the nicking site for initiation of virion-sense DNA replication), and the TATA box and iterons that act as binding sites for the replication-associated protein (Hanley-Bowdoin *et al.*, 1999). By far the most informative of both genomic components is the DNA A that encodes two overlapping virion-sense open reading frames (ORFs) AV1 and AV2, and at least four overlapping complementary-sense ORFs AC1, AC2, AC3 and AC4 (Fauquet *et al.*, 2008; Brown *et al.*, 2012; Hull, 2014). AV1 encodes the coat protein gene and is the determinant of vector transmission (Harrison *et al.*, 2002) in addition to its role in genome encapsidation. As depicted by their names, the complementary-sense genes AC1 through AC4, individually and in concert, are implicated in the replication of CMGs

within the host cell. ORF AC1 encodes a replication-associated protein (Rep), AC2 a transcriptional activator protein (TrAP) and AC3 a replication enhancer protein (REn). ORF AC4 plays a role as a host activation protein, which serves as an important symptom determinant implicated in cell-cycle control, and may also counteract the host response to Rep gene expression (Stanley *et al.*, 2005). The AC4 protein of EACMCV was shown to be a pathogenicity determinant and suppressor of the systemic phase of RNA silencing in *N. benthamiana* (Fondong *et al.*, 2007). The AC2 of ACMV can act as a *trans*-activator of several plant genes (Trinks *et al.*, 2005) in addition to its role as a *trans*-activator of the late viral genes AV1 and BV1 (Sunter and Bisaro, 1992). The TrAP also functions in the suppression of post-transcriptional gene silencing (Vanitharani *et al.*, 2004). A putative ORF, AC5, encoded in the complementary sense and embedded within the coat protein gene, reported for some CMGs (Hong *et al.*, 1993) has not yet been proven to be transcribed and translated. The AV2 ORF, a signature of Old World begomoviruses (Rybicki, 1994), functions as a movement protein. The two ORFs of the DNA B component, BV1 and BC1, encode the nuclear shuttle protein and the movement protein, respectively (Brown *et al.*, 2012). These two ORFs are non-overlapping and code for genes that play a role in intracellular (BV1) and intercellular (BC1) movement of virions within the host plant cell (Hull, 2014).

Since CMGs are often present as mixtures in infected plants, they may exchange genetic material to produce recombinant variants (Deng *et al.*, 1997; Zhou *et al.*, 1998; Fondong *et al.*, 2000; Maruthi *et al.*, 2004; Tiendrébéogo *et al.*, 2012) and are also sometimes associated with satellite DNA molecules (Ndunguru *et al.*, 2008; Patil and Fauquet, 2010) that can modulate disease. The most successful of such recombinant CMGs is EACMV-UG with a DNA A genome composed of 16% ACMV and 84% EACMV (Deng *et al.*, 1997), as a consequence of exchange of genetic material between the two parental viruses (Zhou *et al.*, 1997). Other recombinant CMGs include: SACMV, EACMCV, EACMMV, EACMZV, EACMKV, SLCMV and ICMV, indicating that the genomes of CMGs are highly plastic giving rise to stable recombinant variants capable of independent evolution (Fig. 5.2a). Although the majority of recombinant CMGs, including the recently characterized Cassava mosaic Madagascar virus, have an EACMV lineage, it was not until very recently that an ACMV-like recombinant CMG was reported to occur naturally in cultivated cassava. Genetic analysis of the complete DNA A genome of African cassava mosaic Burkina Faso virus indicates that it arose by interspecific recombination between West African isolates of ACMV as the major parental virus, and minor parents related to Tomato leaf curl Cameroon virus and *Cotton leaf curl Gezira virus* (Tiendrébéogo *et al.*, 2012). Some of the factors that could contribute to molecular diversity among CMGs, especially those with an EACMV lineage, were reviewed by Patil and Fauquet (2009).

Adding to the complexity of the CMD situation is a recent report that disease resistance-breaking satellite DNA molecules have been found associated with CMD in Tanzania (Ndunguru *et al.*, 2008). Many ssDNA satellites of ~1.3 kb have been associated with several begomovirus disease complexes and they are generally of two types: (i) the nanovirus-like DNA 1 or alpha-satellites; and (ii) the DNA B-like DNA β or beta-satellites (Briddon *et al.*, 2008; Nawaz-ul-Rehman and Fauquet, 2009; Hull, 2014). The alpha-satellites are capable of independent replication although they depend on the helper virus-encoded proteins for their movement and encapsidation, whereas the beta-satellites depend on their helper virus for replication, movement and encapsidation (Briddon *et al.*, 2008; Nawaz-ul-Rehman *et al.*, 2009; Hull, 2014). It was recently demonstrated that several CMGs showed contrasting and differential interactions with alpha- and beta-satellites derived from other *Begomovirus* species resulting in the modulation of symptom phenotypes by these satellites in *N. benthamiana* (Patil and Fauquet, 2010). Since the development and deployment of disease resistant cultivars remain the main thrust of CMD management in sub-Saharan Africa, the possibility that disease resistance-breaking satellite DNA molecules exist within the

sub-region could further complicate the CMD situation and pose a significant new threat to cassava production.

5.5 Host Range

All known CMGs and their genetic variants are capable of systemic infection of cassava, their natural host plant. In addition to cultivated cassava, some CMGs have also been reported to occur naturally in *Manihot glaziovii* Müll. Arg., a wild relative of cultivated cassava known also by the common name of Ceará Rubber (Fauquet and Fargette, 1990; Ogbe *et al.*, 2006; Alabi *et al.*, 2008) and in several non-cassava plants including *Jatropha multifida* (Euphorbiaceae; Fauquet and Fargette, 1990), *Senna occidentalis* (Fabaceae; Ogbe *et al.*, 2006; Alabi *et al.*, 2008), *Glycine max* (Fabaceae; Alabi *et al.*, 2008; Mgbechi-Ezeri *et al.*, 2008), *Ricinus communis* (Euphorbiaceae; Shoyinka *et al.*, 2001; Alabi *et al.*, 2008), *Combretum confertum* (Combretaceae; Ogbe *et al.*, 2006; Alabi *et al.*, 2008), *Leucaena leucocephala* (Fabaceae; Alabi *et al.*, 2011), *Centrosema pubescens* (Fabaceae; Monde *et al.*, 2010), *Pueraria phaseoloides* var. *javanica* (Fabaceae; Monde *et al.*, 2010) and *Jatropha curcas* (Euphorbiaceae; Ramkat *et al.*, 2011). These reports are largely based on serological and/or molecular detection and/or characterization of the infecting CMGs from these hosts. Transmission of CMGs from non-cassava hosts back to cassava has not been undertaken, although anecdotal reports indicating that non-cassava host of CMGs also support high whitefly populations (Ogbe *et al.*, 2006; Alabi *et al.*, 2008) implicate them in the epidemiology of CMD as virus reservoirs for vector-mediated transmission of CMGs to cassava. A possible role for these non-cassava hosts in CMD epidemiology was further supported by experimental demonstration of whitefly-mediated transmission of ACMV from cassava to *J. curcas* (Amoatey *et al.*, 2013). It has also been demonstrated that CMGs are capable of infecting additional non-cassava plant species under experimental conditions. For instance, ACMV has been shown to infect members of the family Solanaceae, especially those belonging to the genera *Nicotiana* and *Datura* (Bock and Woods, 1983); additionally, ICMV, SLCMV and EACMV can infect some species in the genus *Nicotiana*, SACMV is capable of infecting *Phaseolus vulgaris* (Fabaceae; Berrie *et al.*, 2001) and *Malva parviflora* (Malvaceae; Berrie *et al.*, 2001), and SLCMV was shown to infect *Ageratum conyzoides* (Asteraceae; Saunders *et al.*, 2002) and *Arabidopsis thaliana* (Brassicaceae; Mittal *et al.*, 2008).

5.6 Transmission and Epidemiology

Cassava is a vegetatively propagated crop, with the consequence that CMGs, their genetic variants and DNA satellite molecules often take advantage of this mode of propagation for their spread over short and long distances. Since CMD symptoms are only apparent on the foliage and local farmers are seldom familiar with the viral aetiology of symptomatic plants, frequent exchange of virus-infected vegetative cuttings is the primary route of CMD spread across most cassava-growing regions. CMGs are also transmissible through grafting (Atiri *et al.*, 2004), via biolistic inoculation using a gene gun (Briddon *et al.*, 1998; Makwarela *et al.*, 2006) and by agroinoculation (Berrie *et al.*, 2001). These procedures have therefore been used for biological characterization of CMG isolates (Fondong *et al.*, 2000) and to conduct virus infectivity assays (Berrie *et al.*, 2001).

From the infection foci generated by the establishment of virus-infected host plants, the whitefly vector can acquire and transmit CMGs to otherwise healthy plants, thus contributing to short distance (within field) and sometimes long-distance (between fields) spread of CMD (Chant, 1958; Dubern, 1994). Epidemiological studies have demonstrated that external sources of inoculum are more important than internal sources, and that there are strong environmental gradients of spread linked to prevailing wind direction (Fargette *et al.*, 1990). Whitefly abundance and seasonal population dynamics determine the rate and pattern of CMG spread within cassava fields, and these populations are in turn driven by local climatic conditions (Fargette *et al.*, 1994). Vector abundance is much more important as a determinant of CMG spread than transmission efficiency (Legg, 2010). However, there is some evidence for virus–vector co-adaptation in the CMD–cassava pathosystem, since cassava *B. tabaci* populations from specific geographical locations transmitted homologous viruses with equivalent efficiency, whereas the same population transmitted CMGs from other regions with significantly lower efficiency (Maruthi *et al.*, 2002). Importantly, while African *B. tabaci* were less efficient in transmitting CMGs from South Asia than in transmitting African CMGs, and Asian *B. tabaci* were similarly better adapted to transmitting their own CMGs, there were no significant differences in transmission of ACMV (from West Africa) and EACMV (from East Africa) by *B. tabaci* from West or East Africa. In addition to *B. tabaci*, other whitefly species such as *B. afer* (Priesner & Hosny) can also transmit CMGs (Palaniswami *et al.*, 1996), albeit at lower efficiencies. Generally, starvation of non-viruliferous whiteflies prior to acquisition feeding on infected cassava accelerated virus acquisition from source plants (Dubern, 1994).

Once acquired, a latent period of about 6–8 hours must lapse before the whitefly is able to transmit the virus, which can thereafter be retained by an infectious whitefly for at least 9 days (Dubern, 1994). Viruliferous whiteflies require a 10–30 minute inoculation access period for virus inoculation into healthy cassava plants. Under experimental conditions, ten viruliferous whiteflies are needed to achieve the optimal rate of transmission when released on a cassava plant (Dubern, 1994), although a single whitefly is capable of virus transmission (Chant, 1958; Dubern, 1994). ACMV is transstadially (Chant, 1958; Dubern, 1994), but not transovarially (Dubern, 1994) transmitted. Of note, CMGs can be transmitted via mechanical inoculation from cassava to herbaceous hosts (Bock and Woods, 1983; Amoatey *et al.*, 2013) albeit with difficulty, but they are not known to be seed-borne or seed-transmitted in cassava or transmitted via dodder (Storey and Nichols, 1938).

Populations of *B. tabaci* occurring on cassava have been shown to be genetically distinct to those that occur on annual crops and herbaceous weeds (Burban *et al.*, 1992; Berry *et al.*, 2004; Sseruwagi *et al.*, 2005). Furthermore, non-cassava genotypes are unable to colonize cassava and die when forced to feed on the plant (Burban *et al.*, 1992; Legg *et al.*, 1994). The ability of cassava-colonizing genotypes to feed on several crop and weed species other than cassava (Sseruwagi *et al.*, 2006) does mean that potential exists for alternative hosts of CMGs to act as reservoirs of these viruses. Since cassava is strongly preferred by cassava-colonizing *B. tabaci* genotypes, and cassava crops are typically available year round, infected cassava plantings are by far the most important source of CMG inoculum. However, non-cassava hosts may harbour other begomoviruses along with CMGs, thus creating potential recombination opportunities that may give rise to new genetic variants and novel CMGs. The recombinant African cassava mosaic Burkina Faso virus (Tiendrébéogo *et al.*, 2012) provides a good example of such an occurrence.

5.7 Diagnostics

The successful purification of CMGs (Bock *et al.*, 1977) paved the way for their antibody-based diagnoses. Polyclonal antibodies have been used for the detection of ACMV in cassava leaf samples by the double antibody sandwich method of ELISA (Sequeira and Harrison, 1982) and immunosorbent electron microscopy (Roberts *et al.*, 1984). The availability of a panel of monoclonal antibodies (Thomas *et al.*, 1986) spurred rapid detection and discrimination of CMGs using triple antibody sandwich-ELISA (Thomas *et al.*, 1986; Harrison *et al.*, 2002). Although diagnosis of CMGs by ELISA is versatile and can be used for large-scale testing of field samples in diagnostic surveys (Ogbe *et al.*, 1997), its major limitation lies in its inability to distinguish different CMGs in mixed virus infections (Thottappilly *et al.*, 2003). In addition, similarities in the coat protein

epitopes of recombinant CMGs such as EACMV-UG and their parental viruses further complicate efforts to differentiate CMGs by ELISA in mixed-infected plants (Thottappilly *et al.*, 2003). Thus, the advent of the PCR technique has advanced molecular diagnosis of CMGs in singleplex (Fondong *et al.*, 2000; Berry and Rey, 2001b; Pita *et al.*, 2001b; Ndunguru *et al.*, 2005; Ogbe *et al.*, 2006; Alabi *et al.*, 2008; Sserubombwe *et al.*, 2008; Monde *et al.*, 2010) and multiplex (Alabi *et al.*, 2008; Abarshi *et al.*, 2012; Aloyce *et al.*, 2013) formats and has contributed to the rapid and reliable assessment of CMGs in epidemiological studies, crop improvement and phytosanitary programs in many sub-Saharan African countries. In most cases, these assays are developed using oligonucleotide primers specific to the DNA A component of CMGs. Amplified DNA fragments are then analyzed using restriction enzymes (Sserubombwe *et al.*, 2008) in heteroduplex mobility assays (Berry and Rey, 2001a) or sequenced for profiling CMGs (Fondong *et al.*, 2000; Berry and Rey, 2001b; Pita *et al.*, 2001b; Ndunguru *et al.*, 2005; Alabi *et al.*, 2008; Sserubombwe *et al.*, 2008; Monde *et al.*, 2010). A multiplex assay has also been manipulated to incorporate the simultaneous detection of RNA viruses associated with CBSD when present in mixed infections with CMGs (Abarshi *et al.*, 2012). Although PCR-based assays allow for sensitive and rapid detection of CMGs, the lack of capacity for molecular diagnostics in many sub-Saharan African countries, coupled with challenges associated with cold chain procurement of molecular reagents, make ELISA the most valuable and affordable diagnostic tool in these countries.

5.8 Distribution

CMD is widely distributed across sub-Saharan Africa and also occurs in cassava-growing regions of India and Sri Lanka (Fig. 5.3). Although cassava originated from South America, there is no report of CMD to date from that region, suggesting that CMGs are indigenous to sub-Saharan Africa and to

Fig. 5.3. Global distribution map of cassava mosaic disease (CMD). Each black circle represents countries where CMD has been reported. The image was produced using the Google Earth software and modified on Adobe Photoshop CS6.

a lesser extent South-East Asia, where it is assumed that they were present in indigenous host plants long before cassava was introduced. For Africa, it has been argued that East Africa may be the centre of origin of CMGs (Ndunguru *et al.*, 2005). Coupled with reports of an African origin for *B. tabaci* (Boykin *et al.*, 2007), the CMGs–cassava pathosystem represents a type of 'new encounter phenomenon' (Buddenhagen, 1977) where a pathogen that co-evolved with indigenous plant species made a host 'jump' to an introduced plant species, with the help of a competent vector, and subsequently becomes an important pathogen (Thresh and Fargette, 2001). The frequent cross-border exchange of cassava vegetative cuttings of unknown virus infection status and windaided long-distance dispersal of viruliferous whiteflies across natural boundaries may have then contributed to the widespread distribution of CMD and CMGs across sub-Saharan Africa. Indeed, a recently conducted evolutionary study concluded that anthropic factors in the spread of CMGs from Africa to the South West Indian Ocean Islands are the principal axes of viral migration corresponding with major routes of human movement

and commercial trade (De Bruyn *et al.*, 2012). While humans may be largely responsible for long-distance movements of CMGs, *B. tabaci* whiteflies have also been shown to be responsible for the spread of CMGs across large areas of East and Central Africa at rates of 20–30 km per year (Legg, 1999).

5.9 Economic Impact

But for a few limited recent reports, the majority of yield loss estimates due to CMD were conducted in the last century (Thresh *et al.*, 1994; Thottappilly *et al.*, 2003; Legg *et al.*, 2004). A synthesis of these yield loss estimates led Thresh *et al.* (1997) to approximate annual CMD-associated tuberous root loss to be between 15% and 24%, an equivalent of 12–23 million tonnes or US$1.2–2.3 billion. Other studies reported a region-wide CMD-associated yield loss in sub-Saharan Africa to be over 30 million tonnes of fresh cassava roots on an annual basis (Legg and Thresh, 2000; Legg *et al.*, 2006). As previously mentioned, such losses are often influenced by the severity of the disease owing to the presence of single or multiple viruses and their genetic variants, and the level of susceptibility of affected cultivars. For instance, Owor *et al.* (2004) reported that a huge variation in CMD-associated losses, ranging from 12%–82%, was a function of the presence of single or mixed virus infections. A regional pandemic of an unusually severe form of CMD further underscored the potential of CMGs to cause not only significant economic problems, but also associated negative social impacts with a magnitude comparable to the infamous potato late blight disease outbreak in the 19th century in Ireland. The severe CMD pandemic began in Uganda in the early to mid-1990s (Gibson *et al.*, 1996; Otim-Nape *et al.*, 1997) on popular and widely cultivated cassava varieties and soon spread to other countries in East Africa, including Kenya and Tanzania (Otim-Nape *et al.*, 1997; Legg, 1999). The CMD pandemic devastated many cassava farms, forced thousands of subsistence farmers to abandon the crop (Otim-Nape *et al.*, 1997), and resulted in famine-related deaths (Otim-Nape *et al.*, 1998).

5.10   Management

A disease as complex as CMD requires a multi-faceted management approach that takes into account the different vertices of the 'disease quadrangle' consisting of CMGs, their natural and alternative host plants, their whitefly vector, and environmental factors influencing/modulating each of these elements. To this end, numerous approaches have been developed for the management of CMD and many of these approaches have been discussed in several review articles (Atiri *et al.*, 2004; Thresh and Cooter, 2005; Vanderschuren *et al.*, 2007). A synopsis of CMD management strategies is discussed below under four broad categories.

5.10.1   Natural and transgenic resistance to cassava mosaic disease

The main thrust of CMD management over the years has been breeding for resistance to CMGs using conventional approaches (Thresh and Cooter, 2005; Dixon *et al.*, 2001, 2010). The primary source of resistance used in these efforts was *M. glaziovii* (Jennings, 1994), although later efforts also focused on cassava landraces with single CMD resistance genes (Fregene *et al.*, 2001). This resulted in the production of a series of resistant materials, notably the tropical *Manihot* species (TMS) derived from experimental crosses, the tropical *Manihot esculenta* (TME) lines that mainly comprised local West African landraces, and various crosses between the two groups (TMS × TME). Combination of polygenic TMS-type resistance (Jennings, 1994) and single gene-derived TME-type resistance (Fregene *et al.*, 2001) gave rise to several varieties that were near immune to CMG infection. These were evaluated and released in countries of East and Central Africa and were hugely successful in controlling the pandemic of severe CMD in the region (Jennings, 1994; Legg *et al.*, 2006). Such materials also show resistance to cassava bacterial blight, caused by *Xanthomonas axonopodis* pv. *manihotis*, and have been widely deployed across sub-Saharan Africa over the course of several

decades (Manyong *et al.*, 2000). Unfortunately, most of the CMD-resistant materials are similar to local varieties in being susceptible to CBSD (Winter *et al.*, 2010). In order to address this problem, efforts are being made to use transgenic approaches based on RNA interference to incorporate CBSD resistance into varieties that already have high levels of conventionally bred CMD resistance and farmer-preferred quality characteristics (Vanderschuren *et al.*, 2012). Furthermore, the use of marker-assisted breeding techniques (Akano *et al.*, 2002; Lokko *et al.*, 2005; Okogbenin *et al.*, 2012) and the genotype-by-sequencing approach (Rabbi *et al.*, 2014) have also been exploited to fast-track the conventional breeding process.

To complement conventional breeding programs, recent global efforts have focused on the development of transgenic resistance to CMD (Vanderschuren *et al.*, 2007; Vanderschuren *et al.*, 2009; Sayre *et al.*, 2011) while preserving consumer-preferred attributes and enhancing other nutritional and agronomic attributes of cassava. Such efforts are now being integrated into the larger goal of the BioCassava Plus initiative, funded by the Bill and Melinda Gates Foundation, with the main objective of reducing malnutrition by delivering improved cassava cultivars that provide complete and balanced nutrition in a readily marketable and higher yielding food crop (https://www.danforthcenter.org/scientists-research/research-institutes/institute-for-international-crop-improvement/crop-improvement-projects/biocassava-plus). The thrust of transgenic resistance breeding efforts in cassava against CMD has been based on RNA interference technology (Sayre *et al.*, 2011). Initially, several targets including the viral non-coding intergenic region and messenger RNAs of Rep (AC1), TrAP (AC2) and REn (AC3) showed promise for increased ACMV resistance (Zhang *et al.*, 2005). Subsequent efforts have led to the production of stable transgenic cassava lines overexpressing hairpin double-stranded RNAs homologous to the non-coding intergenic region of ACMV. Genetically modified cassava lines so produced, though susceptible to ACMV infection, showed an enhanced recovery phenotype (Vanderschuren *et al.*, 2007) relative to their non-transgenic cousins. In addition, transgenic cassava lines generated via a dose-dependent constitutive expression of artificial hairpin double-stranded RNAs homologous to the Rep coding sequence showed immunity to virus infection in an otherwise ACMV-susceptible cultivar (Vanderschuren *et al.*, 2009). The added advantage of a Rep coding sequence-targeted transgenic resistance is that the gene is fairly conserved across geminiviruses, which means that transgenic lines may have broader resistance to multiple CMGs and perhaps other geminiviruses as well (Brunetti *et al.*, 2001). Transgenic cassava lines are still undergoing field trials in Uganda, Kenya and, most recently in Nigeria to evaluate their performance under natural CMD pressure. Hopefully, increased public awareness of the benefits of transgenic cassava, the debunking of myths and falsehoods surrounding the supposed risk of their consumption, and the establishment of enabling legislatures across sub-Saharan African countries will permit large-scale field evaluation of transgenic materials resulting in their accelerated release for commercial production.

5.10.2 Cassava mosaic disease avoidance and cultural control

Among the various cultural methods evaluated for CMD management, planting of virus-free cuttings is the most effective for minimizing spread of CMD in Africa (Fargette *et al.*, 1994; Thresh and Cooter, 2005). Currently, most farmers in sub-Saharan Africa do their own cassava vegetative cutting selection from the preceding season's crop, often at a time when most of the foliage has senesced and CMD symptoms are no longer apparent. Hence, CMD-affected cuttings can be inadvertently selected leading to disease perpetuation between seasons and across fields. With the transition of the cassava enterprise in sub-Saharan Africa from subsistence to a more commercially oriented approach (Nassar and Ortiz, 2010), it is hoped that cassava vegetative cutting production will be undertaken by entities capable of screening such materials for virus infection and mass producing virus-free propagules under controlled conditions to ensure their virus-free status. In addition to clean plants, cultural management strategies employed to tackle the menace of CMD include disease and/or vector avoidance through adjustment of planting dates (Adipala *et al.*, 1998; Adjata *et al.*, 2012), intercropping and varietal mixtures (Sserubombwe *et al.*, 2001; Fondong *et al.*, 2002) and roguing of infected plants especially at early stages of growth (Thresh and Otim-Nape, 1994). The results of these approaches are often variable with some studies reporting them as efficacious (Sserubombwe *et al.*, 2001; Fondong *et al.*, 2002; Thresh and Otim-Nape, 1994) while others dispute their effectiveness (Fargette and Fauquet, 1988; Otim-Nape *et al.*, 1997). Roguing of volunteer cassava plants and alternative hosts of CMGs (Alabi *et al.*, 2008; Bragard *et al.*, 2013) also deserve attention to exclude potential sources of virus inoculum and further ensure the success of CMD cultural control efforts.

5.10.3 Whitefly management

Farmers in sub-Saharan African countries seldom practice chemical control of the whitefly vector of CMGs due to the costs associated with implementing such measures, including the cost of the chemical product as well as the application equipment. In addition, pesticides are relatively less effective in controlling arthropod-borne viruses if the main spread is from external sources and not within crops (Thresh and Cooter, 2005). The negative impact of pesticides on the environment and the risks to beneficial organisms, including pollinators and natural enemies, as well as risks to public health makes pesticide use less appealing. The potential for biological control of the whitefly vector remains to be exploited. Preliminary studies conducted on this theme have highlighted the important role that natural enemies currently play in constraining whitefly populations, but have not yet demonstrated how natural enemies can be effectively utilized to deliver a sufficient level of whitefly control to cause significant reductions in the spread of CMGs (Legg *et al.*, 2014a and references therein). It is believed that the transition of cassava from a largely subsistence crop to a more commercially oriented enterprise will encourage an increased willingness of growers to invest in their crop, thus providing new opportunities for the development and application of integrated whitefly control strategies akin to the successful integrated pest management programmes developed in different regions of the USA (Legg *et al.*, 2014b).

5.10.4 Cassava mosaic disease monitoring and forecasting

Virus disease surveillance programs can help determine the status of the disease, its associated viruses and whitefly vector abundance. Such information can in turn be employed towards the design and implementation of disease management strategies. Several CMD surveys have been, and continue to be, conducted across sub-Saharan Africa over the years (Ndunguru *et al.*, 2005; Bull *et al.*, 2006; Ogbe *et al.*, 2006; Sserubombwe *et al.*, 2008; De Bruyn *et al.*, 2012; Harimalala *et al.*, 2012; Muengula-Manyi *et al.*, 2012; Zinga *et al.*, 2012; Chikoti *et al.*, 2013), and approaches used and results obtained have been reviewed (Legg and Thresh, 2001; Sseruwagi *et al.*, 2004). Surveys have been used to generate CMG distribution maps across the region and have sometimes facilitated the discovery of novel CMGs and their strains. The application of surveillance data has, however, largely been restricted to raising awareness about threats of new spread to previously less-affected areas and to developing course predictions of likely future patterns of epidemic development. Considerable opportunities exist for the more intensive use of these datasets, such as in the design of integrated pest management programmes for disease and vector management similar to several integrated pest management pest information programmes being implemented in the developed countries for management of various economically important diseases of different crops (http://www.ipmpipe.org/).

5.11 Concluding Remarks

In conclusion, cassava mosaic disease will continue to engage the global scientific community for years to come given the complexity of the disease, the successful attributes of CMGs and their whitefly vector and the increasing value of cassava. However, when it comes to effective management of CMD, it will continue to be impracticable to recommend a 'one-size-fits-all' strategy. Rather, an integrated approach that takes into consideration 'traditional' versus 'novel', 'ancient' versus 'modern', and 'conventional' versus 'unconventional' solutions will hold better promise for a comprehensive ecologically friendly and economically sustainable CMD management effort in sub-Saharan Africa. In line with this, largely neglected components of the CMG–cassava pathosystem such as volunteer cassava plants and alternative hosts of CMGs, abundance of natural enemies of *B. tabaci* and the use of transgenic resistance will need to form part of an overall CMD management strategy.

References

Abarshi, M.M., Mohammed, I.U., Jeremiah, S.C., Legg, J.P., Lava-Kumar, P., *et al.* (2012) Multiplex RT-PCR assays for the simultaneous detection of both RNA and DNA viruses infecting cassava and the common occurrence of mixed infections by two cassava brown streak viruses in East Africa. *Journal of Virological Methods* 179, 176–184.

Adipala, E., Byabakama, B.A., Ogenga-Latigo, M.W. and Otim-Nape, G.W. (1998) Effect of planting date and varietal resistance on the development of cassava mosaic virus disease in Uganda. *African Plant Protection* 4, 71–79.

Adjata, K.D., Tchacondo, T., Tchansi, K., Banla, E. and Gumedzoe, Y.M.D. (2012) Cassava mosaic disease transmission by whiteflies (*Bemisia tabaci* Genn.) and its development on some plots of cassava (*Manihot esculenta* Crantz) clones planted at different dates in Togo. *American Journal of Plant Physiology* 7, 200–211.

Akano, A.O., Dixon, A.G.O., Mba, C., Barrera, E. and Fregene, M. (2002) Genetic mapping of a dominant gene conferring resistance to cassava mosaic disease. *Theoretical and Applied Genetics* 105, 521–535.

Akinbade, S.A., Hanna, R., Nguenkam, A., Njukwe, E., Fotso, A., *et al.* (2010) First report of the *East African cassava mosaic virus*-Uganda (EACMV-UG) infecting cassava (*Manihot esculenta*) in Cameroon. *New Disease Reports* 21, 22.

Alabi, O.J., Ogbe, F.O., Bandyopadhyay, R., Kumar, P.L., Dixon, A.G.O., *et al.* (2008) Alternate hosts of *African cassava mosaic virus* and *East African cassava mosaic Cameroon virus* in Nigeria. *Archives of Virology* 153, 1743–1747.

Alabi, O.J., Kumar, P.L. and Naidu, R.A. (2011) Cassava mosaic disease: a curse to food security in sub-Saharan Africa. *APSnet Features*. Available at: http://www.apsnet.org/publications/apsnetfeatures/Pages/cassava.aspx (accessed 30 October 2014).

Aloyce, R.C., Tairo, F., Sseruwagi, P., Rey, M.E. and Ndunguru, J. (2013) A single-tube duplex and multiplex PCR for simultaneous detection of four cassava mosaic begomovirus species in cassava plants. *Journal of Virological Methods* 189, 148–156.

Amoatey, H.M., Appiah, A.S., Danso, K.E., Amiteye, S., Appiah, R., *et al.* (2013) Controlled transmission of *African cassava mosaic virus* (ACMV) by *Bemisia tabaci* from cassava (*Manihot esculenta* Crantz) to seedlings of physic nut (*Jatropha curcas* L.). *African Journal of Biotechnology* 12, 4465–4472.

Atiri, G.I., Ogbe, F.O., Dixon, A.G.O., Winter, S. and Ariyo, O. (2004) Status of cassava mosaic virus diseases and cassava begomoviruses in sub-Saharan Africa. *Journal of Sustainable Agriculture* 24, 5–35.

Berrie, L.C., Palmer, K.E., Rybicki, E.P. and Rey, M.E.C. (1998) Molecular characterization of a distinct South African cassava infecting geminivirus. *Archives of Virology* 143, 2253–2260.

Berrie, L.C., Rybicki, E.P. and Rey, M.E. (2001) Complete nucleotide sequence and host range of *South African cassava mosaic virus*: further evidence for recombination amongst begomoviruses. *Journal of General Virology* 82, 53–58.

Berry, S. and Rey, M.E.C. (2001a) Differentiation of cassava-infecting begomoviruses using heteroduplex mobility assays. *Journal of Virological Methods* 92, 151–163.

Berry, S. and Rey, M.E.C. (2001b) Molecular evidence for diverse populations of cassava-infecting begomoviruses in southern Africa. *Archives of Virology* 146, 1795–1802.

Berry, S.D., Fondong, V., Rey, C., Rogan, D., Fauquet, C.M., *et al.* (2004) Molecular evidence for five distinct *Bemisia tabaci* (Homoptera: Aleyrodidae) geographic haplotypes associated with cassava in sub-Saharan Africa. *Annals of the Entomological Society of America* 97, 852–859.

Bigirimana, S., Barumbanze, P., Obonyo, R. and Legg, J.P. (2004) First evidence for the spread of *East African cassava mosaic virus*-Uganda (EACMV-UG) and the pandemic of severe cassava mosaic disease to Burundi. *Plant Pathology* 53, 231.

Bock, K.R. and Woods, R.D. (1983) Etiology of African cassava mosaic disease. *Plant Disease* 67, 994–995.

Bock, K.R., Guthrie, E.J., Meredith, G. and Barker, H. (1977) RNA and protein components of maize streak and cassava latent viruses. *Annals of Applied Biology* 85, 305–308.

Boykin, L.M., Shatters, R.G. Jr., Rosell, R.C., McKenzie, C.L., Bagnall, R.A., *et al.* (2007) Global relationships of *Bemisia tabaci* (Hemiptera: Aleyrodidae) revealed using Bayesian analysis of mitochondrial *COI* DNA sequences. *Molecular Phylogenetics and Evolution* 44, 1306–1319.

Bragard, C., Caciagli, P., Lemaire, O., Lopez-Moya, J.J., MacFarlane, S., *et al.* (2013) Status and prospects of plant virus control through interference with vector transmission. *Annual Review of Phytopathology* 51, 177–201.

Briddon, R.W., Liu, S., Pinner, M.S. and Markham, P.G. (1998) Infectivity of *African cassava mosaic virus* clones to cassava by biolistic inoculation. *Archives of Virology* 143, 2487–2492.

Briddon, R.W., Brown, J.K., Moriones, E., Stanley, J., Zerbini, M., *et al.* (2008) Recommendations for the classification and nomenclature of the DNA-β satellites of begomoviruses. *Archives of Virology* 153, 763–781.

Brown, J.K., Fauquet, C.M., Briddon, R.W., Zerbini, M., Moriones, E., *et al.* (2012) *Geminiviridae*. In: King, A.M.Q., Adams, M.J., Carstens, E.B. and Lefkowitz, E.J. (eds) *Virus Taxonomy, Ninth Report of the ICTV*. Elsevier/Academic Press, London, pp. 351–373.

Brown, J.K., Zerbini, M., Navas-Castillo, J., Moriones, E., Ramos-Sobrinho, R. *et al.* (2015) Revision of *Begomovirus* taxonomy based on pairwise sequence comparisons. *Archives of Virology* 160, 1593–1619.

Brunetti, A., Tavazza, R., Noris, E., Lucioli, A., Acotto, G.P., *et al.* (2001) Transgenically expressed T-Rep of *Tomato yellow leaf curl Sardinia virus* acts as a trans-dominant-negative mutant, inhibiting viral transcription and replication. *Journal of Virology* 75, 10573–10581.

Buddenhagen, I.W. (1977) Resistance and vulnerability of tropical crops in relation to their evaluation and breeding. *Annals of the New York Academy of Sciences* 287, 309–326.

Bull, S.E., Briddon, R.W., Sserubombwe, W.S., Ngugi, K., Markham, P.G., *et al.* (2006) Genetic diversity and phylogeography of cassava mosaic viruses in Kenya. *Journal of General Virology* 87, 3053–3065.

Burban, C., Fishpool, L.D.C., Fauquet, C., Fargette, D. and Thouvenel, J.-C. (1992) Host-associated bio-types within West African populations of the whitefly *Bemisia tabaci* (Genn.) (Hom. Aleyrodidae). *Journal of Applied Entomology* 113, 416–423.

Carter, S.E., Fresco, L.O., Jones, P.G. and Fairbairn, J.N. (1995) Introduction and diffusion of cassava in Africa. International Institute of Tropical Agriculture (IITA) Research Guide No. 49, IITA, Ibadan, Nigeria.

Chant, S.R. (1958) Studies on the transmission of cassava mosaic virus by *Bemisia* spp. (*Aleyrodidae*). *Annals of Applied Biology* 46, 210–215.

Chikoti, P.C., Ndunguru, J., Melis, R., Tairo, F., Shanahan, P., *et al.* (2013) Cassava mosaic disease and associated viruses in Zambia: occurrence and distribution. *International Journal of Pest Management* 59, 63–72.

De Bruyn, A., Villemot. J., Lefeuvre, P., Villar, E., Hoareau, M., *et al.* (2012) *East African cassava mosaic*-like viruses from Africa to Indian ocean islands: molecular diversity, evolutionary history and geographical dissemination of a bipartite begomovirus. *BMC Evolutionary Biology* 12, 228.

Deng, D., Otim-Nape, W.G., Sangare, A., Ogwal, S., Beachy, R.N., *et al.* (1997) Presence of a new virus closely related to East African cassava mosaic geminivirus, associated with cassava mosaic outbreak in Uganda. *African Journal of Root Tuber Crops* 2, 23–28.

Dixon, A.G.O., Whyte, J.B.A., Mahungu, N.M. and Ng, S.Y.C. (2001) Tackling the cassava mosaic disease (CMD) challenge in Sub-Saharan Africa: the role of plant host resistance and germplasm deployment. In: Taylor, N.J., Ogbe, F. and Fauquet, C.M. (eds) *Cassava, an Ancient Crop for Modern Times: Food, Health, Culture.* Donald Danforth Plant Sciences Center, St. Louis, Missouri, pp. S8–05.

Dixon, A.G.O., Ogbe, F.O. and Okechukwu, R.U. (2010) Cassava mosaic disease in sub-Saharan Africa: a feasible solution for an unsolved problem. *Outlook on Agriculture* 39, 89–94.

Dubern, J. (1994) Transmission of African cassava mosaic geminivirus by the whitefly (*Bemisia tabaci*). *Tropical Science* 34, 82–91.

FAOSTAT (2013) Food and Agriculture Organization (FAO) of the United Nations, Rome, Italy. Available at: faostat.fao.org (accessed 24 October 2013).

Fargette, D. and Fauquet, C. (1988) A preliminary study on the influence of intercropping maize and cas-sava on the spread of *African cassava mosaic virus* by whiteflies. *Aspects of Applied Biology* 17, 195–202.

Fargette, D., Fauquet, C., Grenier, E. and Thresh, J.M. (1990) The spread of *African cassava mosaic virus* into and within cassava fields. *Journal of Phytopathology* 130, 289–302.

Fargette, D., Jeger, M., Fauquet, C. and Fishpool, L.D.C. (1994) Analysis of temporal disease progress of *African cassava mosaic virus*. *Phytopathology* 84, 91–98.

Fauquet, C. and Fargette, D. (1990) *African cassava mosaic virus*: etiology, epidemiology, and control. *Plant Disease* 74, 404–411.

Fauquet, C.M., Briddon, R.W., Brown, J.K., Moriones, E., Stanley, J., *et al.* (2008) Geminivirus strain de-marcation and nomenclature. *Archives of Virology* 153, 783–821.

Fondong, V.N.F., Pita, S., Rey, M.E.C., de Kochko, A., Beachy, R.N., *et al.* (2000) Evidence of synergism between *African cassava mosaic virus* and a new double-recombinant geminivirus infecting cassava in Cameroon. *Journal of General Virology* 81, 287–297.

Fondong, V.N., Thresh, J.M. and Zok, S. (2002) Spatial and temporal spread of cassava mosaic virus dis-ease in cassava grown alone and intercropped with maize and/or cowpea. *Journal of Phytopathology* 150, 365–374.

Fondong, V.N., Reddy, R.V., Lu, C., Hankoua, B., Felton, C., *et al.* (2007) The consensus N-myristoylation motif of a geminivirus AC4 protein is required for membrane binding and pathogenicity. *Molecular Plant-Microbe Interactions* 20, 380–391.

Fregene, M., Okogbenin, E., Mba, C., Angel, F., Suarez, M.C., *et al.* (2001) Genome mapping in cassava improvement: challenges, achievements and opportunities. *Euphytica* 120, 159–165.

Gibson, R.W., Legg, J.P. and Otim-Nape, G.W. (1996) Unusually severe symptoms are a characteristic of the current epidemic of mosaic virus disease of cassava in Uganda. *Annals of Applied Biology* 128, 479–490.

Hanley-Bowdoin, L., Settlage, S.B., Orozco, B.M., Nagar, S. and Robertson, D. (1999) Geminiviruses: models for plant DNA replication, transcription, and cell cycle regulation. *CRC Critical Reviews in Plant Sciences* 18, 71–106.

Harimalala, M., Lefeuvre, P., De Bruyn, A., Tiendrébéogo, F., Hoareau, M., *et al*. (2012) A novel cassava-infecting begomovirus from Madagascar: Cassava mosaic Madagascar virus. *Archives of Virology* 157, 2027–2030.

Harrison, B.D., Liu, Y.L., Zhou, X., Robinson, D.J., Calvert, L., *et al*. (1997) Properties, differentiation and geographical distribution of geminivirus that cause cassava mosaic disease. *African Journal of Root Tuber Crops* 2, 19–22.

Harrison, B.D., Swanson, M.M. and Fargette, D. (2002) Begomovirus coat protein: serology, variations and functions. *Physiological and Molecular Plant Pathology* 60, 257–271.

Hong, Y.G., Robinson, D.J. and Harrison, B.D. (1993) Nucleotide sequence evidence for the occurrence of three distinct whitefly-transmitted geminiviruses in cassava. *Journal of General Virology* 74, 2437–2443.

Hull, R. (2014) Genome composition, organization, and expression. In: *Plant Virology*, 5th edn. Elsevier, Inc., London, pp. 247–339.

Jennings, D.L. (1994) Breeding for resistance to *African cassava mosaic virus* in East Africa. *Tropical Science* 34, 110–122.

Karakacha, H.W., Koerbler, M., Epampuka, P.B., Ayecho, P. and Winter, S. (2001) Characterization and distribution of begomoviruses causing – Uganda variant – cassava mosaic virus disease in East and Central Africa. *Proceedings of the Third Geminivirus Symposium*, John Innes Center, Norwich, UK, 24-28 July 2001, p. 115.

Kumar, P.L., Akinbade, S.A., Dixon, A.G.O., Mahungu, N.M., Mutunda, M.P., *et al*. (2008) First report of the occurrence of *East African cassava mosaic virus*-Uganda (EACMV-UG) in Angola. *New Disease Reports* 18, 20.

Legg, J.P. (1999) Emergence, spread and strategies for controlling the pandemic of cassava mosaic virus disease in east and central Africa. *Crop Protection* 18, 627–637.

Legg, J.P. (2010) Epidemiology of a whitefly-transmitted cassava mosaic geminivirus pandemic in Africa. In: Stansly, P.A. and Naranjo, S.E. (eds) *Bemisia: Bionomics and Management of a Global Pest*. Springer, Dordrecht–Heidelberg–London–New York, pp. 233–257.

Legg, J.P. and Fauquet, C.M. (2004) Cassava mosaic geminiviruses in Africa. *Plant Molecular Biology* 56, 585–599.

Legg, J.P. and Thresh, J.M. (2000) Cassava mosaic virus disease in East Africa: a dynamic disease in a changing environment. *Virus Research* 71, 135–149.

Legg, J.P. and Thresh, J.M. (2001) Cassava virus diseases in Africa. In: Hughes, J.d' A. and Odu, B.O. (eds) *Proceedings of Plant Virology in Sub-Saharan Africa*. International Institute of Tropical Agriculture, Ibadan, Nigeria, pp. 517–552.

Legg, J.P., Gibson, R.W. and Otim-Nape, G.W. (1994) Genetic polymorphism amongst Ugandan populations of *Bemisia tabaci* (Gennadius) (Homoptera: Aleyrodidae), vector of African cassava mosaic geminivirus. *Tropical Science* 34, 73–81.

Legg, J.P., Okao-Okuja, G., Mayala, R. and Muhinyuza, J.B. (2001) Spread into Rwanda of the severe cassava mosaic virus disease pandemic and associated Ugandan variant of *East African cassava mosaic virus* (EACMV-UG). *Plant Pathology* 50, 796.

Legg, J.P., Ndjelassili, F. and Okao-Okuja, G. (2004) First report of cassava mosaic disease and cassava mosaic geminiviruses in Gabon. *Plant Pathollology* 53, 232.

Legg, J.P., Owor, B., Sseruwagi, P. and Ndunguru, J. (2006) Cassava mosaic virus disease in East and Central Africa: epidemiology and management of a regional pandemic. *Advances in Virus Research* 67, 355–418.

Legg, J.P., Shirima, R., Tajebe, L.S., Guastella, D., Simon, B., *et al*. (2014a) Biology and management of *Bemisia* whitefly vectors of cassava virus pandemics in Africa. *Pest Management Science* 70, 1446–1453.

Legg, J.P., Sseruwagi, P., Boniface, S., Okao-Okuja, G., Shirima, R., *et al*. (2014b) Spatio-temporal patterns of genetic change amongst populations of cassava *Bemisia tabaci* whiteflies driving virus pandemics in East and Central Africa. *Virus Research* 186, 61–75.

Lokko, Y., Danquah, E.Y., Offei, S.K., Dixon, A.G.O. and Gedil, M.A. (2005) Molecular markers associated with a new source of resistance to the cassava mosaic disease. *African Journal of Biotechnology* 4, 873–881.

Makwarela, M., Taylor, N.J., Fauquet, C.M. and Rey, M.E.C. (2006) Biolistic inoculation of cassava (*Manihot esculenta* Crantz) with *South African cassava mosaic virus*. *African Journal of Biotechnology* 5, 154–156.

Manyong, V.M., Dixon, A.G.O., Makinde, K.O., Bokanga, M. and Whyte, J. (2000) The contribution of IITA-improved cassava to food security in sub-Saharan Africa: an impact study. International Institute of Tropical Agriculture, Ibadan, Nigeria.

Maruthi, M.N., Colvin, J., Seal, S., Gibson, G. and Cooper, J. (2002) Co-adaptation between cassava mosaic geminiviruses and their local vector populations. *Virus Research* 86, 71–85.

Maruthi, M.N., Seal, S., Colvin, J., Briddon, R.W. and Bull, S.E. (2004) *East African cassava mosaic Zanzibar virus*—a recombinant begomovirus species with a mild phenotype. *Archives of Virology* 149, 2365–2377.

Matthew, A.V. and Muniyappa, V. (1992) Purification and characterization of *Indian cassava mosaic virus*. *Journal of Phytopathology* 135, 299–308.

Mgbechi-Ezeri, J.U., Alabi, O.J., Naidu, R.A. and Kumar, P.L. (2008) First report of the occurrence of *African cassava mosaic virus* in soybean in Nigeria. *Plant Disease* 92, 1709.

Mittal, D., Borah, B.K. and Dasgupta, I. (2008) Agroinfection of cloned Sri Lankan cassava mosaic virus DNA to *Arabidopsis thaliana*, *Nicotiana tabacum* and cassava. *Archives of Virology* 153, 2149–2155.

Monde, G., Walangululu, J., Winter, S. and Bragard, C. (2010) Dual infection by cassava begomoviruses in two leguminous species (Fabaceae) in Yangambi, Northeastern Democratic Republic of Congo. *Archives of Virology* 155, 1865–1869.

Muengula-Manyi, M., Nkongolo, K.K., Bragard, C., Tshilenge-Djim, P., Winter, S., *et al.* (2012) Incidence, severity and gravity of cassava mosaic disease in savannah agro-ecological region of DR-Congo: analysis of agro-environmental factors. *American Journal of Plant Sciences* 3, 512–519.

Nassar, N. and Ortiz, R. (2010) Breeding cassava to feed the poor. *Scientific American*. Available at: www.scientificamerican.com/article.cfm?id=breeding-cassava (accessed 30 October 2014).

Nawaz-ul-Rehman, M.S. and Fauquet, C.M. (2009) Evolution of geminiviruses and their satellites. *FEBS Letters* 583, 1825–1832.

Ndunguru, J., Legg, J.P., Aveling, T.A.S., Thompson, G. and Fauquet, C.M. (2005) Molecular biodiversity of cassava begomoviruses in Tanzania: evolution of cassava geminiviruses in Africa and evidence for East Africa being a center of diversity of cassava geminiviruses. *Virology Journal* 2, 21.

Ndunguru, J., Fofana, B., Legg, J.P., Chellappan, P., Taylor, N., *et al.* (2008) Two novel satellite DNAs associated with bipartite cassava mosaic begomoviruses enhancing symptoms and capable of breaking high virus resistance in a cassava landrace. In: *Book of Abstracts, Global Cassava Partnership–First Scientific Meeting: Cassava Meeting the Challenges of the New Millennium*. Ghent University, Ghent, Belgium, 21–25 July 2008, p. 141.

Neuenschwander, P., Hughes, J.d'A., Ogbe, F., Ngatse, J.M. and Legg, J.P. (2002) The occurrence of the Uganda variant of *East African cassava mosaic virus* (EACMV-Ug) in western Democratic Republic of Congo and the Congo Republic defines the westernmost extent of the CMD pandemic in East/Central Africa. *Plant Pathology* 51, 384.

Ntawuruhunga, P., Okao-Okuja, G., Bembe, A., Obambi, M., Mvila, A.J.C., *et al.* (2007) Incidence, severity and viruses associated with cassava mosaic disease in the Republic of Congo. *African Crop Sciences Journal* 15, 1e9.

Ogbe, F.O., Legg, J., Raya, M.D., Muimba-Kankolongo, A., Theu, M.P., *et al.* (1997) Diagnostic survey of cassava mosaic viruses in Tanzania, Malawi and Zambia. *Roots* 4, 12–15.

Ogbe, F.O., Atiri, G.I., Dixon, A.G.O. and Thottappilly, G. (2003) Symptom severity of cassava mosaic disease in relation to concentration of *African cassava mosaic virus* in different cassava genotypes. *Plant Pathology* 52, 84–91.

Ogbe, F.O., Dixon, A.G.O., Hughes, J.d'A., Alabi, O.J. and Okechukwu, R. (2006) Status of cassava begomoviruses and their new natural hosts in Nigeria. *Plant Disease* 90, 548–553.

Okogbenin, E., Egesi, C.N., Olasanmi, B., Ogundapo, O., Kahya, S., *et al.* (2012) Molecular marker analysis and validation of resistance to cassava mosaic disease in elite cassava genotypes in Nigeria. *Crop Science* 52, 2576–2586.

Otim-Nape, G.W., Bua, A., Thresh, J.M., Baguma, Y., Ogwal, S., *et al.* (1997) *Cassava Mosaic Virus Disease in Uganda: The Current Pandemic and Approaches to Control*. Natural Resources Institute, Chatham, UK.

Otim-Nape, G.W., Thresh, J.M. and Shaw, M.W. (1998) The incidence and severity of cassava mosaic virus disease in Uganda: 1990–1992. *Tropical Science* 38, 25–37.

Owor, B., Legg, J.P., Okao-Okuja, G., Obonyo, R. and Ogenga-Latigo, M.W. (2004) The effect of cassava mosaic geminiviruses on symptom severity, growth and root yield of a cassava mosaic virus disease-susceptible cultivar in Uganda. *Annals of Applied Biology* 145, 331–337.

Palaniswami, M.S., Nair, R.R., Pillai, K.S. and Thankappan, M. (1996) Whiteflies on cassava and its role as vector of cassava mosaic in India. *Journal of Root Crops* 22, 1–8.

Patil, B.L. and Fauquet, C.M. (2009) Cassava mosaic geminiviruses: actual knowledge and perspectives. *Molecular Plant Pathology* 10, 685–701.

Patil, B.L. and Fauquet, C.M. (2010) Differential interaction between cassava mosaic geminiviruses and geminivirus satellites. *Journal of General Virology* 91, 1871–1882.

Pita, J.S., Fondong, V.N., Sangare, A., Otim-Nape, G.W., Ogwal, S., *et al.* (2001a) Recombination, pseudorecombination and synergism of geminiviruses are determinant keys to the epidemic of severe cassava mosaic disease in Uganda. *Journal of General Virology* 82, 655–665.

Pita, J.S., Fondong, V.N., Sangaré, A., Kokora, R.N.N. and Fauquet, C.M. (2001b) Genomic and biological diversity of the African cassava geminiviruses. *Euphytica* 120, 115–125.

Rabbi, I.Y., Hamblin, M.T., Kumar, P.L., Gedil, M.A., Ikpan, A.S., *et al.* (2014) High-resolution mapping of resistance to cassava mosaic geminiviruses in cassava using genotyping-by-sequencing and its implications for breeding. *Virus Research* 186, 87–96.

Ramkat, R.C., Calari, A., Maghuly, F. and Laimer, M. (2011) Biotechnological approaches to determine the impact of viruses in the energy crop plant *Jatropha curcas*. *Virology Journal* 8, 386.

Roberts, I.M., Robinson, D.J. and Harrison, B.D. (1984) Serological relationships and genome homologies among geminiviruses. *Journal of General Virology* 65, 1723–1730.

Rybicki, E.P. (1994) A phylogenetic and evolutionary justification for three genera of *Geminiviridae*. *Archives of Virology* 139, 49–77.

Saunders, K., Nazeera, S., Mali, V.R., Malathi, V.G., Briddon, R., *et al.* (2002) Characterization of *Sri Lankan cassava mosaic virus* and *Indian cassava mosaic virus*: evidence for acquisition of a DNA B component by a monopartite begomovirus. *Virology* 293, 63–74.

Sayre, R., Beeching, J.R., Cahoon, E.B., Egesi, C., Fauquet, C., *et al.* (2011) The BioCassava Plus Program: biofortification of cassava for sub-Saharan Africa. *Annual Review of Plant Biology* 62, 251–272.

Sequeira, J.C. and Harrison, B.D. (1982) Serological studies on cassava latent virus. *Annals of Applied Biology* 101, 33–42.

Shoyinka, S.A., Thottappilly, G., McGrath, F.F. and Harrison, B.D. (2001) Detection, relationships and properties of cassava mosaic geminivirus in naturally infected castor oil plant, *Ricinus communis* L. in Nigeria. In: *The Fifth International Meeting of the Cassava Biotechnology Network*. Donald Danforth Plant Science Center, St. Louis, Missouri, 4–9 November 2001, pp. S8–S20.

Sserubombwe, W.S., Thresh, J.M., Otim-Nape, G.W. and Osiru, D.O.S. (2001) Progress of cassava mosaic virus disease and whitefly vector populations in single and mixed stands of four cassava varieties grown under epidemic conditions in Uganda. *Annals of Applied Biology* 138, 161–170.

Sseruwagi, P., Sserubombwe, W.S., Legg, J.P., Ndunguru, J. and Thresh, J.M. (2004) Methods of surveying the incidence and severity of cassava mosaic disease and whitefly vector populations on cassava in Africa: a review. *Virus Research* 100, 129–142.

Sseruwagi, P., Legg, J.P., Maruthi, M.N., Colvin, J., Rey, M.E.C., *et al.* (2005) Genetic diversity of *Bemisia tabaci* (Gennadius) (Hemiptera: Aleyrodidae) populations and presence of the B biotype and a non-B biotype that can induce silver leaf symptoms in squash, in Uganda. *Annals of Applied Biology* 147, 253–265.

Sseruwagi, P., Maruthi, M.N., Colvin, J., Rey, M.E.C., Brown, J.K., *et al.* (2006) Colonisation of non-cassava plant species by cassava whiteflies (*Bemisia tabaci*) (Gennadius) (Hemiptera: Aleyrodidae) in Uganda. *Entomologia Experimentalis et Applicata* 119, 145–153.

Sserubombwe, W.S., Briddon, R.W., Baguma, Y.K., Ssemakula, G.N., Bull, S.E., *et al.* (2008) Diversity of begomoviruses associated with mosaic disease of cultivated cassava (*Manihot esculenta* Crantz) and its wild relative (*Manihot glaziovii* Mull. Arg.) in Uganda. *Journal of General Virology* 89, 1759–1769.

Stanley, J. and Gay, M.R. (1983) Nucleotide sequences of cassava latent virus DNA. *Nature* 301, 260–262.

Stanley, J., Bisaro, D.M., Briddon, R.W., Brown, J.K., Fauquet, C.M., *et al.* (2005) Geminiviridae. In: Fauquet, C.M., Mayo, M.A., Maniloff, J., Desselberger, U. and Ball, L.A. (eds) *Virus Taxonomy, VIIIth Report of the ICTV*. Elsevier/Academic Press, London, UK, pp. 301–326.

Storey, H.H. and Nichols, R.F.W. (1938) Studies of the mosaic disease of cassava. *Annals of Applied Biology* 25, 790–806.

Sunter, G. and Bisaro, D.M. (1992) Transactivation of geminivirus AR1 and BR1 gene expression by the viral AL2 gene product occurs at the level of transcription. *The Plant Cell* 4, 1321–1331.

Thomas, J.E., Massalski, P.R. and Harrison, B.D. (1986) Production of monoclonal antibodies to *African cassava mosaic virus* and differences in their reactivity with other whitefly-transmitted geminiviruses. *Journal of General Virology* 67, 2739–2748.

Thottappilly, G., Thresh, J.M., Calvert, L.A. and Winter, S. (2003) Cassava. In: Loebenstein, G. and Thottappilly, G. (eds) *Virus and Virus-Like Diseases of Major Crops in Developing Countries*. Kluwer Academic Publishers, Dordrecht, The Netherlands, pp. 107–165.

Thresh, J.M. and Cooter, R.J. (2005) Strategies for controlling cassava mosaic virus disease in Africa. *Plant Pathology* 54, 587–614.

Thresh, J.M. and Fargette, D. (2001) The epidemiology of African plant viruses: basic principles and concepts. In: Hughes, J.d'A. and Odu, B.O. (eds) *Proceedings of Plant Virology in Sub-Saharan Africa.* International Institute of Tropical Agriculture, Ibadan, Nigeria, pp. 61–111.

Thresh, J.M. and Otim-Nape, G.W. (1994) Strategies for controlling African cassava mosaic geminivirus. *Advances in Disease Vector Research* 10, 215–236.

Thresh, J.M., Fargette, D. and Otim-Nape, G.W. (1994) Effects of African cassava mosaic geminivirus on the yield of cassava. *Tropical Science* 34, 26–42.

Thresh, J.M., Otim-Nape, G.W., Legg, J.P. and Fargette, D. (1997) *African cassava mosaic virus* disease: The magnitude of the problem. *African Journal of Root Tuber Crops* 2, 13–18.

Tiendrébéogo, F., Lefeuvre, P., Hoareau, M., Traoré, E.V.S., Barro, N., *et al.* (2009) Occurrence of *East African cassava mosaic virus*-Uganda (EACMV-UG) in Burkina Faso. *Plant Pathology* 58, 783.

Tiendrébéogo, F., Lefeuvre, P., Hoareau, M., Harimalala, M.A., De Bruyn, A., *et al.* (2012) Evolution of *African cassava mosaic virus* by recombination between bipartite and monopartite begomoviruses. *Virology Journal* 9, 67.

Trinks, D., Rajeswaran, R., Shivaprasad, P.V., Akbergenov, R., Oakeley, E.J., *et al.* (2005) Suppression of RNA silencing by a geminivirus nuclear protein, AC2, correlates with transactivation of host genes. *Journal of Virology* 79, 2517–2527.

Vanderschuren, H., Stupak, M., Futterer, J., Gruissem, W. and Zhang, P. (2007) Engineering resistance to geminiviruses – review and perspectives. *Plant Biotechnology Journal* 5, 207–220.

Vanderschuren, H., Alder, A., Zhang, P. and Gruissem, W. (2009) Dose-dependent RNAi-mediated geminivirus resistance in the tropical root crop cassava. *Plant Molecular Biology* 70, 265–272.

Vanderschuren, H., Moreno, I., Anjanappa, R.B., Zainuddin, I.M. and Gruissem, W. (2012) Exploiting the combination of natural and genetically engineered resistance to cassava mosaic and cassava brown streak viruses impacting cassava production in Africa. *PLoS One* 7, e45277.

Vanitharani, R., Chellappan, P., Pita, J.S. and Fauquet, C.M. (2004) Differential roles of AC2 and AC4 of cassava geminiviruses in mediating synergism and suppression of posttranscriptional gene silencing. *Journal of Virology* 78, 9487–9498.

Warburg, O. (1894) Die kulturpflanzen usambaras. *Mittcilungenaus den Deutschen Schutzgebieten* 7, 131.

Winter, S., Koerbler, M., Stein, B., Pietruszka, A., Paape, M., *et al.* (2010) Analysis of cassava brown streak viruses reveals the presence of distinct virus species causing cassava brown streak disease in East Africa. *Journal of General Virology* 91, 1365–1372.

Zhang, P., Vanderschuren, H., Fütterer, J. and Gruissem, W. (2005) Resistance to cassava mosaic disease in transgenic cassava expressing antisense RNAs targeting virus replication genes. *Plant Biotechnology Journal* 3, 385–397.

Zhou, X., Liu, Y., Calvert, L., Munoz, C., Otim-Nape, G.W., *et al.* (1997) Evidence that DNA A of a geminivirus associated with severe cassava mosaic disease in Uganda has arisen by interspecific recombination. *Journal of General Virology* 78, 2101–2111.

Zhou, X., Robinson, D.J. and Harrison, B.D. (1998) Types of variation in DNA A among isolates of *East African cassava mosaic virus* from Kenya, Malawi and Tanzania. *Journal of General Virology* 79, 2835–2840.

Zinga, I., Harimalala, M., De Bruyn, A., Hoareau, M., Mandakombo, N., *et al.* (2012). *East African cassava mosaic virus*-Uganda (EACMV-UG) and *African cassava mosaic virus* (ACMV) reported for the first time in Central African Republic and Chad. *New Disease Reports* 26, 17.

Zinga, I., Chiroleu, F., Legg, J., Lefeuvre, P., Komba, E.K., *et al.* (2013) Epidemiological assessment of cassava mosaic disease in Central African Republic reveals the importance of mixed viral infection and poor health of plant cuttings. *Crop Protection* 44, 6–12.

6 Cucumber Mosaic

Masarapu Hema,[1] Pothur Sreenivasulu[2] and P. Lava Kumar[3]*

[1]Department of Virology, Sri Venkateswara University, Tirupati, India; [2]Formerly Department of Virology, Sri Venkateswara University, Tirupati, India; [3]International Institute of Tropical Agriculture, Ibadan, Nigeria

6.1 Introduction

Cucumber mosaic virus (CMV) causes significant economic losses in several agricultural and horticultural crops worldwide (Jacquemond, 2012). The virus was first reported as the causal agent of diseases inflicting cucumber (*Cucumis sativus*) and muskmelon (*Cucumis melo*) in Michigan and cucumber in New York in 1916 (Palukaitis *et al.*, 1992). It has since been listed as a virus of greatest economic importance in cucurbits (*Cucurbita* spp.), pepper (*Capsicum annuum, C. frutescens*), tomato (*Solanum lycopersicum*), celery (*Apium graveolens*), cowpea (*Vigna unguiculata*), lettuce (*Lactuca sativa*) and banana (*Musa* spp.). Forage legumes and ornamentals are also affected by CMV (Gallitelli, 2000, 2002). Economic losses in crops are highest in field-grown vegetables and ornamentals, and pasture legumes (García-Arenal and Palukaitis, 2008; Jacquemond, 2012; Makkouk *et al.*, 2012; Moury and Verdin, 2012; Lecoq and Desbiez, 2012).

There is considerable variation in the expression of foliar symptoms in CMV-infected plants, and yield fluctuations differ from year to year between locations and are difficult to quantify, especially when mixed infections are involved. Nonetheless, some values associated with the direct effects of CMV on crop losses have been reported; for example, yield losses of 25–50% were reported in tomato plants in China (Tien and Wu, 1991), and 60% of melon plants and up to 80% of pepper plants in Spain (Avilla *et al.*, 1997; Luis-Arteaga *et al.*, 1998). When a necrogenic satellite RNA is present, the recorded losses in Spain and Italy were 80% of tomato plants in 70% of the growing regions (Jordá *et al.*, 1992; Gallitelli, 2000, 2002).

The isolates of CMV are distributed worldwide in both temperate and tropical areas (Scholthof *et al.*, 2011; Rybicki, 2015). It appears to be the most important virus of some annual crops in Argentina, Eastern China, Croatia, France, Egypt, Greece, Israel, Italy, Japan, Poland, Portugal, Spain, Sweden and in the north-eastern regions of the USA. In other countries, CMV ranks second or third in importance (Tomlinson, 1987). A number of reports indicate that CMV is well established in the Mediterranean region where it is frequently found in mixed infections with viruses, such as *Alfalfa mosaic virus*, *Tomato spotted wilt virus* and several potyviruses.

CMV is one of the most extensively studied viruses both as a plant pathogen and also as a model virus in virology. Many of its characteristics (i.e. transmission by mechanical sap inoculation to a range of herbaceous hosts, high virus titres in host plants, high degree of stability *in planta* and *in vitro*, tripartite positive sense RNA genome, and a wide host range) have made this virus a model system in plant virus research.

*E-mail: l.kumar@cgiar.org

Additionally, the CMV coat protein (CP) has been used as a platform for foreign epitope presentation (Natilla *et al.*, 2004, 2006; Nuzzaci *et al.*, 2010; Vitti *et al.*, 2014). Several reviews on CMV have been published over the past two decades or so (Palukaitis *et al.*, 1992; Palukaitis and García-Arenal, 2003a,b; García-Arenal and Palukaitis 2008; Jacquemond, 2012). This chapter presents an overview of the features of the virus, its economic importance and disease management strategies that are applicable to various crops it affects.

6.2   Virion and genome properties

6.2.1   Virions

The virions of CMV are non-enveloped, isometric particles measuring ~29 nm in diameter. They are composed of 180 subunits of single CP subunits arranged in pentamer–hexamer clusters with T=3 quasi-symmetry, and with a three-component single-stranded RNA genome (18%). The genome is packaged in three different icosahedral particles, which sediment at the same rate. Virions of CMV are stabilized by protein–RNA interactions and the nucleic acid is essential for assembly. The sedimentation coefficient (*S*) is circa 98*S* and

the particle weight is $(5.8–6.7) \times 10^{-6}$ Da. RNA1 and RNA2 are encapsidated in different particles, whereas RNA3 and RNA4 are packaged together in the same particle; some particles may contain three molecules of RNA4. In some isolates, virus particle preparations are shown to contain low levels of the other RNA species such as RNA4a, RNA5 and RNA6. There is a limit to the size of the encapsidated RNAs; those larger than RNA1 are not encapsidated *in vivo* (Palukaitis and García-Arenal, 2003a,b; ICTVdB Management, 2006; García-Arenal and Palukaitis, 2008; Jacquemond, 2012).

6.2.2   Genome

The CMV genome is composed of three single-stranded (+) RNAs designated as RNA1 (*circa* 3.3 kb), RNA2 (*circa* 3.0 kb) and RNA3 (*circa* 2.2 kb). A schematic illustration of the genome organization is given in Fig. 6.1. Monocistronic RNA1 codes for protein 1a, which possesses methyltransferase and helicase activities. The RNA2 is bicistronic, coding for a large protein 2a, that possesses RNA-dependent RNA polymerase activity, and a small protein 2b, expressed from a subgenomic RNA4a, determines virulence and inhibits the RNA interference (RNAi)

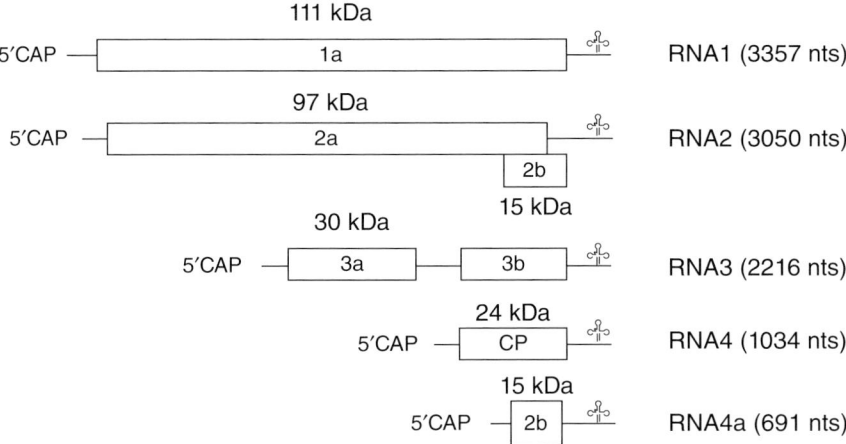

Fig. 6.1. The three genomic RNAs (RNA1, RNA2 and RNA3), and the two sub-genomic RNAs (RNA4 and RNA4a) of CMV. The open reading frames are indicated as boxes, and the RNA UTRs as solid horizontal lines. All the RNAs are capped with 7-methyl guanosine at the 5′ end, and possess a tRNA (⌀)-like structure at the 3′ end.

pathway in the host (Masuta and Shimura, 2013). This protein is primarily localized in the nucleus of the infected cells and interacts with the host protein involved in protein localization to the nucleus (Wang *et al.*, 2004b). The role of CMV protein 2b in counteracting host defences has been reviewed by Jacquemond (2012). Bicistronic RNA3 encodes the 3a protein or movement protein from the virus sense RNA3, while the second protein, the 3b protein or CP, is expressed from a sub-genomic RNA4. The size of genomic and sub-genomic RNAs may differ slightly according to the strain, but each open reading frame (ORF) of different strains has a similar size except for ORF 2a and ORF 2b of subgroup II isolates. Each of the RNAs has a cap structure at its 5′ end and a tRNA-like structure at its 3′ hydroxylated end. About 150 nucleotides (nts) at the 3′ terminal end are highly conserved among the different RNAs of a strain (Palukaitis *et al.*, 1992; Palukaitis and García-Arenal, 2003a,b; Mochizuki and Ohki, 2012).

Among the small RNAs encapsidated in virus particles, RNA5 and satellite RNAs (satRNAs) are well characterized. RNA5 is frequently detected with subgroup II CMV strains and in *Tomato aspermy virus* (genus *Cucumovirus*). It is approximately 300 nts long and consists of a mixture of the 3′ termini of RNA2 and RNA3. It is not capped and no polypeptides are associated with its presence (Jacquemond, 2012). CMV also supports non-coding satRNAs varying in size from 333 to 405 nts in length that share almost no sequence similarity. These satRNAs are dependent upon the helper CMV for both their replication and encapsidation (reviewed by Jacquemond, 2012; Palukaitis and García-Arenal, 2003a,b). More than 100 satellite variants have been associated with over 65 isolates of CMV from both subgroups of the virus. These satRNAs usually decrease the accumulation of the helper virus, and in most hosts, also reduce the virulence of CMV. However, satRNAs in selected hosts can enhance the disease induced by CMV (Kouadio *et al.*, 2013). For example, infection of tomato plants with CMV and certain satellites lead to systemic necrosis in the fields in several Mediterranean countries. This necrosis

is actually caused by sequences of the complementary-sense satRNA produced in large quantities during satellite replication. Structural and functional analysis of CMV satRNAs in RNA silencing was conducted in order to determine whether the satRNA-induced symptoms are due to a down regulation of the target host gene expression (Shimura and Masuta, 2012). Like all plant viruses and subviral agents, replication of viral satRNAs was found associated with the accumulation of 21–24 nts viral small interfering RNA derived from the whole region of a satRNA genome in both plus and minus strands. These satRNA-derived small interfering RNAs have recently been shown to play an important role in the trilateral interactions among host plants, helper viruses and satRNAs (Fang *et al.*, 2015). Defective RNA3 has occasionally been identified in some CMV progenies. These RNAs possess deletions of a few hundred nucleotides in either the movement protein or CP gene, or both. Their origin is unclear, as they occur under different experimental conditions (Jacquemond, 2012).

6.3 Taxonomy

CMV is the type member of the genus *Cucumovirus* of the family *Bromoviridae*. Other members of this genus include *Peanut stunt virus* (PSV), *Tomato aspermy virus* and *Gayfeather mild mottle virus* (Jacquemond, 2012). Serological tests suggest that *Peanut stunt virus* and *Tomato aspermy virus* are distantly related to CMV. Nucleotide sequence similarity among these three cucumoviruses ranges from 60% to 65% (García-Arenal and Palukaitis, 2008; Mochizuki and Ohki, 2012).

CMV isolates have been classified based on serological typing, peptide mapping of the CP, and sequence similarity of their genomic RNAs into two major subgroups, named subgroup I and subgroup II (Palukaitis *et al.*, 1992; Kumari *et al.*, 2013). The percentage identity in the nucleotide sequence between pairs of isolates belonging to each of these subgroups ranges from 69% to 77%, depending on the pair of isolates and the RNA segment compared (dissimilarity being highest for RNA2). Nucleotide sequence identity among isolates within a subgroup is above 88% for subgroup I and above 96% for subgroup II, indicating a higher heterogeneity of subgroup I. Overall, nucleotide sequence similarity among CMV strains is approximately 70–98%. Subgroup I has been further divided into IA and IB groups by phylogenetic analysis of the 5' non-coding region of RNA3 (Roossinck *et al.*, 1999). This grouping is also supported by phylogenetic analysis of the CP ORF. Isolates of subgroup II are frequently noticed in temperate regions. Most isolates of subgroup IB are reported from East Asia, which is presumed to be the place of origin of this group, and also in the Mediterranean region, California, Brazil and Australia (García-Arenal and Palukaitis, 2008; Jacquemond, 2012).

6.4 Diversity

The occurrence of several CMV strains that have a very broad host range suggests that the multicomponent nature of the virus has minimal or no effect on its ecological success or epidemiological competence. Viable reassortments have been prepared *in vitro* and used to assign specific functions to individual segments of the genome (Palukaitis *et al.*, 1992). True recombination (RNA-RNA recombination) and the production of mixed subunit capsids from two different cucumoviruses were shown to affect insect mediated-transmission of CMV (Chen *et al.*, 1995). But reassortment or recombination are rare in natural populations of this virus. Jacquemond (2012) reviewed the role of recombination and genetic drift in the evolution of CMV. Surveys of naturally infected crops suggest that subgroup I strains are more frequent than those of subgroup II and sometimes they represent more than 80% of all isolates (Gallitelli, 2000). Genetic variation of 32 CMV field isolates as well as historic isolates collected from different regions suggests low nucleotide diversity. Phylogenetic and computational analysis confirmed recombination between subgroups I and II, as well as between IA and IB in RNA3 (Nouri *et al.*, 2014). Pseudorecombination between the

two distinct strains, CMV-209 isolated from *Glycine soja* and CMV-Fny, using infectious cDNA clones was shown to enhance symptom severity (Phan *et al.*, 2014b). Although CMV is known to be prone to recombination and sequence data of various isolates indicate high genetic diversity, this has not been reported with populations within crop plants, field or wild plants (Jacquemond, 2012).

CMV isolate identification is based on the nucleotide diversity of the CP encoding gene. But phylogenetic trees constructed with proteins other than CP do not completely support the grouping previously proposed based on CP. Subgroup IA is more heterogeneous than subgroup IB. Jacquemond (2012) reviewed the phylogenetic analysis of various isolates of CMV and suggested that analysis of at least a part of each genome segment is necessary for confident genotyping of CMV isolates.

Studies by Davino *et al.* (2012) on the genetic variation and evolution of CMV from aromatic, medicinal and ornamental plants in northern Italy using sequence analysis and comparison of the movement protein with equivalent sequences of isolates from other countries suggested that long-distance migration plays a role in the evolution and determination of the genetic structure and diversity of CMV in northern Italy and other areas. Kim *et al.* (2014b) analyzed 252 pepper (*C. annuum*) samples through reverse transcription (RT)-PCR analysis and found that the *2b* gene of CMV is under weaker purifying selection than the other genes. Based on the phylogenetic analysis of RNA1, the CMV isolates from pepper were divided into three clusters in subgroup I. Full-genome sequence-based molecular analysis of the Korean CMV population suggest that the subpopulations of CMV have been geographically localized in pepper fields in that country (Kim *et al.*, 2014b).

6.5 Symptomatology

CMV particles can be found in the cytoplasm of all parts of the host plant. Inclusion bodies found in the cytoplasm of infected cells are crystals that are often rhomboidal, hexagonal or roughly spherical, and may appear as solid hollow structures containing virus particles. Inclusions of diagnostic value can be viewed by an optical microscope upon staining with Azure A (Palukaitis and García-Arenal, 2003a,b). Symptoms induced by CMV in various plant species vary with the viral strain or isolate, the host genotype, age of the plant at the time of infection, co-infections and environmental conditions under which the plants are grown. The presence of satRNAs can affect the level of CMV replication and also its pathogenicity (Palukaitis and García-Arenal, 2003a,b). Brief descriptions of the symptoms incited by CMV in selected crops are given below.

6.5.1 Cucurbits

Almost all cucurbit plants are susceptible to CMV and develop symptoms varying in severity. In cucumber, melon and squash (*Cucurbita* spp.), CMV causes prominent foliar yellow mosaic, malformation, and drastic reduction of leaf size and stem internodes, along with severe plant stunting. During the early stages of growth, leaves develop prominent downward leaf curling, mosaic, and are smaller in size. Flowers of severely affected plants are malformed and petals remain green in colour. The most severe symptoms occur in summer squash, some pumpkins (*C. maxima*), and different types of melons. Symptoms are less severe in cucumber, winter squash, and watermelon (*Citrullus lanatus*). Fruits from infected plants are distorted, often discolored, and usually remain small. Seed production in severely affected fruits is negligible (Provvidenti, 1996).

6.5.2 Pepper

CMV causes mild or no symptoms on inoculated pepper leaves followed by mosaic symptom development on uninoculated leaves, and necrotic symptoms on inoculated and uninoculated leaves. Foliar symptoms of pepper plants vary with the stage of infection.

Initial symptoms include chlorosis of young leaves over the basal portion of the leaf lamina or over the entire leaf. Oak leaf and ring spot patterns may develop on leaves of older plants. As new leaves emerge, these leaves typically develop chlorotic mosaic patterns that tend to cover the entire leaf surface. Leaves that develop subsequently may exhibit varied degrees of deformation, including sunken interveinal lamina with protruding primary veins. CMV infected plants also tend to be stunted. Pepper fruit may develop ring spotting, irregular ripening and roughness (bumpy appearance) and are unmarketable. Necrotic spots form on the fruit of some pepper varieties (Murphy, 2003).

6.5.3 Tomato

During the early stages of infection with CMV, tomato plants are typically yellow, bushy and considerably stunted. The leaves may show mottle patterns similar to those caused by *Tobacco mosaic virus* (TMV). The most characteristic symptom of CMV infection is the development of shoestring-like leaf blades. This is often times confused with symptoms of tomato mosaic that are referred to as fern leaf. Severely affected plants produce few fruits that are usually small and show delayed maturity (Zitter, 1991).

6.5.4 Bean

Symptoms of CMV infection in bean (*Phaseolus vulgaris*) crops consist of leaf curling, green mottle, blistering and a zipper-like roughness along the main veins involving only a few leaves. Foliar symptoms are most obvious, and pod infection and loss are greatest when plants are infected before bloom. Early infected plants may yield few or no pods because CMV causes flower abortion and abnormal development. Pods that do develop are mostly curved, mottled and reduced in size. Symptoms can vary among varieties and are generally confused with those caused by *Bean common mosaic virus*.

6.5.5   Chickpea and Cowpea

Desi chickpea (*Cicer arietinum*) develop leaf chlorosis, reddening and stunting of the plants, while cowpea show large or small brown lesions on inoculated leaves. Only some subgroup I isolates are systemic in the host and able to cause mild mosaic; generally these induce chlorotic lesions on inoculated leaves.

6.5.6   Banana

Symptoms in banana depend on the strain of CMV, ambient temperatures and probably the genotype of the host. Leaf symptoms are usually obvious along the veins as light green or yellow striping, which later develop into chlorotic streaks. Other symptoms include yellow, light green and dark green stripes, generally along the veins, or a more general mosaic. Affected areas may become necrotic as time progresses. Leaf veins may be prominent and abnormally thickened. Chlorotic streaks, mosaic and distortion tend to develop on fruits. Bunches produced on infected plants consist of small fruits or few fruits. When ambient temperatures fall below 24°C, which is common during the winter period in the subtropics or at higher elevations in the tropics, the additional stress results in the leaves, heart leaf and central cylinder of a CMV-infected plant becoming necrotic. In extreme cases, death of plants ensues. A mild strain of CMV was reported in plantains in Honduras. Symptoms are usually masked, appearing only during cool winter months (Ploetz, 1994).

6.5.7   Co-infections

CMV is known to occur in mixed infections. Co-infection with viruses from other genera such as *Crinivirus*, *Potexvirus*, *Potyvirus* and *Tobamovirus* can intensify disease symptoms (synergy), particularly in cucurbits, legumes and solanaceous hosts and result in an increase in the accumulation of CMV. For instance, mixed infections of CMV and *Turnip mosaic*

virus (genus *Potyvirus*) were shown to induce more severe symptoms on *Nicotiana benthamiana* than either single infection (Takeshita *et al.*, 2012).

Synergy has also been shown to occur in plants that are tolerant to one of the two viruses by restoring a deficient function (Palukaitis and García-Arenal, 2003a,b). In cucumbers resistant to CMV, co-infection with *Zucchini yellow mosaic virus* (genus *Potyvirus*) results in increased CMV accumulation without an increase in the severity of symptoms (Wang *et al.*, 2004a).

Tomato plants exhibit more severe symptom expression when co-infected with *Potato virus Y* (genus *Potyvirus*) and CMV. This synergy is reported to be accompanied by an increasing level of CMV and a decrease in *Potato virus Y*, essentially through the action of the CMV 2b protein. CMV with deletions in the *2b* gene is unable to move systemically, but its defect is compensated for when plants are co-infected with *Potato virus Y* (Mascia *et al.*, 2010). Malformation of young tobacco leaves co-infected with CMV and TMV was observed in transgenic plants expressing CMV *2b* (Siddiqui *et al.*, 2011). These authors showed that the synergistic effect resulted from the joint action of the two viral silencing suppressors (protein 2b and TMV replicase). Chen *et al.* (2014) found that *Turnip crinkle virus* provides effective resistance to infection by CMV in *Arabidopsis* plants co-infected by both viruses and this antagonistic effect is much weaker than when both viruses are inoculated on different leaves of the same plant. These studies suggest that CMV interacts with a number of viruses and in most cases it increases the titres and accentuates symptoms in the co-infecting virus.

6.6   Host Range

CMV is probably one of the plant viruses that has the broadest host range. Isolates have been reported to naturally infect various agricultural and horticultural crops as well as numerous weed species; the latter act as reservoirs of the virus and its vector species (Table 6.1).

Table 6.1. Examples of crop and weed plants reported to be naturally infected with *Cucumber mosaic virus*.

Field crops	**Herbaceous foliage and flowering ornamentals**
Lucerne (*Medicago sativa*)	*Amaranthus* spp.
Beet (*Beta vulgaris*)	*Anemone* spp.
Chickpea (*Cicer arietinum*)	*Aquilegia* spp.
Clovers (*Trifolium* spp.)	*Begonia* tuber hybrid
Cowpea (*Vigna unguiculata*)	*Calendula* spp.
Faba bean (*Vicia faba*)	*Canna* x *generalis*
French bean (*Phaseolus vulgaris*)	*Dahlia* spp.
Groundnut (*Arachis hypogaea*)	*Delphinium* spp.
Lentil (*Lens culinaris*)	*Dieffenbachia seguine*
Lupins (*Lupinus* spp.)	*Eliokarmos thyrsoides*
Maize (*Zea mays*)	*Geranium* spp.
Mungbean (*Vigna radiata*)	*Gladiolus* spp.
Mustard (*Brassica* and *Sinapis* spp.)	*Hydrangea macrophylla*
Pea (*Pisum sativum*)	*Iberis* spp.
Rape seed (*Brassica napus*)	*Iris* spp.
Safflower (*Carthamus tinctorius*)	*Lilium* spp.
Soybean (*Glycine max*)	*Narcissus tazetta*
Sunflower (*Helianthus annuus* and hybrids)	*Pelargonium* spp.
Tobacco (*Nicotiana* spp.)	*Peperomia magnoliifolia*
Urdbean (*Vigna mungo*)	*Petunia* spp. and hybrids
Vetches (*Vicia*, *Lathyrus* and *Astragalus* spp.)	*Phlox* spp.
	Rosmarinus officinalis
Horticultural crops	*Salvia splendens*
Artichoke (*Cynara cardunculus* var. scolymus)	*Tagetes* spp.
Banana and plantains (*Musa* spp., polyploids and hybrids)	*Viola* spp.
Carrot (*Daucus carota*)	*Zinnia* spp.
Celery (*Apium graveolens* Dulce group)	
Chili pepper (*Capsicum frutescens*)	**Medicinal plants**
Bell pepper (*Capsicum annuum*)	*Catharanthus roseus*
Cucurbits (*Cucumis*, *Cucurbita*, *Citrullus*, *Momordica* spp.)	
Aubergine (*Solanum melongena*)	**Weeds**
Lettuce (*Lactuca* spp.)	*Stellaria media*
Papaya (*Carica papaya*)	*Spergula arvensis*
Potato (*Solanum tuberosum*)	*Echinocystis lobata*
Spinach (*Spinacia oleracea*)	*Lamium purpureum*
Sweet potato (*Ipomoea batatas*)	*Commelina diffusa*
Tomato (*Solanum lycopersicum*)	
	Woody and semi-woody plants
Plantation crops	*Ixora* spp.
Black pepper (*Piper nigrum*)	*Passiflora edulis*
Vanilla (*Vanilla* spp.)	

The host range of CMV exceeds 1300 species in more than 500 genera of over 100 families of monocots and dicots, including many vegetables, ornamentals and woody and semi-woody plants (Lawson, 1985; Palukaitis *et al.*, 1992; Palukaitis and García-Arenal, 2003a,b; Afreen *et al.*, 2009; Kim *et al.*, 2011; Hosseinzadeh *et al.*, 2012; Jacquemond, 2012; Lecoq and Desbiez, 2012; Makkouk *et al.*, 2012; Moury and Verdin, 2012). New hosts are being reported at regular intervals (Afreen *et al.*, 2009; Kim *et al.*, 2011; Miura *et al.*, 2013; Kim *et al.*, 2014a). Some strains (e.g. CMV-209 isolated from wild soybean, *G. soja*) have been shown to have a restricted host range and to have lost the ability to infect several of the typical hosts of CMV (Phan *et al.*, 2014a). Traditionally, CMV isolates infect plant species of the families Cucurbitaceae, Fabaceae and Solanaceae including *Arachis hypogaea, Capsicum annuum, Cicer arietinum, Citrullus lanatus, Commelina diffusa, Cucumis sativus, C. melo, Cucurbita pepo, C. maxima, Dioscorea alata, D. rotundata, Ipomoea batatas, Lens culinaris, Solanum lycopersicum, Musa* spp. polyploids and hybrids, *M. textilis, N. tabacum, Passiflora edulis, Phaseolus* spp., *Pisum sativum, Psophocarpus tetragonolobus, Trifolium* spp., *Vicia faba, Vigna angularis, V. radiata* and *V. unguiculata.*

Plants such as *C. sativus, N. clevelandii, N. glutinosa* and *N. tabacum* are the major virus propagation hosts. *Chenopodium amaranticolor, C. quinoa* and *V. unguiculata* respond with local lesions after sap inoculation. *C. amaranticolor, C. quinoa* (chlorotic lesions), *C. sativus* (systemic mosaic), *S. lycopersicum, N. edwardsonii, N. glutinosa* and *N. tabacum* (variable symptoms depending on strain) are useful as diagnostic hosts for CMV isolates.

All viral-encoded proteins are known to be involved in pathogenesis (Jacquemond, 2012). However, new insights into virus disease development involving the silencing of physiologically important host genes were recently revealed using next-generation sequencing methods. For instance, two studies elucidated the mechanism involved in the induction of bright chlorosis in tobacco by a satRNA. Y-SatRNA (satellite RNA of CMV-Y strain) has a 22 nucleotide sequence complementary to the mRNA of the tobacco magnesium protoporphyrin chelatase unit I (*ChlI*), a gene essential in chlorophyll synthesis (Smith *et al.*, 2011; Shimura and Masuta, 2012). The small RNAs derived from the Y-SatRNA induced the development of chlorosis in tobacco by down regulating the *ChlI* gene through RNA silencing.

Necrotic symptoms can occur either on inoculated leaves or as a systemic syndrome leading to plant death in some cases. The hypersensitive response in local lesion hosts was associated with protein 1a, 2a or CP, depending on the host–virus strain combination (Palukaitis and García-Arenal, 2003a,b). The hormone, spermine, was shown to act as a signal for a hypersensitive defence response to

CMV in *A. thaliana* (Mitsuya *et al.*, 2009). The mechanism of systemic necrotic induction has been reviewed by Jacquemond (2012).

6.7 Transmission

CMV is graft and mechanically sap transmissible under experimental inoculations. Natural spread of CMV is through seeds of some host plant species, through propagules of vegetatively propagated plants, and by more than 80 aphid species (Hemiptera: Aphididae) in a non-persistent manner (Palukaitis and García-Arenal, 2003a,b). *Myzus persicae* and *Aphis gossypii* are the most important aphid species involved in transmitting the virus in vegetable crops. They are also widely used for experimental inoculations under laboratory conditions. A study on relative importance of aphid species in the transmission of CMV in snap bean crops in the USA identified 25 aphid species in different fields; however, only six were recognized as major contributors to CMV transmission (Gildow *et al.*, 2008). *A. gossypii, A. glycines, Acyrthosiphon pisum* and *Therioaphis trifolii* proved to be the most efficient vectors and were also the most prevalent in the field (Gildow *et al.*, 2008). CP is the sole viral determinant of CMV transmissibility by aphids. Studies have revealed that specific viral CP domains and amino acids affect transmissibility and vector specificity (Ng *et al.*, 2005; Pierrugues *et al.*, 2007).

Transmission efficiency depends on virus accumulation. This has been demonstrated in studies with CMV isolates and their associated satRNAs. Satellite RNA drastically reduces the replication of the helper viral genome especially in solanaceous hosts, and such plants are poor sources for the acquisition of the CMV by aphids. In such cases, the density of the vector population determines the initiation of CMV epidemics. Various epidemiological models are used to explain the likelihood that damaging epidemics will develop for a particular pathosystem (Escriu *et al.*, 2003).

Mauck *et al.* (2010) demonstrated that infections with CMV can induce volatile emissions that affect the attractiveness of the plants for insect vectors. In the case of squash infected with CMV, it was shown that the population density of *A. gossypii* appeared to be less important on CMV-infected plants than on healthy ones, and that winged aphids less frequently colonized infected plants. In contrast, infection greatly increased attractiveness of plants for both *A. gossypii* and *M. persicae*, by inducing increased emissions of a mixture of volatiles by the plants (Mauck *et al.*, 2010).

Transmission of viruses through seeds of either cultivated crops or wild species is an important parameter for the study of disease epidemics, as plants grown from infected seeds constitute a primary source of virus inoculum which can be efficiently disseminated by insect vectors. Based on biological, serological and molecular techniques, transmission of CMV through seeds has been demonstrated for several plant species. True transmission of CMV through seeds (via embryo) has been established for crops such as bean, groundnut, lentil, spinach, lupin and pepper. CMV detection in seed lots, or in plants germinated from seeds, has also been documented for common bean, cowpea, green gram (*V. radiata*), pea (*P. sativum*), faba bean (*V. faba*), chickpea, vetch (*V. sativa*), tomato, lucerne (*Medicago sativa*) and pumpkin (*C. maxima*) (Jacquemond, 2012). Transmission rates are usually low (less than 2.5%), but sufficient to successfully serve as disease foci and initiate epidemics under conditions favourable to vectors. Higher rates were reported for lentil (*L. culinaris*) (up to 9.5%), tomato (8%), spinach (*Spinacia oleracea*) (15%) and cowpea (21%). CMV isolates are also vertically transmitted through the propagules of vegetatively propagated plants such as banana, sweet potato, black pepper and certain ornamentals (Gallitelli, 2000, 2002; Jacquemond, 2012).

6.8 Detection and Diagnosis

Successful control of CMV infections requires the availability of strain-specific, reliable and rapid methods of virus detection. Biological, serological and molecular techniques have been adapted and successfully applied to the diagnosis of CMV infections. In some laboratories, sap inoculation onto diagnostic hosts is initially used to tentatively identify the virus. Subsequent tests employ both polyclonal and monoclonal antibodies that have been generated against various isolates of CMV (Hsu *et al.*, 2000; Zein and Miyatake, 2009; Berniak *et al.*, 2010). Polyclonal antibodies are typically raised against recombinant CMV CP expressed in *Escherichia coli* (Salem *et al.*, 2013; Kapoor *et al.*, 2014). Antibodies to CMV are available commercially (http://ztmbpd.iari.res.in/elisa-kit-detection-cucumber-mosaic-virus).

Among several serological techniques available, enzyme immunoassays (direct and indirect ELISAs, dot-immunobinding ELISA, Western blotting) are routinely used for the detection of CMV in different plant species (Kiranmai *et al.*, 1996; Zein *et al.*, 2007; Berniak *et al.*, 2009; Zein and Miyatake, 2009). Lateral flow immunoassay was developed for detecting TMV and CMV in tomato (Lamptey *et al.*, 2013). Double-stranded RNA analysis from suspected plants is also used to identify CMV in crops such as banana (Kiranmai *et al.*, 1996). Both radioactive and non-radioactive probes are used for CMV detection by various forms of hybridization (Gallitelli and Saldarelli 1996; Kiranmai *et al.*, 1998). Several variants of PCR are also used for its sensitive detection in different plants (Sharman *et al.*, 2000; Berniak *et al.*, 2009). Primers specific to different regions of the tripartite genome and satRNAs are employed and used successfully for the detection of several cucumoviruses (Seo *et al.*, 2014). RT-PCR followed by restriction fragment length polymorphism analysis of amplicons has proved useful in differentiating CMV strains into subgroups (Raj *et al.*, 1998; Niimi *et al.*, 2003; Kumar *et al.*, 2005; Berniak *et al.*, 2009). But precise identification of CMV isolates is achieved by sequence analysis of PCR amplicons. A detection system based on a multiplex RT-PCR has been developed to simultaneously identify multiple viruses including CMV in various plant species (Chen *et al.*, 2011; Dai *et al.*, 2012; Panno *et al.*, 2012; Kwon *et al.*, 2013, 2014). Essentially molecular techniques have provided a simple way to investigate the dynamics of CMV populations in nature. More accurate population analysis and identification of distinct satRNA variants can be addressed by RNase protection assays (Aranda *et al.*, 1995). Loop-mediated isothermal amplification assays are available for detecting the virus in banana, tobacco and black pepper crops (Peng *et al.*, 2012; Zhao *et al.*, 2012; Bhat *et al.*, 2013).

Other technologies have been employed for the analysis of whole genome expression in different host plants responding to CMV infection in order to understand the molecular basis of pathogenesis and the resistant response. Whole genome microarray has been used to study the *A. thaliana* resistome in response to CMV infection (Marathe *et al.*, 2004). Deyong *et al.* (2005) developed microarrays to detect and differentiate CMV serogroups and subgroups using oligonucleotide probes. Lang *et al.* (2011) profiled CMV-responsive mRNAs in tomato using micro paraflo microfluidics microarrays that resulted in identification of a wide range of genes involved in regulation of host susceptibility to CMV. Transcriptome analysis of *N. tabacum* infected with CMV through next-generation sequencing identified differentially expressed genes involved in many biological processes such as photosynthesis, pigment metabolism and plant pathogen interaction (Lu *et al.*, 2012). MicroRNAs have been recognized to play an important regulatory role in plant development and stress responses. Small RNA libraries from CMV-infected and mock-inoculated tomato revealed that the defence response and photosynthesis-related genes were most affected in CMV-infected tomatoes (Feng *et al.*, 2014). It is conceivable that application of next-generation sequencing tools will result in precise mapping of host regulatory response to CMV and identification of candidate genes for developing transgenic resistance or genetic modification using gene editing tools.

6.9   Ecology and Epidemiology

Combinations of natural plant communities harbouring CMV and aphid vectors create complex pathosystems. CMV strains survive in nature due to their wide host range, including annuals and perennials, and vertical transmission through seed and vegetative propagules. Nearly 80 aphid species are known to be involved in horizontal transmission of CMV. Additionally, the virus overwinters or oversummers in a number of weed species, many of which are perennials. CMV has caused severe epidemics in many crops in different geographical regions (Gallitelli, 2000, 2002; García-Arenal *et al.*, 2000). The planting of infected seed (or infected propagules) or inoculum from sources outside the crop such as other crops in the vicinity, volunteer plants, self-sown plants or weeds are known to serve as primary sources of inoculum for a newly sown crop with aphids serving as vectors. Various studies have demonstrated the ability of CMV to induce changes in its host that make the virus-infected plants more attractive to aphids (alighting, settling and probing), leading to enhanced transmission (Carmo-Sousa *et al.*, 2014). A study by Mauck *et al.* (2014) showed that CMV reduced host–plant quality of *Cucurbita pepo* causing aphids to rapidly leave infected plants, but increased the attractiveness of infected plants to aphids through elevated emissions of plant volatiles that are similar to those emitted by healthy plants.

Epidemiological models have been developed to forecast the likelihood of CMV epidemics. Thackray *et al.* (2004) forecasted epidemics of aphids and CMV in lupin crops in the Mediterranean-type of environment. Based on this pathosystem, they proposed a model that could be useful for similar virus/aphid vector combinations. Betancourt *et al.* (2013) modelled the evolution of CMV virulence in two hosts that differ in susceptibility (virus multiplication rate and virulence). Their model allowed for inoculum flow between hosts and for the co-infection of host plants with competing virus genotypes, as well as competition which affects transmission rates to new hosts. Emergence of highly virulent genotypes was predicted as mixed infections favoured high vector densities.

6.10 Control

The broad host range and large number of aphid species that transmit CMV make the virus a serious threat to the health of certain crops and their related economies. An essential prerequisite for the control of CMV is the availability of healthy propagules (seeds or seedlings), although for many crops this is very difficult to achieve. In the absence of totally virus-free material, threshold levels of CMV infection should be established for each crop in order to minimize the risk of epidemics (Gallitelli, 2000, 2002). However, such information in reality is seldom available.

As the virus is vertically transmitted through the seed and vegetative propagules of several CMV-susceptible crops, the primary focus is on the selection and planting of virus-free propagules in agroecosystems where the virus was not reported in the previous seasons. Since the virus is worldwide in distribution and has a broad host range, quarantine inspections are directed to seed germplasm and new accessions of plant species in which the virus may be seed-borne. These inspections are also necessary to limit the introduction of new isolates into geographical areas where such isolates have not been previously recorded. Ornamental plants propagated through bulbs, rhizomes, tubers or stem pieces may be virus reservoirs, and such plants are rigorously screened in quarantine stations. Thus, certification of quarantined plant propagules is essential in limiting the introduction of CMV and its strains into new areas (Gallitelli, 2002).

Removal of reservoir hosts and roguing of volunteer plants and previous crop debris in and around the field are necessary. Cropping practices (e.g. adjustment of planting dates, planting in isolation, border cropping, mulching, intercropping, crop density, crop rotation) have to be manipulated to avoid all possible sources of inoculum and vector populations. Biological (e.g. border crops, botanicals, predators), physical (e.g. mulching) and chemical (e.g. pesticide spray) control of aphid vectors to minimize the spread of the CMV have been practiced (Hooks and Fereres, 2006). The effects of two aphidophagous predators (the larvae of *Chrysoperla carnea* and adults of *Adalia bipunctata*) on the spread of CMV transmitted in a non-persistent manner by the cotton aphid *A. gossypii* were studied under semifield conditions. Both natural enemies significantly reduced the number of aphids on the CMV-source cucumber plant after 5 days, but not after 1 day. The CMV transmission rate was generally low, especially after 1 day, due to the limited movement of aphids from the central CMV-source plant. This increased slightly after 5 days. Infected plants were mainly located around the central virus-infected source plant and the percentage of the aphid occupation and CMV-infected plants did not differ significantly in absence or presence of natural enemies (Garzón *et al.*, 2015).

Planting of virus-tolerant or -resistant crops is ideal. Some CMV-tolerant or -resistant varieties of tomato, pepper, legumes and some cucurbits are available (Gallitelli, 2000, 2002; Jacquemond, 2012). Several natural CMV-resistant genes of either cultivated crops or related wild species have been discussed by Palukaitis and García-Arenal (2003a,b) and Jacquemond (2012). As breeding for resistance is a continuous process, recent attempts have been made to identify sources of resistance to CMV in different plants (Cai *et al.*, 2003; Akhtar *et al.*, 2010). In spite of these efforts, breeding for resistance in crops like tomato is often hampered by difficulties in obtaining viable and/or fertile hybrids between tomato and its wild relatives. Some resistance genes especially from *A. thaliana* have been characterized (Jacquemond, 2012). It appears that several basal defence mechanisms to CMV infections have been deployed by host plants (Palukaitis and Carr, 2008). Molecular resistance reactions of *A. thaliana* to CMV infection, for example, were investigated by various groups and allowed for the identification of the genes and their products involved in defence reactions (Bouché *et al.*, 2006; Du *et al.*, 2007; Cillo *et al.*, 2009; Wang *et al.*, 2010; Harvey *et al.*, 2011).

Finally, work with antiviral microbial sources holds promise to protect against CMV infections. In Egypt, antiviral-producing *Streptomyces* spp. were isolated from the soil rhizosphere. It was found that five *Streptomyces* species were non-phytotoxic and effective in local as well as systemic control of CMV infection in *C. amaranticolor* (El-Dougdoug *et al.*, 2012). Elsharkawy *et al.* (2012) also demonstrated the induction of systemic resistance against CMV by the plant growth-promoting fungus, *Penicillium simplicissimum* GP17-2 in *A. thaliana* and tobacco plants, and suggested the involvement of multiple defence pathways including an increased expression of regulatory and defence genes involved in the salicylic and jasmonic acid/ethylene signalling pathways.

6.10.1 Transgenic approaches

Several groups have investigated the feasibility of transgenic resistance for the management of diseases induced by CMV. Morroni *et al.* (2008) reviewed CMV-derived transgenic resistance (e.g. CP-, satRNA-, antisense RNA-, replicase-based strategies) and non-pathogen derived transgenic resistance (e.g. strategies involving transgenes such as RNase, ribozyme, pathogeneis-related proteins, plantibodies or ribosome-inactivating protein from *Phytolacca americana* [pokeweed antiviral protein, PAP] or *Trichosanthes kirilowii*) in plant species such a tobacco, pepper, tomato, potato, cucumber, melon and squash. CP-transgenic plants were shown to confer broad spectrum resistance in several investigations (Jacquemond, 2012). Over-expression of a coiled-coil nucleotide site leucine-rich repeat-type resistance gene, *RCYI*, conferred resistance to a yellow strain of CMV (CMV-Y) in *A. thaliana* (Sekine *et al.*, 2008). RNAi-based constructs targeting CMV protein 2b, a strong suppressor of post-transcriptional gene silencing in tobacco was shown to offer complete resistance in 30% plants, whereas 30% showed delayed symptom development (Kavosipour *et al.*, 2012). Expression of artificial microRNAs targeting the 2b mRNA of CMV also efficiently inhibited gene expression and protein suppressor function, and thus conferred effective resistance to CMV infection in transgenic tobacco plants (Qu *et al.*, 2007). Ntui *et al.* (2014) generated a construct consisting of an inverted repeat

of a 1793 base pair fragment from a defective CMV replicase gene derived from the RNA2 of CMV-O. Of the four transgenic lines inoculated with CMV-O or CMV-Y *in vitro* and *ex vivo*, three lines showed immunity to both strains of CMV as no symptoms were detected, whereas one line exhibited high resistance. These plants expressed mild symptoms limited to the regions of the plant that were inoculated.

6.10.2 Crop-specific control measures

Cucurbits

CMV in cucurbits is mainly managed using resistant cultivars or by controlling vectors. CMV resistance was identified in a few cucurbit species but not in all. For instance, the CMV resistance in cucumber was derived from 'Chinese Long' and 'Tokyo Long Green' (Zitter and Murphy, 2009). High levels of resistance to CMV are reported in a number of wild squash species, of which *Cucurbita okeechobeensis* subsp. *martinezii* and *C. ecuadorensis* are extensively used in interspecific crosses with *C. pepo* (Gong *et al.*, 2013). Resistance or tolerance to CMV is also described in *C. maxima* lines from South America and in a *C. moschata* line from Nigeria. Genetic resistance for CMV in muskmelon is derived from oriental melons (Zitter and Murphy, 2009).

To increase resistance against CMV, the CP of the virus has been introduced into cucumber, melon and squash, and has conferred good levels of resistance against several strains of CMV (Provvidenti, 1996). Transgenic watermelons resistant to multiple virus infections were developed using a single chimeric transgene comprising a silencer DNA from the partial N gene of *Watermelon silver mottle virus* fused to partial CP sequences of CMV, *Cucumber green mottle mosaic virus* (CGMMV) and *Watermelon mosaic virus* (WMV) (Lin *et al.*, 2012). Transgenic watermelon R(0) plants were individually challenged with CMV, CGMMV or WMV, or with a mixture of the three viruses. Two lines were identified as resistant to CMV, CGMMV and WMV individually, and to mixed inoculations with the three viruses. The R(1) progeny of the two resistant R(0) lines showed resistance to CMV and WMV, but not to CGMMV. Low-level accumulation of trans-gene transcripts in resistant plants and small interfering siRNAs specific to CMV and WMV were readily detected in the resistant R(1) plants by Northern blot analysis, indicating that the resistance was established via RNA-mediated PTGS.

Insecticides, reflective mulches and mineral oils (Simons and Zitter, 1980) are widely used for the control of CMV. Eradication of weed hosts is often an impossible task, because of the extensive host range of the virus. Long-lasting insecticide-treated nets (LLITNs) constitute a novel alternative that combines physical and chemical tactics to prevent insect access and the spread of insect-transmitted plant viruses in protected enclosures. This approach is based on a slow-release insecticide-treated net with large hole sizes that allow improved ventilation of greenhouses. The efficacy of a wide range of LLITNs was tested under laboratory conditions against *M. persicae* and *A. gossypii*. Two nets were selected for field tests under a high insect infestation pressure in the presence of plants infected with CMV and *Cucurbit aphid-borne yellows virus* (family *Luteoviridae*, genus *Polerovirus*). LLITNs resulted in high mortality of aphids, although their efficacy decreased over time because of exposure to the sun. Nets effectively blocked the invasion of aphids and reduced the incidence of viruses in the field. LLITNs of appropriate mesh size are a very valuable IPM tool, in combination with biocontrol agents, for protecting against aphid vectors of CMV (Dáder *et al.*, 2014).

Spinach and lettuce

CMV resistance in spinach is controlled by a single dominant gene in the variety 'Virginia Savoy', and this has been incorporated into many current spinach varieties. This resistance, however, is not complete and may break down at temperatures greater than 28°C. Virus resistance to CMV in lettuce was identified in *Lactuca saligna* PI 26153 from Portugal, a distantly related species of lettuce (*L. sativa*). However, this resistance is strain specific. An accession from *L. serriola* proved tolerant to three strains of CMV, but this tolerance has not been transferred to any of the current commercial varieties (Zitter and Murphy, 2009).

Pepper

Extensive effort has been deployed to the development of pepper varieties with resistance to CMV. Case in point, the pepper variety 'Peacework' was developed and exhibits high levels of virus resistance (Mazourek *et al.*, 2009). However, several *Capsicum* species show polygenic resistance against CMV during the initial steps of pathogenesis. In most cases, the major resistance gene remains unknown and thus unavailable for practical deployment (Jacquemond, 2012). Tolerance has been described in some varieties such as 'Perennial', but it has not been utilized in any commercial varieties. Protection is extended to only a small number of strains or isolates of the virus.

Lee *et al.* (2009) developed a transformation system of pepper using *Agrobacterium* and the CP gene, CMVP0-CP, with the aim of developing a new CMVP1-resistant pepper line. A large number of transgenic pepper plants were screened for CMVP1 tolerance under greenhouse and field conditions. Three independent pepper lines were found highly tolerant to CMVP1 as well as CMVP0. The transgenic plants did not develop symptoms of stunting and fruits attained the expected size.

Cultural practices with this crop include the use of reflective mulches to deter aphid vectors, elimination of weeds in the vicinity of the crop; and timely planting to avoid exposing young plants to high aphid populations in the field and migrating aphid populations. Live mulches or border crops that are not susceptible to CMV are often used as depositories for the virus by viruliferous aphids entering the crop. The combined effect of oil sprays and rapid-acting insecticides (e.g. pyrethroids) may reduce losses if applied in a timely manner (Gallitelli, 2000, 2002). Alternative approaches to managing CMV such as cross-protection, involving the inoculation of young plants with a mild strain of the virus or with the virus and an associated satRNA, have shown promise in pepper (Murphy, 2003).

Tomato

In tomato, early sowing or late transplanting, as well as avoiding overlapping cycles with susceptible crops, have proved effective in preventing diseases caused by CMV (Gallitelli, 2000, 2002). Sources of resistance and tolerance to CMV have been reported in wild tomato relatives, but not in *S. lycopersicum*. No tomato varieties with CMV resistance are available.

Expression of the CMV CP gene in a local popular tomato cultivar 'L4783', transgenic tomato line R8, has showed consistent CMV resistance from R(0) through to R(8). The allergenicity of the CMV-CP expressed in transgenic tomato R8 was assessed by analysis of the expression of the transgene source of the protein, sequence similarity with known allergens and resistance to pepsin hydrolysis. Following the most recent Food and Agricultural Organization/World Health Organization decision tree, all results indicate that the CMV-CP was a protein with low allergenic properties and that the transgenic tomato R8 should be considered as safe as the non-transformed tomato (Lin *et al.*, 2010). The novel approach of conferring resistance through the expression of single-chain variable fragment antibody has also been examined. Di Carli *et al.* (2010) compared the proteome of genetically engineered tomato immunoprotected (single-chain variable fragment) against CMV and wild-type tomato, and showed that CMV is restricted to inoculated leaves and also identified key proteins involved in antibody-mediated resistance.

Banana

Typically, CMV infection does not have a major impact on banana and plantain production, but it can cause significant losses in new plantings where intensive production practices are used (such as under plastic greenhouses in Morocco), and in smallholder situations. Severe or heart-rot isolates of CMV, which are far more damaging than common isolates of the virus, do not occur in all banana-producing areas. It is therefore important to avoid the introduction of severe CMV isolates into new areas where they can cause significant damage (Bouhida and Lockhart, 1990; Lockhart and Jones, 2000).

Simple management practices including the use of virus-free suckers (derived from field plants or tissue culture), avoiding interplanting with susceptible cucurbitaceous and solanaceous crops, the destruction of CMV-susceptible weeds (e.g. *Commelina diffusa*) in and around production fields, the immediate removal and destruction of suspected infected plants and the subsequent disinfection of machetes with sodium hypochlorite or heat can reduce or eliminate CMV infection. The use of insecticides to control the aphid vectors of CMV is viewed as an alternative disease control measure (Niblett et al., 1994; Lockhart and Jones, 2000; Thomas et al., 2003). Heat treatment in combination with meristem tip culture has been successful for CMV eradication in banana (Gupta, 1986). Cryopreservation of meristematic tissues has been successfully used to eradicate a CMV isolate and naturally infected *Banana streak virus* (Helliot *et al.*, 2002). The possibility of obtaining mosaic-resistant banana clones by incorporating the CP gene of CMV into the banana genome has been suggested by Fauquet and Beachy (1993), and cross protection using mild strains by Wu *et al.* (1997), as cited by Lockhart and Jones (2000).

Forage legumes

Nutter *et al.* (1999) reported that CMV infection greatly decreased seed size in narrow-leafed lupin. Therefore, sieving seed lots to remove small seed may be beneficial in: (i) significantly reducing the risk of introducing CMV-infected lupin seeds into growers fields; (ii) increasing the probability of detecting the virus in subsequent tests on the seeds; and (iii) speeding up analyses (as only small fractions of seed need to be tested) during quarantine checks. This approach may be useful with other species in which CMV is also seed-borne (Gallitelli, 2000, 2002).

Ornamentals

As most of the ornamental plants are vegetatively propagated, selection and planting of virus-free propagules is an ideal approach to disease management. Virus-free plants can be generated by tissue culture technologies, combined with thermotherapy and chemotherapy. Transgenic approaches have been used to develop CMV resistance in *Gladiolus* (Kamo *et al.*, 2010) and *Lilium* (Azadi *et al.*, 2011).

6.11   Concluding Remarks

Worldwide, CMV has been reported to naturally infect several economically important agricultural and horticultural crops, forage legumes, medicinal plants and wild plants including weeds, and cause significant yield losses. The virus continues to be a threat to the production of food crops in various agroecosystems as it is transmitted vertically through the propagules of several plant species, and very efficiently horizontally by several aphid species in a non-persistent manner. Several of the characteristics of this virus have made it one of the model systems in plant virus research. The role of CMV-encoded gene products and RNAi in pathogenesis is well understood in plant systems, tobacco and *Arabidopsis*. Molecular parasitism of CMV with satRNA is well established. Sequences of satRNA-modulating symptom expression by the helper virus CMV have been identified. For its detection and identification, indicator and diagnostic hosts, enzyme immunoassays, peptide profiling, double-stranded RNA analysis, nucleic acid hybridization, microarray, variants of PCR and next-generation sequencing have been applied. The genome of several CMV isolates collected from different plant species in varied geographical locations is partially sequenced and their phylogenetic relatedness (subgroups IA, IB and II) has been established. Workable practices for the management of CMV diseases in several

crops (e.g. cucurbits, pepper, tomato, legumes, banana) have been formulated for different agroecosystems and used with varying degrees of success. These include the selection and planting of CMV-free seeds and vegetative propagules, manipulation of crop cultural practices, and the control of aphid vectors with insecticides and oil sprays. Nonetheless, alternate environmentally friendly approaches instead of chemical control need to be promoted for the control of aphid vectors. Additionally, tissue culture technologies should be exploited to produce CMV-free plants that are propagated through vegetative propagules. The adoption of CMV tolerant or resistant crop cultivars developed either by conventional breeding and/or by transgenics is an option in some regions. It is ideal to stack the *R* genes to control co-infections involving CMV and other viruses of economic importance.

As CMV continues to be an important pathogen of several crop plants in the fast-changing farming systems, consistent vigilance is required in order to avoid its introduction through propagules into virus-free agroecosystems where severe isolates of the virus have not been previously reported. Even though CMV is worldwide in distribution, the virus is still of quarantine importance. CMV emergence in developing countries as a major virus disease agent is a significant risk because of agriculture intensification. Timely identification of new virus strains that can result in severe symptom phenotypes is an important requirement to avoid CMV outbreaks that can impair the sustainability of modern farming systems. In developed as well as in some developing countries, CMV may become less of a disease problem if novel control strategies are successfully implemented.

Acknowledgements

M. Hema acknowledges Council of Scientific and Industrial Research (CSIR) and Department of Biotechnology (DBT), New Delhi, for providing financial support.

References

Afreen, B., Khan, A.A., Naqvi, Q.A., Kumar, S., Pratap, D., *et al.* (2009) Molecular identification of a *Cucumber mosaic virus* subgroup II isolate from carrot. *Journal of Plant Diseases and Protection* 116, 193–199.

Akhtar, K.P., Saleem, M.Y., Asghar, M., Ahmad, M. and Sarwar, N. (2010) Resistance of *Solanum* species to *Cucumber mosaic virus* subgroup IA and its vector *Myzus persicae. European Journal of Plant Pathology* 128, 435–450.

Aranda, M.A., Fraile, A., García-Arenal, F. and Malpica, J.M. (1995) Experimental evaluation of the ribonuclease protection assay method for the assessment of genetic heterogeneity in populations of RNA viruses. *Archives of Virology* 140, 1373–1383.

Avilla, C., Collar, J.L., Duque, M. and Fereres, A. (1997) Yield of bell pepper (*Capsicum annuum*) inoculated with CMV and/or PVY at different time intervals. *Journal of Plant Disease and Protection* 104, 1–8.

Azadi, P., Otang, N.V., Supaporn, H., Khan, R.S., Chin, D.P., *et al.* (2011) Increased resistance to *Cucumber mosaic virus* (CMV) in *Lilium* transformed with a defective CMV replicase gene. *Biotechnology Letters* 33, 1249–1255.

Berniak, H., Malinowski, T. and Kaminska, M. (2009) Comparison of ELISA and RT-PCR assays for detection and identification of *Cucumber mosaic virus* (CMV) isolates infecting horticultural crops in Poland. *Journal of Fruit and Ornamental Plant Research* 17, 5–20.

Berniak, H., Malinowski, T. and Kaminska, M. (2010) Characterisation of polyclonal antibodies raised against two isolates of *Cucumber mosaic virus. Journal of Plant Pathology* 92, 231–234.

Betancourt, M., Escriu, F., Fraile, A. and García-Arenal, F. (2013) Virulence evolution of a generalist plant virus in a heterogeneous host system. *Evolutionary Applications* 6, 875–890.

Bhat, A.I., Siljo, A. and Deeshma, K.P. (2013) Rapid detection of *Piper yellow mottle virus* and *Cucumber mosaic virus* infecting black pepper (*Piper nigrum*) by loop-mediated isothermal amplification (LAMP). *Journal of Virological Methods* 193, 190–196.

Bouché, N., Lauressergues, D., Gasciolli, V. and Vaucheret, H. (2006) An antagonistic function for *Arabidopsis* DCL2 in development and a new function for DCL4 in generating viral siRNAs. *EMBO Journal* 25, 3347–3356.

Bouhida, M. and Lockhart, B.E. (1990) Increase in importance of *Cucumber mosaic virus* infection in greenhouse grown bananas in Morocco. *Phytopathology* 80, 981.

Cai, W.Q., Fang, R.X., Shang, H.S., Wang, X., Zhang, F.L., *et al.* (2003) Development of CMV and TMV resistant chili pepper: field performance and biosafety assessment. *Molecular Breeding* 11, 25–35.

Carmo-Sousa, M., Moreno, A., Garzo, E. and Fereres, A. (2014) A non-persistently transmitted-virus induces a pull-push strategy in its aphid vector to optimize transmission and spread. *Virus Research* 186, 38–46.

Chen, B., Randles, J.W. and Francki, R.I. (1995) Mixed-subunit capsids can be assembled *in vitro* with coat protein subunits from two cucumoviruses. *Journal of General Virology* 76, 971–973.

Chen, S., Gu, H., Wang, X., Chen, J. and Zhu, W. (2011) Multiplex RT-PCR detection of *Cucumber mosaic virus* subgroups and tobamoviruses infecting tomato using 18S rRNA as an internal control. *Acta Biochimica et Biophysica Sinica* 43, 465–471.

Chen, Y.J., Zhang, J., Liu, J., Deng, X.G., Zhang, P., *et al.* (2014) The capsid protein p38 of *Turnip crinkle virus* is associated with the suppression of *Cucumber mosaic virus* in *Arabidopsis thaliana* co-infected with *Cucumber mosaic virus* and *Turnip crinkle virus. Virology* 462, 71–80.

Cillo, F., Mascia, T., Pasciuto, M.M. and Gallitelli, D. (2009) Differential effects of mild and severe *Cucumber mosaic virus* strains in the perturbation of microRNA-regulated gene expression in tomato map to the 3'sequence of RNA2. *Molecular Plant-Microbe Interactions* 22, 1239–1249.

Dáder, B., Legarrea, S., Moreno, A., Plaza, M., Carmo-Sousa, M., *et al.* (2014) Control of insect vectors and plant viruses in protected crops by novel pyrethroid-treated nets. *Pest Management Science* doi: 10.1002/ps.3942.

Dai, J., Cheng, J., Huang, T., Zheng, X. and Wu, Y.A. (2012) Multiplex reverse transcription PCR assay for simultaneous detection of five tobacco viruses in tobacco plants. *Journal of Virological Methods* 183, 57–62.

Davino, S., Panno, S., Rangel, E.A., Davino, M., Bellardi, M.G., *et al.* (2012) Population genetics of *Cucumber mosaic virus* infecting medicinal, aromatic and ornamental plants from northern Italy. *Archives of Virology* 157, 739–745.

Deyong, Z., Willingmann, P., Heinze, C., Adam, G., Pfunder, M., *et al.* (2005) Differentiation of *Cucumber mosaic virus* isolates by hybridization to oligonucleotides in a microarray format. *Journal of Virological Methods* 123, 101–108.

Di Carli, M., Villani, M.E., Bianco, L., Lombardi, R., Perrotta, G., *et al.* (2010) Proteomic analysis of the plant-virus interaction in *Cucumber mosaic virus* (CMV) resistant transgenic tomato. *Journal of Proteome Research* 9, 5684–5697.

Du, Z.-Y., Chen, F.-F., Liao, Q.-S., Zhang, H.-R., Chen, Y.-F., *et al.* (2007) *2b* ORFs encoded by subgroup IB strains of *Cucumber mosaic virus* induce differential virulence on *Nicotiana* species. *Journal of General Virology* 88, 2596–2604.

El-Dougdoug, Kh.A., Ghaly, M.F. and Taha, M.A. (2012) Biological control of *Cucumber mosaic virus* by certain local *Streptomyces* isolates: inhibitory effects of selected five Egyptian isolates. *International Journal of Virology* 8, 151–164.

Elsharkawy, M.M., Shimizu, M., Takahashi, H. and Hyakumachi, M. (2012) Induction of systemic resistance against *Cucumber mosaic virus* by *Penicillium simplicissimum* GP17-2 in *Arabidopsis* and tobacco. *Plant Pathology* 61, 964–976.

Escriu, F., Fraile, A. and García-Arenal, F. (2003) The evolution of virulence in a plant virus. *Evolution* 57, 755–765.

Fang, Y.Y., Smith, N.A., Zhao, J.H., Lee, J.R., Guo, H.S., *et al.* (2015) Cloning and profiling of small RNAs from *Cucumber mosaic virus* satellite RNA. *Methods in Molecular Biology* 1236, 99–109.

Fauquet, C.M. and Beachy, R.N. (1993) Status of the coat protein-mediated resistance and its potential application for banana viruses. In: Proceedings of the workshop on *Biotechnology Applications for Banana and Plantain Improvement* held in San Jose, Costa Rica, 27–31 January 1992. INIBAP, Mountpellier, France, pp. 69–84.

Feng, J., Liu, S., Wang, M., Lang, Q. and Jin, C. (2014) Identification of microRNAs and their targets in tomato infected with *Cucumber mosaic virus* based on deep sequencing. *Planta* 240, 1335–1352.

Gallitelli, D. (2000) The ecology of *Cucumber mosaic virus* and sustainable agriculture. *Virus Research* 71, 9–21.

Gallitelli, D. (2002) Present status of controlling *Cucumber mosaic virus*. In: Hadidi, A., Khetarpal, R.K. and Koganezawa, H. (eds), *Plant Virus Disease Control,* APS Press, St. Paul, Minnesota, pp. 507–523.

Gallitelli, D. and Saldarelli, P. (1996) Molecular identification of phytopathogenic viruses. *Methods in Molecular Biology* 50, 57–79.

García-Arenal, F. and Palukaitis, P. (2008) *Cucumber mosaic virus*. In: Mahy, B.W.J. and van Regenmortel, M.H.V. (eds), *Desk Encyclopedia of Plant and Fungal Virology*, Academic Press-Elsevier, Oxford, UK, pp. 171–176.

García-Arenal, F., Escriu, F., Aranda, M.A., Alonso-Prados, J.L., Malpica, J.M., *et al.* (2000) Molecular epidemiology of *Cucumber mosaic virus* and its satellite RNA. *Virus Research* 71, 1–8.

Garzón, A., Budia, F., Medina, P., Morales, I., Fereres, A., *et al.* (2015) The effect of *Chrysoperla carnea* (Neuroptera: *Chrysopidae*) and *Adalia bipunctata* (Coleoptera: *Coccinellidae*) on the spread of *Cucumber mosaic virus* (CMV) by *Aphis gossypii* (Hemiptera: *Aphididae*). *Bulletin of Entomological Research* 105, 13–22.

Gildow, F.E., Shah, D.A., Sackett, W.M., Butzler, T., Nault, B.A., *et al.* (2008) Transmission efficiency of *Cucumber mosaic virus* by aphids associated with virus epidemics in snap bean. *Phytopathology* 98, 1233–1241.

Gong, L., Paris, H.S., Stift, G., Pachner, M., Vollmann, J., *et al.* (2013) Genetic relationships and evolution in *Cucurbita* as viewed with simple sequence repeat polymorphisms: the centrality of *C. okeechobeensis*. *Genetic Resources and Crop Evolution* 60, 1531–1546.

Gupta, P.P. (1986) Eradication of mosaic disease and rapid clone multiplication of bananas and plantains through meristem tip culture. *Plant Cell, Tissue and Organ Culture*, 6, 33–39.

Harvey, J.J., Lewsey, M.G., Patel, K., Westwood, J., Heimstädt, S., *et al.* (2011) An antiviral defence role of AGO2 in plants. *PLoS One* 6, 14639.

Helliot, B., Panis, B., Poumay, Y., Swennen, R., Lepoivre, P., *et al.* (2002) Cryopreservation for the elimination of cucumber mosaic and banana streak viruses from banana. *Plant Cell Reports* 20, 1117–1122.

Hooks, C.R.R. and Fereres, A. (2006) Protecting crops from non-persistently aphid-transmitted viruses: a review on the use of barrier plants as a management tool. *Virus Research* 120, 1–16.

Hosseinzadeh, H., Nasrollanejad, S. and Khateri, H. (2012) First report of *Cucumber mosaic virus* subgroups I and II on soybean, pea, and eggplant in Iran. *Acta Virologica* 56, 145–148.

Hsu, H.T., Barzuna, L., Hsu, Y.H., Bliss, W. and Perry, K.L. (2000) Identification and subgrouping of *Cucumber mosaic virus* with mouse monoclonal antibodies. *Phytopathology* 90, 615–620.

ICTVdB Management (2006) 00.010.0.04.001. *Cucumber mosaic virus*. In: Büchen-Osmond, C. (Ed), *ICTVdB - The Universal Virus Database, version 4*. Columbia University, New York (http://www.ncbi. nlm.nih.gov/ICTVdb/ICTVdB/00.010.0.04.001.htm).

Jacquemond, M. (2012) *Cucumber mosaic virus*. *Advances in Virus Research* 84, 439–504.

Jordá, C., Afar, A., Aranda, M.A., Moriones, E. and García-Arenal, F. (1992) An epidemic of *Cucumber mosaic virus* plus satellite RNA in tomatoes in eastern Spain. *Plant Disease* 76, 363–366.

Kamo, K., Jordan, R., Guaranga, M.A., Hsu, H.T. and Ueng, P. (2010) Resistance to *Cucumber mosaic virus* in gladiolus plants transformed with either a defective replicase or coat protein subgroup II gene from *Cucumber mosaic virus*. *Plant Cell Reports* 29, 695–704.

Kapoor, R., Mandal, B., Paul, P.K., Chigurupati, P. and Jain, R.K. (2014) Production of cocktail polyclonal antibodies using bacterial expressed recombinant protein for multiple virus detection. *Journal of Virological methods* 196, 7–14.

Kavosipour, S., Niazi, A., Izadpanah, K., Afsharifar, A. and Yasaie, M. (2012) Induction of resistance to *Cucumber mosaic virus* (CMV) using hairpin construct of *2B* gene. *Iranian Journal of Plant Pathology* 48, 65–67.

Kim, M.K., Jeong, R.D., Kwak, H.R., Lee, S.H., Kim, J.S., *et al.* (2011) Characteristics of *Cucumber mosaic virus* isolated from *Zea mays* in Korea. *Plant Pathology Journal* 27(4), 372–377.

Kim, M.K., Jeong, R.D., Kwak, H.R., Lee, S.H., Kim, J.S., *et al.* (2014a) First report of *Cucumber mosaic virus* isolated from wild *Vigna angularis* var. *nipponensis* in Korea. *Plant Pathology Journal* 30, 200–207.

Kim, M.K., Seo, J.K., Kwak, H.R., Kim, J.S., Kim, K.H., *et al.* (2014b) Molecular genetic analysis of *Cucumber mosaic virus* populations infecting pepper suggests unique patterns of evolution in Korea. *Phytopathology* 104, 993–1000.

Kiranmai, G., Sreenivasulu, P. and Nayudu, M.V. (1996) Comparison of three different tests for detection of Cucumber mosaic cucumovirus in banana (*Musa paradisiaca*). *Current Science* 71, 764–767.

Kiranmai, G., Satyanarayana, T. and Sreenivasulu, P. (1998) Molecular cloning and detection of Cucumber mosaic cucumovirus causing infectious chlorosis disease of banana using DNA probes. *Current Science* 74, 356–359.

Kouadio, O.T., De Clerck, C., Agneroh, T.H., Parisi, O., Lepoivre, P., *et al.* (2013) Role of satellite RNAs in *Cucumber mosaic virus*-host plant interactions: a review. *Biotechnologie, Agronomie, Société et Environnement* 17, 1780–4507.

Kumar, S., Srivastava, A. and Raj, S.K. (2005) Molecular diagnosis of *Cucumber mosaic virus* in *Chrysanthemum*. *Indian Journal of Virology* 16, 16.

Kumari, R., Bhardwaj, P., Singh, L., Zaidi, A.A. and Hallan, V. (2013) Biological and molecular characterization of *Cucumber mosaic virus* subgroup II isolate causing severe mosaic in cucumber. *Indian Journal of Virology* 24, 27–34.

Kwon, J.Y., Ryu, K.H. and Choi, S.H. (2013) Reverse Transcription Polymerase Chain Reaction-based system for simultaneous detection of multiple lily-infecting viruses. *Plant Pathology Journal* 29, 338–343.

Kwon, J.Y., Hong, J.S., Kim, M.J., Choi, S.H., Min, B.E., *et al.* (2014) Simultaneous multiplex PCR detection of seven cucurbit-infecting viruses. *Journal of Virological Methods* 206, 133–139.

Lamptey, J.N.L., Osei, M.K., Mochiah, M.B., Osei, K., Berchie, J.N., *et al.* (2013) Serological detection of *Tobacco mosaic virus* and *Cucumber mosaic virus* infecting tomato (*Solanum lycopersicum*) using a lateral flow immunoassay. *Journal of Agricultural Studies* 1, dx.doi.org/10.5296/jas.v1i2.3768.

Lang, Q., Jin, C., Gao, X., Feng, J., Chen, S., *et al.* (2011) Profiling of *Cucumber mosaic virus* responsive mRNAs in tomato using micro paraflo microfluidics microarrays. *Journal of Nanoscience and Nanotechnology* 11, 3115–3125.

Lawson, R.H. (1985) Viruses and virus diseases. In: Strider, D.L. (Ed) *Diseases of Floral Crops*, Vol. I. Praeger Scientific Publisher, New York, pp. 253–294.

Lecoq, H. and Desbiez, C. (2012) Viruses of cucurbit crops in the Mediterranean region: an ever-changing picture. *Advances in Virus Research* 84, 67–126.

Lee, Y.H., Jung, M., Shin, S.H., Lee, J.H., Choi, S.H., *et al.* (2009) Transgenic peppers that are highly tolerant to a new CMV pathotype. *Plant Cell Reports* 28, 223–232.

Lin, C.H., Sheu, F., Lin, H.T. and Pan, T.M. (2010) Allergenicity assessment of genetically modified *Cucumber mosaic virus* (CMV) resistant tomato (*Solanum lycopersicon*). *Journal of Agricultural and Food Chemistry* 58, 2302–2306.

Lin, C.Y., Ku, H.M., Chiang, Y.H., Ho, H.Y., Yu, T.A., *et al.* (2012) Development of transgenic watermelon resistant to *Cucumber mosaic virus* and *Watermelon mosaic virus* by using a single chimeric transgene construct. *Transgenic Research* 21, 983–993.

Lockhart, B.E.L. and Jones, D.R. (2000) Banana Mosaic. In: Jones, D.R. (ed.), *Diseases of Banana, Abaca and Enset*, CABI Publishing, Wallingford, UK, pp. 256–263.

Lu, J., Du, Z.X., Kong, J., Chen, L.N., Qiu, Y.H., *et al.* (2012) Transcriptome analysis of *Nicotiana tabacum* infected by *Cucumber mosaic virus* during systemic symptom development. *PLoS One* 7, e43447.

Luis-Arteaga, M., Alvarez, J.M., Alonso-Prados, J.L., Bernal, J.J., García-Arenal F., *et al.* (1998) Occurrence, distribution and relative incidence of mosaic viruses infecting field-grown melon in Spain. *Plant Disease* 82, 979–982.

Makkouk, K., Pappu, H. and Kumari, S.G. (2012) Virus diseases of peas, beans, and faba bean in the Mediterranean region. *Advances in Virus Research* 84, 367–402.

Marathe, R., Guan, Z., Anandalakshmi, R., Zhao, H. and Dinesh-Kumar, S.P. (2004) Study of *Arabidopsis thaliana* resistome in response to *Cucumber mosaic virus* infection using whole genome microarray. *Plant Molecular Biology* 55, 501–520.

Mascia, T., Cillo, F., Fanelli, V., Finetti-Sialer, M.M., De Stradis, A., *et al.* (2010) Characterization of the interactions between *Cucumber mosaic virus* and *Potato virus Y* in mixed infections in tomato. *Molecular Plant-Microbe Interactions* 23, 1514–1524.

Masuta, C. and Shimura, H. (2013) RNA silencing against viruses: molecular arms race between *Cucumber mosaic virus* and its host. *Journal of General Plant Pathology* 79, 227–232.

Mauck, K.E., De Moraes, C.M. and Mescher, M.C. (2010) Deceptive chemical signals induced by a plant virus attract insect vectors to inferior hosts. *Proceedings of the National Academy of Sciences USA* 107, 3600–3605.

Mauck, K.E., De Moraes, C.M. and Mescher, M.C. (2014) Biochemical and physiological mechanisms underlying effects of *Cucumber mosaic virus* on host-plant traits that mediate transmission by aphid vectors. *Plant Cell & Environment* 37, 1427–1439.

Mazourek, M., Moriarty, G., Glos, M., Fink, M., Kreitinger, M., *et al.* (2009) 'Peacework': A *Cucumber mosaic virus*-resistant early red bell pepper for organic systems. *HortScience* 44, 1464–1467.

Miura, N.S., Beriam, L.O.S. and Rivas, E.B. (2013) Detection of *Cucumber mosaic virus* in commercial anthurium crops and genotypes evaluation. *Horticultura Brasileira* 31, 322–327.

Mitsuya, Y., Takahashi, Y., Berberich, T., Miyazaki, A., Matsumura, H., *et al.* (2009) Spermine signaling plays a significant role in the defense response of *Arabidopsis thaliana* to *Cucumber mosaic virus*. *Journal of Plant Physiology* 166, 626–643.

Mochizuki, T. and Ohki, S.T. (2012) *Cucumber mosaic virus*: viral genes as virulence determinants. *Molecular Plant Pathology* 13, 217–225.

Morroni, M., Thompson, J.R. and Tepfer, M. (2008) Twenty years of transgenic plants resistant to *Cucumber mosaic virus*. *Molecular Plant-Microbe Interactions* 21, 675–684.

Moury, B. and Verdin, E. (2012) Viruses of pepper crops in the Mediterranean basin: a remarkable stasis. *Advances in Virus Research* 84, 127–162.

Murphy, J.F. (2003) *Cucumber mosaic virus*. In: Pernezny, K.L., Roberts, P.D., Murphy, J.F. and Goldberg, N.P. (Eds.) *Compendium of pepper diseases*. APS Press, St. Paul, Minnesota, pp. 29–31.

Natilla, A., Piazzolla, G., Nuzzaci, M., Saldarelli, P., Tortorella, C., *et al.* (2004) *Cucumber mosaic virus* as carrier of a hepatitis C virus-derived epitope. *Archives of Virology* 149, 137–154.

Natilla, A., Hammond, R.W. and Nemchinov, L.G. (2006) Epitope presentation system based on *Cucumber mosaic virus* coat protein expressed from a *Potato virus X*-based vector. *Archives of Virology* 151, 1373–1386.

Ng, J.C., Josefsson, C., Clark, A.J., Franz, A.W. and Perry, K.L. (2005) Virion stability and aphid vector transmissibility of *Cucumber mosaic virus* mutants. *Virology* 332, 397–405.

Niblett, C.L., Pappu, S.S., Bird, J. and Lastra, R. (1994) Infectious chlorosis, mosaic and heart rot. In: Ploetz, R.C., Zentmyer, G.A., Nishijima, W.T., Rohrback, K.G. and Ohr, H.D. (eds) *Compendium of tropical fruit diseases*. APS Press, St. Paul, Minnesota, pp. 18–19.

Niimi, Y., Han, D.S., Mori, S. and Kobayashi, H. (2003) Detection of *Cucumber mosaic virus*, *Lily symptomless virus* and *Lily mottle virus* in *Lilium* species by RT-PCR technique. *Scientia Horticulturae* 97, 57–63.

Nouri, S., Arevalo, R., Falk, B.V.V. and Groves, R.L. (2014) Genetic structure and molecular variability of *Cucumber mosaic virus* isolates in the USA. *PLoS One* 9, e96582.

Ntui, V.O., Kynet, K., Khan, R.S., Ohara, M., Goto, Y., *et al.* (2014) Transgenic tobacco lines expressing defective CMV replicase-derived dsRNA are resistant to CMV-O and CMV-Y. *Molecular Biotechnology* 56, 50–63.

Nutter, F.W., Alberts, E. and Graetz, D. (1999) Quantifying the relationship between seed size in nar-row-leafed lupin and incidence of *Cucumber mosaic virus*. In: *Plant Virus Epidemiology: Current Status and Future Prospects*. VII International Plant Virus Epidemiology Symposium, 11–16 April, Almeria, Spain, pp. 97–98.

Nuzzaci, A., Vitti, A., Condelli, V., Lanorte, M.T., Tortorella, C., *et al.* (2010) *In vitro* stability of *Cucumber mosaic virus* nanoparticles carrying a *Hepatitis C virus*-derived epitope under simulated gastrointestinal conditions and *in vivo* efficacy of an edible vaccine. *Journal of Virological Methods* 165, 211–215.

Palukaitis, P. and Carr, J.P. (2008) Plant resistance responses to viruses. *Journal of Plant Pathology* 90, 153–171.

Palukaitis, P. and García-Arenal, F. (2003a) Cucumoviruses. *Advances in Virus Research* 62, 241–323.

Palukaitis, P. and García-Arenal, F. (2003b) *Cucumber mosaic virus*. In: Antoniw, J. and Adams, M. (eds) *Description of Plant viruses*. DPV400, Association of Applied Biologists, Rothamstead Research, UK. http://www.dpvweb.net/dpv/showdpv.php?dpvno=400 (accessed September 2007).

Palukaitis, P., Roossinck, M.J., Dietzgen, R.G. and Francki, R.I.B. (1992) *Cucumber mosaic virus*. *Advances in Virus Research* 41, 281–348.

Panno, S., Davino, S., Rubio, L., Rangel, E., Davino, M., *et al.* (2012) Simultaneous detection of the seven main tomato-infecting RNA viruses by two multiplex reverse transcription polymerase chain reactions. *Journal of Virological Methods* 186, 152–156.

Peng, J., Shi, M., Xia, Z., Huang, J. and Fan, Z. (2012) Detection of *Cucumber mosaic virus* isolates from banana by one-step reverse transcription loop-mediated isothermal amplification. *Archives of Virology* 157, 2213–2217.

Phan, M.S., Seo, J.K., Choi, H.S., Lee, S.H. and Kim, K.H. (2014a) Pseudorecombination between two distinct strains of *Cucumber mosaic virus* results in enhancement of symptom severity. *Plant Pathology Journal* 30, 316–322.

Phan, M.S., Seo, J.K., Choi, H.S., Lee, S.H. and Kim, K.H. (2014b) Molecular and biological characterization of an isolate of *Cucumber mosaic virus* from *Glycine soja* by generating its infectious full-genome cDNA clones. *Plant Pathology Journal* 30, 159–167.

Pierrugues, O., Guilbaud, L., Fernandez-Delmond, I., Fabre, F., Tepfer, M., *et al.* (2007) Biological properties and relative fitness of inter-subgroup *Cucumber mosaic virus* RNA 3 recombinants produced *in vitro*. *Journal of General Virology* 88, 2852–2861.

Ploetz, R.C. (1994) Bract mosaic. In: Ploetz, R.C., Zentmyer, G.A., Nishijima, W.T., Rohrbach, K.G. and Ohr, H.D. (eds) *Compendium of Tropical Plant Diseases*. APS Press, St. Paul, Minnesota, p. 20.

Provvidenti, R. (1996) Cucumber mosaic. In: Zitter, T.A., Hopkins, D.L. and Thomas, C.E. (eds) *Compendium of Cucurbit Diseases*. APS Press, St. Paul, Minnesota, pp. 38–39.

Qu, J., Ye, J. and Fang, R. (2007) Artificial microRNA-mediated virus resistance in plants. *Journal of Virology* 81, 6690–6699.

Raj, S.K., Saxena, S., Hallan, V. and Singh, B.P. (1998) Reverse transcription-polymerase chain reaction (RT-PCR) for direct detection of *Cucumber mosaic virus* in gladiolus. *Biochemistry and Molecular Biology International* 44, 89–95.

Roossinck, M.J., Zhang, L. and Hellwald, K.-H. (1999) Rearrangements in the 5′ nontranslated region and phylogenetic analyses of *Cucumber mosaic virus* RNA3 indicate radial evolution of three subgroups. *Journal of Virology* 73, 6752–6758.

Rybicki, E.P. (2015) A top ten list for economically important plant viruses. *Archives of Virology* 160, 17–20.

Salem, R.E., Salama, M.I. and Nour-El-Din, H.A. (2013) Expression of *Cucumber mosaic virus* coat protein gene for the production of recombinant polyclonal antibodies. *Arab Journal of Biotechnology* 16, 243–252.

Scholthof, K.B.G., Adkins, S., Czosnek, H., Palukaitis, P., Jacquot, E., *et al.* (2011) Top 10 plant viruses in molecular plant pathology. *Molecular Plant Pathology* 12, 938–954.

Sekine, K.T., Kawakami, S., Hase, S., Kubota, M., Ichinose, Y., *et al.* (2008) High level expression of a virus resistance gene, *RCY1*, confers extreme resistance to *Cucumber mosaic virus* in *Arabidopsis thaliana*. *Molecular Plant-Microbe Interactions* 21, 1398–1407.

Seo, J.K., Lee, Y.J., Kim, M.K., Lee, S.H., Kim, K.H., *et al.* (2014) A novel set of polyvalent primers that detect members of the genera *Bromovirus* and *Cucumovirus*. *Journal of Virological Methods* 203, 112–115.

Sharman, M., Thomas, J.E. and Dietzen, R.G. (2000) Development of a multiplex immunocapture PCR with colourimetric detection for viruses of banana. *Journal of Virological Methods* 89, 75–88.

Shimura, H. and Masuta, C. (2012) Structural and functional analysis of CMV satellite RNAs in RNA silencing. *Methods in Molecular Biology* 894, 273–286.

Siddiqui, S.A., Valkonen, J.P., Rajamäki, M.L. and Lehto, K. (2011) The 2b silencing suppressor of a mild strain of *Cucumber mosaic virus* alone is sufficient for synergistic interaction with *Tobacco mosaic virus* and induction of severe leaf malformation in 2b-transgenic tobacco plants. *Molecular Plant-Microbe Interactions* 24, 685–693.

Simons, J.N. and Zitter, T.A. (1980) Use of oils to control aphid-borne viruses. *Plant Disease* 64, 542–546.

Smith, N.A., Eamens, A.L. and Wang, M.-B. (2011) Viral small interfering RNAs target host genes to mediate disease symptoms in plants. *PloS Pathogens* 7, e1002022.

Takeshita, M., Koizumi, E., Noguchi, M., Sueda, K., *et al.* (2012) Infection dynamics in viral spread and interference under the synergism between *Cucumber mosaic virus* and *Turnip mosaic virus*. *Molecular Plant-Microbe Interactions* 25, 18–27.

Thackray, D.J., Diggle, A.J., Berlandier, F.A. and Jones, R.A.C. (2004) Forecasting aphid outbreaks and epidemics of *Cucumber mosaic virus* in lupin crops in a Mediterranean-type environment. *Virus Research* 100, 67–82.

Thomas, J.E., Geering, A.D.W., Dahal, G., Lockhart, B.E.L. and Thottappilly, G. (2003) Banana and Plantain. In: Lobenstein, G. and Thottappilly, G. (eds) *Viruses and Virus-like Diseases of Major Crops in Developing Countries*. Kluwer Academic Publishers, Dordrecht, The Netherlands, pp. 477–494.

Tien, P. and Wu, G. (1991) Satellite RNA for the biocontrol of plant disease. *Advances in Virus Research* 39, 321–339.

Tomlinson, J.A. (1987) Epidemiology and control of virus diseases of vegetables. *Annals of Applied Biology* 66, 381–386.

Vitti, A., Nuzzaci, M., Condelli, V. and Piazzolla, P. (2014) Simulated digestion for testing the stability of edible vaccine based on *Cucumber mosaic virus* (CMV) chimeric particle display *Hepatitis C virus* (HCV) peptide. *Methods in Molecular Biology* 1108, 41–56.

Wang, X.B., Wu, Q., Ito, T., Cillo, F., Li, W.X., *et al.* (2010) RNAi-mediated viral immunity requires amplification of virus-derived siRNAs in *Arabidopsis thaliana*. *Proceedings of the National Academy of Sciences USA* 107, 484–489.

Wang, Y., Lee, K.C., Gaba, V., Wong, S.M., Palukaitis, P., *et al.* (2004a) Breakage of resistance to *Cucumber mosaic virus* by co-infection with *Zucchini yellow mosaic virus*: enhancement of CMV accumulation independent of symptom expression. *Archives of Virology* 149, 379–396.

Wang, Y., Tzfira, T., Gaba, V., Citovsky, V., Palukaitis, P., *et al.* (2004b) Functional analysis of the *Cucumber mosaic virus* 2b protein: pathogenicity and nuclear localization. *Journal of General Virology* 85, 3135–1347.

Wu, J.L., Hung, T.H. and Su, H.J. (1997) Strain differentiation of *Cucumber mosaic virus* associated with banana mosaic disease in Taiwan. *Annals of the Phytopathological Society of Japan* 63, 176–178.

Zein, H.S. and Miyatake, K. (2009) Development of rapid, specific and sensitive detection of *Cucumber mosaic virus*. *African Journal of Biotechnology* 8, 751–759.

Zein, H.S., Nakazawa, M., Ueda, M. and Miyatake, K. (2007) Development of serological procedures for rapid and reliable detection of *Cucumber mosaic virus* with Dot-immunobinding assay. *World Journal of Agricultural Sciences* 3, 430–439.

Zhao, L., Cheng, J., Hao, X., Tian, X. and Wu, Y. (2012) Rapid detection of tobacco viruses by reverse transcription loop-mediated isothermal amplification. *Archives of Virology* 157, 2291–2298.

Zitter, T.A. (1991) Cucumber mosaic. In: Jones, J.B., Jones, J.P., Stall, R.E. and Zitter, T.A. (eds) *Compendium of Tomato Diseases*. APS Press, St. Paul, Minnesota, pp. 35–36.

Zitter, T.A. and Murphy, J.F. (2009) Cucumber mosaic. *The Plant Health Instructor*. doi: 10.1094/PHI-I-2009-0518-01.

7 Potato Mosaic and Tuber Necrosis

Mohamad Chikh-Ali and Alexander V. Karasev*

*Department of Plant, Soil and Entomological Sciences,
University of Idaho, Moscow, Idaho, USA*

7.1 Introduction: The Aetiologic Agent, Disease Symptoms, Distribution and Economic Importance

Potato virus Y (PVY), the aetiological agent of potato mosaic or potato tuber necrotic ringspot disease (PTNRD), is the most economically important and devastating virus infecting potato crops worldwide (Singh *et al.*, 2008; Gray *et al.*, 2010; Karasev and Gray, 2013b). PVY is the type species of the genus *Potyvirus*, family *Potyviridae*, the second largest family of plant viruses after *Geminiviridae*. The virus has a single-stranded positive-sense RNA genome of about 9.7 kb with a covalently linked VPg protein at the 5′ terminus and a poly-A tail at the 3′ terminus (Adams *et al.*, 2012). The genome RNA of PVY has two non-translated regions, 5′ and 3′, flanking a single open reading frame that encodes for a large polyprotein. Upon translation, the polyprotein is co- and/or post-translationally cleaved by three viral-specific proteases, P1, HC-Pro and NIa-Pro, to produce ten mature viral proteins (Adams *et al.*, 2012). Another small protein, PIPO, was reported to be encoded by potyvirus genomes from a small open reading frame embedded within the P3 region and to be putatively translated via a +2 translational frameshift (Chung *et al.*, 2008). The PVY genome is encapsidated by about 2000 copies of a single capsid protein of 267 amino acids to form filamentous, non-enveloped flexuous rods, with helical symmetry, measuring 730–740 nm in length and 11 nm in diameter.

PVY has a moderately wide host range and it infects a large number of species in the family Solanaceae. This includes important crop species like potato, pepper, tomato and tobacco. Here, we focus on the effects of PVY on potato. Infections of other hosts of PVY have been addressed elsewhere (Kerlan, 2006; Quenouille *et al.*, 2013). PVY infection in potato (*Solanum tuberosum*) results in distinct foliage and tuber symptoms. The type of foliage symptoms and their severity vary depending on several factors which include PVY strain, potato cultivar, environmental conditions and the infection type, that is, primary (or current season) and secondary (or tuber-borne). In general, PVY induces one or more of the following symptoms in foliage: mosaic or mottle of variable severity, crinkle, rugosity, stunting, necrosis of variable types such as necrotic spots/rings and veinal necrosis, and lower leaf drop when infected plants look like palm trees. The severity of PVY infection increases in mixed infections with other viruses like the tymovirales *Potato virus X* (family *Alphaflexiviridae*, genus *Potexvirus*) or *Potato virus S* (family *Betaflexiviridae*, genus *Carlavirus*) (Kerlan, 2006; Valkonen, 2007). Some PVY strains such as PVY[N-Wi] and PVY[N:O] cause mild symptoms in most potato cultivars, which might contribute to

---

*E-mail: akarasev@uidaho.edu

the increased incidence in potato-producing areas. Other potato cultivars such as 'Russet Norkotah' and 'Shepody' are asymptomatic or show very mild symptoms when infected with PVY, and yet exhibit the same yield reductions as PVY infected 'Russet Burbank' that shows severe foliar symptoms. These asymptomatic cultivars are presumably responsible for the recent increase in PVY inoculum in potato areas (Karasev and Gray, 2013b). Tuber symptoms also vary according to PVY strain, potato cultivars and environmental conditions. Some strains of PVY, mainly PVY^{NTN}, cause potato tuber necrotic ringspot disease (Fig. 7.1), which appears at harvest time or later during storage as pink protruded rings and arches that later become necrotic and sunken.

The yield reduction caused by PVY infection varies according to the infection type and time. PVY infection may occur during the growing season by aphid vectors that transmit PVY from infected to healthy potato plants in a non-persistent manner (primary infection). The yield losses and proportion of progeny tubers infected are greater when the primary infection occurs at early stages of the growing season compared to infections at the later stages of development. At the end of the growing season, potato plants become less susceptible to virus infection and translocation to progeny tubers, the phenomenon known as mature plant resistance. Therefore, PVY infection that originated from infected seed tubers (secondary infection) causes the highest yield reduction and is considered the most damaging type of PVY infection.

There are many reports on yield reduction caused by PVY infections with different cultivars and different PVY strains under varying environmental and infection sources (for details see Valkonen, 2007). In a study on the effect of seed-borne PVY infection on three potato cultivars, 'Russet Burbank', 'Russet Norkotah' and 'Shepody', the yield decreased with the increase in virus incidence by about 200 kg/ha for each 1% increase in PVY (Nolte et al., 2004). PVY also affects potato tuber quality by inducing PTNRD that dramatically reduces the marketability of potato tubers and causes significant losses to ware (large tubers meant for consumption) and seed potato producers. Potato cultivars susceptible to PTNRD, such as 'Yukon Gold', develop tuber necrosis of variable severity when infected with most recombinant PVY strains. As a result, this cultivar is no longer produced in some countries, including France (Valkonen, 2007). In addition, PVY is the main obstacle for seed potato production because

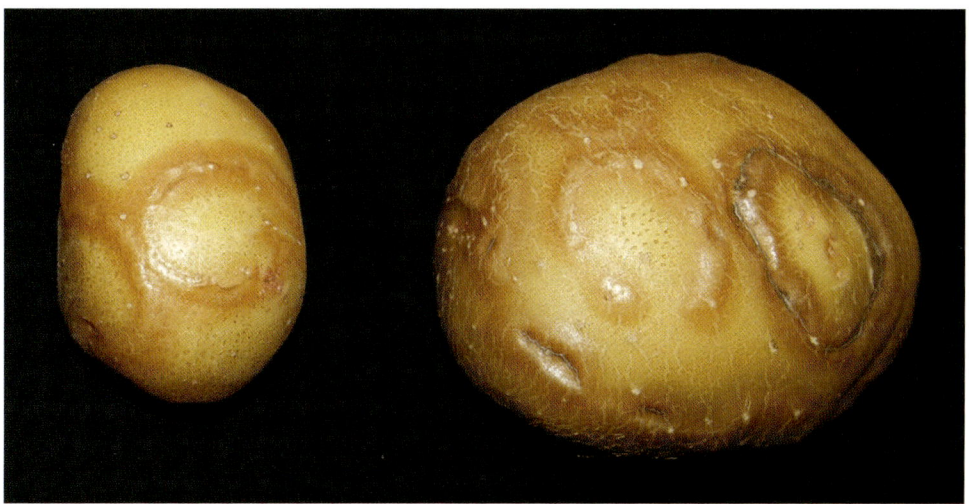

Fig. 7.1. Tuber necrotic ringspot disease. Potato tubers of 'Yukon Gold' showing necrotic arcs and rings of variable development stages.

it is responsible for seed potato degeneration that requires regular replacement of seed potatoes with infection exceeding acceptable levels. High levels of PVY are responsible for the rejection of many seed lots as certified seed and a shortage of certified seed, especially of susceptible potato cultivars. To obtain certified seed potatoes, seed growers and certification programmes implement multiple control measures, including visual inspection, removal of symptomatic plants, postharvest indexing, chemical spraying, early vine killing, among others. These practices generally lead to increases in the cost of seed potato production.

7.2 Host Range and Transmission

PVY has a moderately wide natural and experimental host range. In nature, PVY has been reported from nine plant families such as the family Solanaceae which includes important crops such as potato, tobacco (*Nicotiana* spp.), tomato (*Solanum lycopersicum* L.), pepper (*Capsicum* spp.) and tree tomato (*Solanum betaceum*), as well as solanaceous weeds including *Physalis* spp., *Solanum dulcamara* and *S. nigrum* (Jeffries, 1998). PVY is reported to naturally infect other weed hosts such as *Hyoscyamus desertorum*, *Portulaca oleracea* (Portulacaceae), *Solanum elaeagnifolium*, *S. dulcamara*, *S. nigrum* and *S. villosum* (Boukhris-Bouhachem *et al.*, 2007 and references therein; Chikh-Ali *et al.*, 2008). Ornamental hosts such as *Petunia* spp. and *Dahlia* spp. (Asteraceae) are also reported susceptible (Jeffries, 1998). The experimental host range consists of about 500 plant species in 31 families (Kerlan, 2006). Incidentally, a recombinant strain PVY[NTN] that causes PTNRD was detected in petunias imported to the UK, which indicates that PVY is circulating worldwide not only in infected potato tubers, but also in other plant species. A correlation was also found between weed hosts of PVY and aphid preferences. A report from the Pacific Northwest in the US revealed a preference of *Myzus persicae* and *Macrosiphum euphorbiae*, the two most efficient vectors of PVY, for hairy nightshade

(*S. sarrachoides*) over potato and for infected plants over non-infected plants of both species. This suggests that PVY-infected plants of hairy nightshade could enhance the current season spread of PVY in the field (Alvarez *et al.*, 2008).

PVY is transmitted from one generation/season to another via infected seed tubers that are the main long-distance vehicles of PVY. This mode of transmission is responsible for the global circulation of PVY strains. Most potato-producing areas import their seed potatoes from other areas. In all cases, seed potatoes have a certain tolerance for PVY infection that can be relatively high or low. This explains the close relationships between PVY strains of certain types in different countries and continents, particularly the newly emerged recombinants.

PVY is also transmitted during the growing season from infected to healthy plants by aphids in a non-persistent manner. Both potato colonizing and non-colonizing aphid species can transmit PVY, though the non-colonizing species are believed to transmit PVY at higher efficiencies due to the unrest behaviour they display as frequent flights and probing (Sigvald, 1984; Harrington *et al.*, 1986; De Bokx and Piron, 1990). The most efficient aphid vector of PVY varies from one ecosystem to another (Harrington *et al.*, 1986). The differences in transmission efficiency of PVY strains might be related to the increased fitness and spread of the newly emerged PVY strains. A recent study found that PVY[N], PVY[NTN] and PVY[N-Wi] isolates are transmitted with similar efficiencies by *M. persicae*, but several other non-colonizing aphid species transmitted PVY[N-Wi] at higher efficiencies than PVY[N] or PVY[NTN] (Verbeek *et al.*, 2010). Differences in transmission efficiency were also found among various isolates of the same PVY strains (Verbeek *et al.*, 2010). PVY[NTN] was transmitted more frequently than either PVY[O] or PVY[N:O] when it was present in mixed infections (Srinivasan *et al.*, 2012). Another report, however, did not find differences in transmission efficiency between PVY[O] and PVY[N-Wi] (Mello *et al.*, 2011). The variability of these reports is probably due to the differences in PVY strain/isolate, aphid species/clone and the method used for the efficiency assessment (Karasev and Gray, 2013b).

7.3 Classification of *Potato virus Y* Strains

Traditionally, PVY has been divided into five strain groups (PVY^O, PVY^C, PVY^N, PVY^Z and PVY^E) based on reactions of potato cultivars carrying the resistance genes Ny_{tbr}, Nc_{tbr} and Nz_{tbr} and tobacco (*Nicotiana tabacum* L.) (Jones, 1990; Kerlan *et al.*, 1999, 2011; Singh *et al.*, 2008; Galvino-Costa *et al.*, 2012a; Karasev and Gray, 2013a,b). PVY^O and PVY^C comprise PVY isolates that induce the hypersensitive reaction in potato cultivars carrying the Ny_{tbr} and Nc_{tbr} resistance genes, respectively, and both strains induce only symptoms of mottle in tobacco. PVY^N overcomes Ny_{tbr} and Nc_{tbr} resistance genes and induces veinal necrosis in tobacco. Phylogenetic analysis of PVY^O, PVY^C and PVY^N reveal a distinct evolutionary lineage for each of these strains, which were assumed to be non-recombinant. The PVY^O lineage has multiple sub-lineages, at least six, including PVY^O and PVY^O-O5 (Karasev *et al.*, 2011; Ogawa *et al.*, 2012), which indicates a significant diversity within the PVY^O strain. The PVY^O-O5 is distinct from PVY^O not only in phylogeny, but also biologically – it causes more severe symptoms on potato cultivars carrying the Ny_{tbr} gene and may also react with a PVY^N-specific monoclonal antibody (MAb) 1F5 (Ellis *et al.*, 1997; Karasev *et al.*, 2011). In the same way, PVY^N has at least two sub-lineages, including the PVY^N and PVY^{NA-N} clades (Ogawa *et al.*, 2008, 2012; Karasev *et al.*, 2011). The PVY^Z strain, which was first proposed for two PVY isolates from the UK, overcame the potato resistance genes Ny_{tbr} and Nc_{tbr}, but induced a hypersensitive reaction in potato cultivars 'Maris Bard' and 'Pentland Ivory' carrying the resistance gene Nz_{tbr} (Jones, 1990; Singh *et al.*, 2008; Chikh-Ali *et al.*, 2014). The PVY^Z strain was found to have the typical PVY^{NTN} genome structure that represents a recombination between sections of parental PVY^N and PVY^O genomes with three to four recombination junctions (RJs) (Kerlan *et al.*, 2011; Galvino-Costa *et al.*, 2012a; Karasev and

Gray, 2013b). The fifth strain, PVY^E, is assigned to PVY isolates that are capable of overcoming the three potato resistance genes, Ny_{tbr}, Nc_{tbr} and Nz_{tbr} along with inducing symptoms of mosaic, mottle and vein-clearing in tobacco (Kerlan *et al.*, 1999). Recently, genomic analysis of two PVY^E isolates collected in Brazil, PVY-AGA and PVY-MON, revealed a recombinant genome between two parental strains, PVY^{NTN} and PVY-NE11 (Galvino-Costa *et al.*, 2012a; Fig. 7.2).

In the last three decades, the emergence of recombinant PVY strains/variants has led to a significant shift in the composition of PVY strains in most potato-producing areas. These recombinants are presently the predominating PVY population in most potato-production areas at the expense of the ordinary strain, PVY^O. Several types of PVY recombinants have been identified including PVY^{NTN} (comprising NTN-A and NTN-B with non-recombinant and recombinant P1 regions, respectively), PVY^{N-Wi}, PVY$^{N:O}$, PVY-T13, PVY-261-4 and PVY^{NTN-NW} (Beczner *et al.*, 1984; Chrzanowska, 1991; McDonald and Singh, 1996a,b; Piche *et al.*, 2004; Crosslin *et al.*, 2006; Schubert *et al.*, 2007; Singh *et al.*, 2008; Hu *et al.*, 2009a,b; Gray *et al.*, 2010; Chikh-Ali *et al.*, 2010a; Karasev *et al.*, 2011; Ogawa *et al.*, 2012; Karasev and Gray, 2013a,b; Fig. 7.2). All these recombinants were found to have mosaic genomes built of fragments, mainly of parental PVY^O and PVY^N sequences (Fig. 7.2). Additional PVY genotypes were described that are sufficiently different from other simple O/N recombinants, such as PVY-NE11, with most of the genome sections derived from an unidentified parent (Lorenzen *et al.*, 2008; Fig. 7.2). The phylogenetic relationship of the recombinant PVY strains differ based on the origin of the part of the genome used to construct the phylogenetic trees. When the phylogenetic tree is constructed based on the whole genome, the recombinant strains occupy intermediate positions between the parental lineages (Lorenzen *et al.*, 2006a, 2008; Chikh-Ali *et al.*, 2007a; Karasev *et al.*, 2011). It is worth noting that the possession of similar recombination patterns does not always imply that the strains have descended from the same parental lineages. For instance,

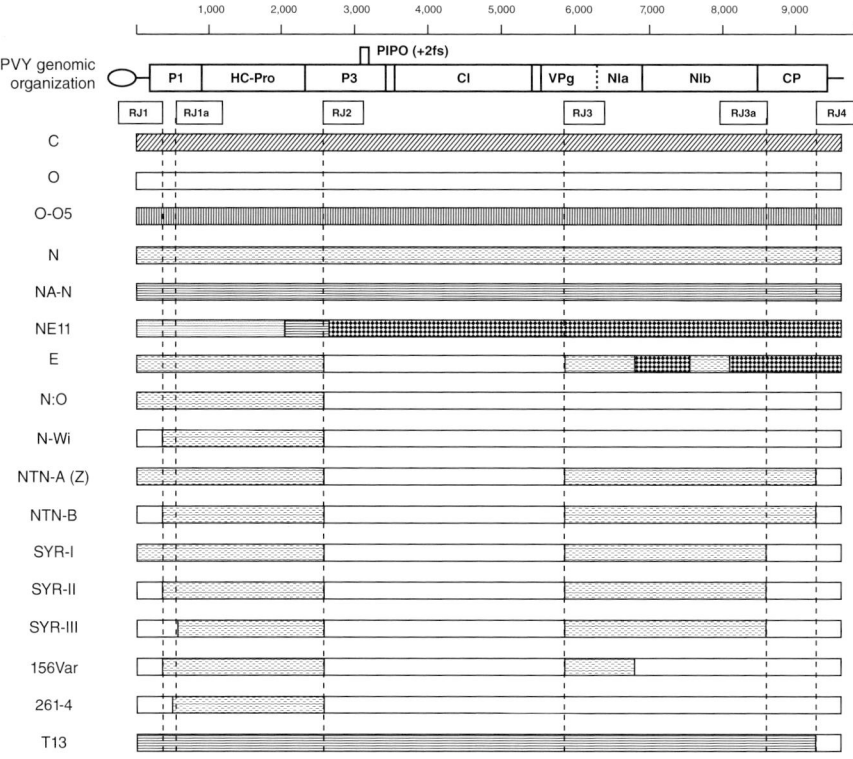

Fig. 7.2. Genomic structure of *Potato virus Y* strains.

PVY[N-Wi] and PVY[N:O] share the recombination junction 2 (RJ2), but PVY[N-Wi] has an additional RJ (RJ1, Fig. 7.2). Phylogenetic analysis suggest that these two recombinants likely originated through different recombination events in which the O sequences in PVY[N-Wi] and PVY[N:O] were donated by different PVY[O] sub-lineages (Karasev *et al.*, 2011), while PVY[NTN]-A (Fig. 7.2) most likely originated from PVY[N:O] through additional recombination events (Karasev *et al.*, 2011). Some PVY recombinants arose through recombination events between recombinant parents. These include PVY[NTN-NW] and PVY[E], of which the former resulted from recombination events between PVY[NTN] and PVY[N-Wi], and the latter arose from recombination between PVY[NTN] and PVY-NE11 (Chikh-Ali *et al.*, 2010a; Galvino-Costa *et al.*, 2012a). There are some more rare variants within all strains that can be differentiated based on biological properties. For example, there are isolates of

PVY[N:O] and PVY[N-Wi] that do not cause tobacco vein necrosis that are referred to as PVY[N:O]-minus and PVY[N-Wi]-minus, respectively (Gray *et al.*, 2010). There are isolates of PVY[NTN] that do not cause tuber necrosis (Chikh-Ali *et al.*, 2013b), and there are isolates of PVY[N], PVY[O], PVY[N:O] and PVY[N-Wi] that can induce tuber necrosis in certain cultivars (Piche *et al.*, 2004; Schubert *et al.*, 2007; Gray *et al.*, 2010; Rigotti *et al.*, 2011).

The shared properties of the recombinant strains with the parental strains complicate the identification and classification of these recombinant strains, and make it inappropriate to associate these recombinants with any of the parental strains. Karasev and Gray (2013b) classified PVY into nine strains (PVY[O], PVY[C], PVY[N], PVY[Z]-NTN, PVY[E], PVY[N-Wi], PVY[N:O], PVY[NA-N] and PVY-NE11) based on the interaction with the potato resistance genes Ny_{tbr}, Nc_{tbr} and Nz_{tbr}, and their genome structures. The convergent evolution of PVY[N:O] and PVY^{N-Wi} placed them in two different strains. Considering the phenotype in tobacco as subordinate to the potato phenotype following infection by diverse strains of the virus, PVY$^{N:O}$, PVY^{N-Wi} and PVY^{NTN} were separated from PVY^N. Due to the similar phenotype in potato, PVY^Z and PVY^{NTN} were merged into one strain, and the name PVY^Z-NTN was thus proposed (Kerlan *et al.*, 2011; Karasev and Gray, 2013b).

7.4 Geographic Distribution of *Potato virus Y* Strains and Population Structure

PVY strains differ in their geographic distribution, pathogenicity, control strategies required and the economic loss to potato production. Therefore, it is very important to identify the main strains circulating in a given potato-production area to develop an efficient control program. To this end, the structure of PVY populations has been studied intensively in many potato-production areas. The identification of PVY strains has been carried out by a multitude of methods, including the reaction in potato indicators, symptoms in tobacco, serological typing using MAbs, reverse transcription (RT)-PCR, restriction fragment length polymorphism, as well as partial and complete sequence analysis. Due to the different approaches used in the identification of PVY strains and the absence of an internationally recognized protocol, the comparison between the results of these studies is not always meaningful or even possible. However, the main trend of PVY populations is that recombinant strains such as PVY^{N-Wi}, PVY^{NTN} and PVY$^{N:O}$ are expanding at the expense of the non-recombinant strains PVY^C, PVY^O and PVY^N in most potato-producing regions. Increasing numbers of theories attempt to explain the fast expansion of recombinant PVY strains. First, the mild symptoms they cause in most potato cultivars facilitated escape from visual inspections, and thus allowed for the build-up of their populations. Second, the absence, until recently, of accurate identification methods of these strains hindered the tracking and control of recombinant strains. For example,

PVY^{NTN} and PVY-NE11 share the same serotype as PVY^N whereas PVY^{N-Wi}, PVY$^{N:O}$ and PVY^{NTN-NW} have the same serotype as PVY^O, and hence they cannot be differentiated using serological methods. On the other hand, PVY^{N-Wi}, PVY$^{N:O}$, PVY^{NTN}, PVY^{NTN-NW} and PVY^N induce similar vein necrotic reactions in tobacco, and therefore they cannot be distinguished using this method. Third, the differences in the efficiency in aphid transmission might be responsible for the expansion in recombinant strains, since recent reports show that in certain cases these strains are transmitted at higher efficiencies (Srinivasan *et al.*, 2012). PVY^C isolates from potato have been found in a few potato-production areas (Kerlan, 2006), but never found in potato-growing regions like North America, Brazil and Japan (Hataya *et al.*, 1994; Ellis *et al.*, 1997; Galvino-Costa *et al.*, 2012b; Chikh-Ali *et al.*, 2013b). The limited distribution of the PVY^C strain could be explained by the lack of aphid transmissibility of some PVY^C isolates (Kerlan, 2006) and by the hypersensitive reaction it causes in most potato cultivars (McDonald and Singh, 2006a).

In North America, PVY^O (including PVY^O-O5), PVY^{N-Wi}, PVY$^{N:O}$, PVY^N, PVY^{NTN}, PVY^{NA-N}, PVY-NE11 and PVY^{N-Wi}-minus have been reported from potato fields (McDonald and Singh 1996a,b; Crosslin *et al.*, 2002; Piche *et al.*, 2004; Baldauf *et al.*, 2006; Crosslin *et al.*, 2006; Lorenzen *et al.*, 2006a, 2008; Gray *et al.*, 2010). PVY^C has not been found in potato crops. Though its name was derived from North America, PVY^{NA-N} is hardly found in North America these days (Nanayakkara *et al.*, 2012; Chikh-Ali and Karasev, personal observations), but it is quite common in Japan (Ogawa *et al.*, 2012; Chikh-Ali *et al.*, 2013b) where it caused a tuber necrosis outbreak three decades ago (Ohshima *et al.*, 2000). PTNRD is associated with most of these strains in the USA and Canada in certain susceptible potato cultivars such as 'Yukon Gold' (Gray *et al.*, 2010). PVY^O and PVY^O-O5 are still the predominant strains, but they are being replaced by recombinant PVY strains such as PVY^{N-Wi}, PVY$^{N:O}$ and PVY^{NTN}. All PVY^{NTN} isolates found in the USA belong to the PVY^{NTN}-A type with a non-recombinant P1 region (see

Fig. 7.2), whereas PVY^{NTN}-B has not been found so far. PVY-NE11 was found in certain areas of the USA, but it has not been reported elsewhere.

In Brazil, PVY has a diverse population structure with at least eight strains/variants including PVY^O, PVY^Z, PVY^E, PVY^{N-Wi}, PVY$^{N:O}$, PVY^{NTN}, PVY^{NA-N} (Galvino-Costa *et al.*, 2012a,b). In agreement with the main trend, the vast majority of PVY strains found in Brazil are recombinants with few isolates belonging to the non-recombinant strain PVY^O (Galvino-Costa *et al.*, 2012a,b).

In the UK, PVY^O, PVY^C, PVY^N, PVY^{NTN}, PVY$^{N:O}$, PVY^Z and PVY^{NA-N} have been reported (Jones, 1990; Boonham *et al.*, 2002a; Barker *et al.*, 2009). In an early study on PVY population structure in France using serological and biological typing, Kerlan *et al.* (1999) found that in addition to PVY^O and PVY^N, there was a significant incidence of PVY^{N-Wi} for the first time in the country. Only one PVY^C isolate was detected in a potato sample as part of a mixed infection with PVY^O; single isolates of PVY^Z and PVY^E were also found. In Spain, PVY^O, PVY^N, PVY^{NTN}, PVY^{N-Wi} and PVY^E have been reported (Blanco-Urgoiti *et al.*, 1998; Kerlan *et al.*, 1999). In Belgium, PVY^{NTN} was the most common strain followed by PVY^{N-Wi}, whereas PVY^O and PVY^N strains were less frequent. Novel isolates were also noted, whereas no PVY^C infection has been found (Kamangar *et al.*, 2014). A similar shift in PVY population structure was reported recently from the Netherlands where PVY^{NTN} and PVY^{N-Wi} are the dominant strains (Van der Vlugt *et al.*, 2008). In the same way, PVY populations in Swiss seed potatoes is now dominated by PVY^{NTN} (Rigotti *et al.*, 2011). PVY^N, PVY^{N-Wi} and PVY$^{N:O}$ are common, whereas the incidence of PVY^O has decreased significantly over time, and PVY^C was not found (Rigotti *et al.*, 2011). In Germany, studies of PVY population over a period of 25 years revealed that the importance of PVY^O has declined and that the strain was replaced by PVY^{NTN} and PVY^{N-Wi} (Lindner, 2008). In addition to these strains, novel recombinants such as PVY-156, PVY-156var, PVY-261-4 and PVY-Nicola have also been reported (Schubert *et al.*, 2007; see Fig. 7.2). In the Czech Republic, a shift

in PVY population from PVY^O dominant to PVY^{N-Wi} and PVY^N was reported (Dědič *et al.*, 2008). It is worth mentioning, however, that the methodology used by Dědič *et al.* (2008) would type PVY^{NTN} as PVY^N since they used serology and tobacco indicators to type PVY strains. In Poland, the PVY population is dominated by recombinant PVY strains, mainly PVY^{N-Wi} followed by PVY^{NTN} and PVY$^{N:O}$, in addition to PVY isolates with novel recombination structures such as PVY-Gr99 and PVY^O is less frequent (Chrzanowska, 1991; Schubert *et al.*, 2007; Yin *et al.*, 2012). In Tunisia, the PVY population was dominated by PVY^{NTN} with few exceptions belonging to PVY^N and PVY^O (Boukhris-Bouhachem *et al.*, 2010; Djilani-Khouadja *et al.*, 2010). In Syria, PVY populations consist solely of recombinant strains without any non-recombinant strains. The majority of these recombinants belong to the PVY^{NTN-NW} type which comprises three recombination patterns, SYR-I, SYR-II and SYR-III. In addition, isolates of PVY^{N-Wi} and PVY^{NTN} were common (Chikh-Ali *et al.*, 2007a, 2010a). A recent survey of potato fields in Jordan revealed that PVY^{NTN-NW} and PVY^{NTN} are common in that region (Anfoka *et al.*, 2014).

The latest survey in northern Japan, prompted by the recent outbreak of PVY in potato in two foundation seed potato stations in Hokkaido Island, revealed a higher occurrence of the recombinant PVY^{NTN} strain than the traditional strains PVY^O and PVY^{NA-N}. The latter strains were the dominant strains for several years prior (Chikh-Ali *et al.*, 2013b). PVY^{NTN} was also found in southern Japan (Ogawa *et al.*, 2012; Chikh-Ali *et al.*, 2013b). In China, sequence analysis of PVY isolates collected from potato and tobacco showed that these isolates were recombinants with recombination patterns resembling those of PVY^{NTN}, PVY^{N-Wi} and PVY^{NTN-NW}, in addition to recombinant isolates possessing unique recombination patterns (Hu *et al.*, 2009a; Tian *et al.*, 2011; Wang *et al.*, 2012). In South Africa, the recombinant PVY^{N-Wi} and PVY^{NTN} were found to be common along with PVY^N and PVY^O (Visser and Bellstedt, 2009). Unlike PVY populations in other potato-production areas, only PVY^O and PVY^N were found in

New Zealand and no recombinant strains were reported which was attributed to the isolated geographic region and efficient plant quarantine measures (Fomitcheva *et al.*, 2009).

7.5 Diagnostic Methods and Management Strategies

Positive diagnosis for PVY can be achieved by various methods which rely on the biological, serological and genomic properties of the virus. ELISA, using MAbs and polyclonal (PAbs) antibodies, is the most commonly used method for the detection of PVY in all laboratories and by all agencies involved in seed potato production and certification. Due to its sensitivity, reliability and simplicity, ELISA is the method of choice for large-scale testing of PVY, which is required by seed inspection agencies and also for international potato trade. Millions of potato samples are tested using ELISA for PVY and other major potato viruses in specialized laboratories every year. Several MAbs and PAbs are available commercially for the reliable detection of PVY (Gugerli and Fries, 1983; Ohshima *et al.*, 1990; Ellis *et al.*, 1996; Karasev *et al.*, 2010).

Although it is a valuable tool for the detection of PVY, ELISA is limited in its ability to differentiate between PVY strains, mainly due to the high level of amino acid sequence identity among strains. ELISA using MAbs can only identify two main serotypes, O and N. That is, it is able to distinguish between capsid proteins coming from two potential parental sequences, PVY^O and PVY^N. Many MAbs have been developed to differentiate between PVY serotypes and most are available commercially (Gugerli and Fries, 1983; Ohshima *et al.*, 1990; Ellis *et al.*, 1996; Karasev *et al.*, 2010). In general, the epitopes targeted by most of these MAbs are located at the N terminus of the PVY capsid protein (Chikh-Ali *et al.*, 2007b; Nikolaeva *et al.*, 2012). A combination of PAbs and MAbs are being used by seed potato certification agencies to track specific serotypes such as the N serotype linked to PVY^{NTN} that is the main cause of PTNRD. However, the identification of a serotype of PVY has proved to be insufficient for accurate strain typing, since different strains share the same serotype, like PVY^{N-Wi}, PVY$^{N:O}$ and PVY^{NTN-NW} that have O serotypes; on the other hand, PVY^{NTN} and PVY-NE11 share the same serotype with PVY^N. The inability to identify PVY strains has complicated all efforts to eradicate and control any specific strain (Gray *et al.*, 2010).

The RT-PCR is another sensitive and reliable method for the detection of PVY, particularly in dormant tubers where ELISA has had limited success. Prior to the use of RT-PCR, the identification of some PVY strains including PVY^O, PVY^N, PVY^C, PVY^Z and PVY^E was a complex process involving a combination of multiple identification methods such as ELISA using MAbs, and bioassays employing *N. tabacum* and a set of potato cultivars carrying the resistance genes Ny_{tbr}, Nc_{tbr} and Nz_{tbr} (Jones 1990; Singh *et al.*, 2008; Karasev and Gray, 2013a,b). Nevertheless, some strains of PVY cannot be differentiated even when all these combined methods are used. PVY^{NTN}, for instance, cannot be differentiated from PVY^N using MAbs or tobacco indicators since they share the same serotype and phenotype in tobacco. Even the ability to induce PTNRD was found unreliable, since some PVY^{NTN} induced PTNRD in the field but not under greenhouse conditions (Boonham *et al.*, 2002b). In addition, other PVY strains may induce PTNRD of various severities in susceptible potato cultivars (Piche *et al.*, 2004; Gray *et al.*, 2010); even worse, some PVY^{NTN} isolates may not induce PTNRD (Chikh-Ali *et al.*, 2013b). As stated above, PVY^{N-Wi} and PVY$^{N:O}$ cannot be differentiated from PVY^O using MAbs and cannot be differentiated from PVY^N, PVY^{NTN}, PVY-NE11 using tobacco indicator plants. PVY^{N-Wi} and PVY$^{N:O}$ themselves are not separately identified using serological and/or biological methods. Differentiation between PVY-NE11 and strains of PVY^{NTN} is also not achieved by these methods, since these strains share the same serotype and phenotype in tobacco, and both induce PTNRD (Piche *et al.*, 2004; Lorenzen *et al.*, 2008).

The reliability of the RT-PCR for the detection of PVY depends on the quality of the primers used. Primers are generally designed based on conserved genomic regions among PVY strains such as the NIb and the central part of the capsid protein region. However, the use of RT-PCR assays to identify PVY strains has evolved concurrently with the spread of PVY recombinants. RT-PCR is an indispensable tool for the detection and identification of PVY strain complexes and has proven to be a reliable and sensitive alternative for the identification of some PVY strains, and as the only tool for the identification of many other strains. Most of these RT-PCR assays target nucleotide polymorphisms around the main RJs reported in the PVY genome. Some of these methods target one or two strains, mainly PVY^{NTN} and/or PVY^{N-Wi} (Moravec et al., 2003; Nie and Singh 2003; Glais et al., 2005), whereas other methods have a wider range of specificity allowing for the distinction between two and five strains (Boonham et al., 2002b; Nie and Singh 2002a,b; Rigotti and Gugerli, 2007; Lorenzen et al., 2006b). Recently, a more advanced multiplex RT-PCR was developed to detect and identify nine established strains and five more unclassified recombinants of PVY simultaneously, including PVY^O, PVY^N, PVY^{NA-N}, PVY^{NTN}-A, PVY^{NTN}-B, PVY^E, PVY-NE11, PVY^{N-Wi}, PVY$^{N:O}$, PVY^{NTN-NW} (SYR-I, SYR-II and SYR-III), and more rare types like PVY-261-4 (Chikh-Ali et al., 2010b, 2013a; see Fig. 7.2). Unlike previously reported RT-PCR methods, which probed one or two RJs of PVY genome, this assay targets five RJs using 12 primers which ensures the flexibility and efficacy of this assay and the detection of even the non-target PVY strains, PVY-NE11, PVY^E and PVY-261-4. In overcoming the limited number of samples that can be handled when conducting RT-PCR and the laborious and time consuming RNA extraction step, conventional RNA extraction was replaced with an immunocapture step in this RT-PCR assay (Chikh-Ali et al., 2013a). This facilitated running the RT-PCR side-by-side with ELISA using the same sample extracts. The test is highly suitable for large-scale testing (Chikh-Ali et al., 2013a).

Other detection methods of PVY have been reported such as PCR-microplate hybridization, which is a nucleic acid-based ELISA-like diagnostic method (Hataya et al., 1994), as well as a cDNA macroarray that

was developed for the simultaneous detection of 12 potato viruses (Maoka et al., 2010). Lateral flow assay is an easy, rapid and sensitive detection method of plant viruses (Tsuda et al., 1992). Commercial lateral flow kits in the form of immunostrips and chips are available for the detection of many potato viruses, including PVY.

As with other plant viruses, PVY-infected plants cannot be cured by chemical treatments, and the non-persistent transmission mode of PVY by aphids minimizes the effect of insecticides on PVY transmission and spread. Therefore, the use of certified seed potatoes is a fundamental step to managing potato viruses including PVY. Certification of seed potatoes and flushing out seed potatoes after a fixed number of years of field growth minimizes the build-up of PVY infection in potato crops and reduces virus infection sources. In order to certify seed potatoes, multiple field inspections and virus tests are carried out during the growing season to confirm the health status of seed potatoes. In addition, due to the late season infection and the asymptomatic infection, winter grow out testing is needed. The importance of laboratory virus tests has increased substantially at the expense of visual inspection. There are many factors limiting the efficiency of visual inspection, including the use of potato cultivars that are asymptomatic or produce mild symptoms when infected with PVY like 'Russet Norkotah' and 'Shepody', as well as the prevalence of PVY strains that induce mild symptoms such as PVY^{N-Wi} and PVY$^{N:O}$ (Karasev and Gray, 2013b).

There are many management approaches taken by seed potato farmers to reduce PVY infection. Most of these approaches are aimed at reducing sources of virus inoculum and spread including the use of certified seed potatoes, roguing, controlling volunteer potatoes, planting border crops, mineral oil and insecticide sprays, and early haulm kill, among others (Gray et al., 2010). The effectiveness of these approaches would increase if they were taken based on educated decisions regarding what approach should be taken and when. For instance, the knowledge of aphid flights and population dynamics during the growing season in a given area is crucial to determine the good timing for roguing, chemical and/or mineral oil sprays, early haulm killing, etc. Another instance is the previous knowledge of symptoms of PVY strains in potato cultivars would help seed farmers and field inspectors identify diseased plants during roguing and visual field inspection. The use of genetic control using N resistance genes is an efficient way of controlling PVY, but this requires knowledge of the prevalent PVY strains in the area since N genes are strain specific.

reliable virus detection and identification methods that are modified every so often in order to keep up with PVY's evolution. The classification system of PVY strains also needs to be revised periodically to enable the accurate classification of the newly emerging strains and correct the classification of the old ones. The search for N resistance genes for PVY in potato cultivars and germplasm, and the incorporation of these N genes into commercial potato cultivars through breeding programs, should be given high priority in order to control the newly emerging and spreading PVY strains.

7.6 Concluding Remarks

PVY is a rapidly evolving virus. Various lines of evidence suggest that recombination is a pivotal process in the emergence of new strains and as a result there is the continuous challenge in the identification, classification and control of the virus. Seed certification programs are aimed at managing PVY and other diseases through the regular supply of seed potatoes with minimal virus infection rates and flushing out seed stocks with high infection rates, which in turn minimizes virus infection sources for the new potato crops. Seed certification programs attempt to implement

Acknowledgments

The research cited here was funded in part through the following grants: the United States Department of Agriculture (USDA)-NIFA-NRI (2009-35600-05025), USDA-NIFA-SCRI (2009-51181-05894), USDA-ARS (58-5354-7-540, 58-5354-2-345 and 58-1907-8-870), Idaho State Department of Agriculture, Idaho Potato Commission, Washington State Potato Commission, Oregon Potato Commission, U.S. Potato Board, and by the Idaho Agricultural Experiment Station.

References

Adams, M.J., Zerbini, F.M., French, R., Rabenstein, F., Stenger, D.C., *et al.* (2012) Potyviridae. In: King, A.M.Q., Adams, M.J., Carstens, E.B. and Lefkowitz, E.J. (eds) *Ninth Report of the International Committee on Taxonomy of Viruses*. Elsevier Academic Press, San Diego, California and London, pp. 1069–1089.

Alvarez, J.M., Srinivasan, R., Cervantes, F., Eigenbrode, S., Bosque-Perez, N., *et al.* (2008) Importance of alternative plant hosts in the epidemiology of PLRV and PVY. Report of the 13th European association for potato research virology section meeting, Coylumbridge, Scotland. *Potato Research* 51, 190.

Anfoka, G., Haj-Ahmad, F., Altaleb, M., Abadi, M., Abubaker, S., *et al.* (2014) First report of recombinant *Potato virus Y* strains infecting potato in Jordan. *Plant Disease* 98, 1017.

Baldauf, P.M., Gray, S.M. and Perry, K.L. (2006) Biological and serological properties of *Potato virus Y* isolates in northeastern United States potato. *Plant Disease* 90, 559–566.

Barker, H., McGeachy, K.D., Toplak, N., Gruden, K., Žel, J., *et al.* (2009) Comparison of genome sequence of PVY isolates with biological properties. *American Journal of Potato Research* 86, 227–238.

Beczner, L., Horvath, H., Romhanyi, L. and Foster, H. (1984) Etiology of tuber ringspot disease in potato. *Potato Research* 27, 339–352.

Blanco-Urgoiti, B., Tribodet, M., Leclere, S., Ponz, F., Perez de San Roman, C., *et al.* (1998) Characterization of Potato potyvirus Y (PVY) isolates from seed potato batches. Situation of the NTN, Wilga and Z isolates. *European Journal of Plant Pathology* 104, 811–819.

Boonham, N., Walsh, K., Hims, M., Preston, S., North, J., *et al.* (2002a) Biological and sequence comparisons of *Potato virus Y* isolates associated with potato tuber necrotic ringspot disease. *Plant Pathology* 51, 117–126.

Boonham, N., Walsh, K., Preston, S., North, J., Smith, P., *et al.* (2002b) The detection of tuber necrotic isolates of *Potato virus Y*, and the accurate discrimination of PVY[O], PVY[N] and PVY[C] strains using RT-PCR. *Journal of Virological Methods* 102, 103–112.

Boukhris-Bouhachem, S., Hullé, M., Rouzé-Jouan, J., Glais, L. and Kerlan, C. (2007) *Solanum elaeagnifolium*, a potential source of Potato virus Y (PVY) propagation. *OEPP/EPPO Bulletin* 37, 125–128.

Boukhris-Bouhachem, S., Djilani-Khouadja, F., Fakhfakh, H., Glais, L., Tribodet, M., *et al.* (2010) Incidence and characterization of *Potato virus Y* in seed potatoes in Tunisia. *Potato Research* 53, 151–166.

Chikh-Ali, M., Maoka, T. and Natsuaki, K.T. (2007a) The occurrence and characterization of new recombinant isolates of PVY displaying shared properties of PVY[N]W and PVY[NTN]. *Journal of Phytopathology* 155, 409–415.

Chikh-Ali, M., Maoka, T. and Natsuaki, K.T. (2007b) A point mutation changes the serotype of a *Potato virus Y* isolate; genomic determination of the serotype of PVY strains. *Virus Genes* 35, 359–367.

Chikh-Ali, M., Katayama, K., Maoka, T. and Natsuaki, K.T. (2008) Significance of weed hosts for *Potato virus Y* protection in Syria. *OEPP/EPPO Bulletin* 83, 226–232.

Chikh-Ali, M., Maoka, T., Natsuaki, T. and Natsuaki, K.T. (2010a) PVY[NTN-NW], a novel recombinant strain of *Potato virus Y* predominating in potato fields in Syria. *Plant Pathology* 59, 31–41.

Chikh-Ali, M., Maoka, T., Natsuaki, K.T. and Natsuaki, T. (2010b) The simultaneous differentiation of *Potato virus Y* strains including the newly described strain PVY[NTN-NW] by multiplex PCR assay. *Journal of Virological Methods* 165, 15–20.

Chikh-Ali, M., Gray, S.M. and Karasev, A.V. (2013a) An improved multiplex IC-RT-PCR assay distinguishes nine strains of *Potato virus Y*. *Plant Disease* 97, 1370–1374.

Chikh-Ali, M., Karasev, A.V., Furutani, N., Taniguchi, M., Kano, Y., *et al.* (2013b) Occurrence of *Potato virus Y* strain PVY[NTN] in foundation seed potatoes in Japan, and screening for symptoms caused by PVY[NTN] in 62 Japanese potato cultivars. *Plant Pathology* 62, 1157–1165.

Chikh-Ali, M., Rowley, J.S., Kuhl, J., Gray, S.M. and Karasev, A.V. (2014) Evidence of a monogenic nature of the *Nz* gene conferring resistance against *Potato virus Y* strain Z (PVY[Z]) in Potato. *American Journal of Potato Research* 91, 649–654.

Chrzanowska, M. (1991) New isolates of the necrotic strain of *Potato virus Y* (PVY[N]) found recently in Poland. *Potato Research* 34, 179–182.

Chung, B.Y.T., Miller, W.A., Athins, J.F. and Firth, A.E. (2008) An overlapping essential gene in the *Potyviridae*. *Proceedings of the National Academy of Sciences USA* 105, 5897–5902.

Crosslin, J.M., Hamm, P.B., Eastwell, K.C., Thornton, R.E., Brown, C.R., *et al.* (2002) First report of the necrotic strain of *Potato virus Y* (PVY[N]) potyvirus on potatoes in the northwestern United States. *Plant Disease* 86, 1177.

Crosslin, J.M., Hamm, P.B., Hane, D.C., Jaeger, J., Brown, C.R., *et al.* (2006) The occurrence of PVY[O], PVY[N], and PVY[N:O] strains of *Potato virus Y* in certified potato seed lot trials in Washington and Oregon. *Plant Disease* 90, 1102–1105.

De Bokx, J.A. and Piron, P.G.M. (1990) Relative efficiency of a number of aphid species in the transmission of potato virus Y[N] in the Netherlands. *European Journal of Plant Pathology* 96, 237–246.

Dědič, P., Ptáček, J. and Čeřovská, N. (2008) A shift of PVY strain spectrum in potatoes in the Czech Republic in past years. Report of the 13th European association for Potato research virology section meeting, Coylumbridge, Scotland. *Potato Research* 51, 198–199.

Djilani-Khouadja, F., Glais, L., Tribodet, M., Kerlan, C. and Fakhfakh, H. (2010) Incidence and characterization of *Potato virus Y* variability in late season planted potato crops in Northern Tunisia. *European Journal of Plant Pathology* 126, 479–488.

Ellis, P., Stace-Smith, R., Bowler, G. and Mackenzie, D.J. (1996) Production of monoclonal antibodies for detection and identification of strains of *Potato virus Y*. *Canadian Journal of Plant Pathology* 18, 64–70.

Ellis, P., Stace-Smith, R. and de Villiers, G. (1997) Identification and geographic distribution of serotypes of *Potato virus Y*. *Plant Disease* 81, 481–484.

Fomitcheva, W.F., Fletcher, J.D. and Schubert, J. (2009) *Potato virus Y* strain spectrum in New Zealand - Absence of Recombinant N:O strains. *Journal of Phytopathology* 157, 507–510.

Galvino-Costa, S.B.F., dos Reis Figueira, A., Camargos, V.V., Geraldino, P.S., Hu, X.-J., *et al.* (2012a) A novel type of *Potato virus Y* recombinant genome, determined for the genetic strain PVY[E]. *Plant Pathology* 61, 388–398.

Galvino-Costa, S.B.F., Figueira, A.R., Rabelo-Filho, F.A.C., Moraes, F.H.R., Nikolaeva, O.V., *et al.* (2012b) Molecular and serological typing of *Potato virus Y* isolates from Brazil reveals a diverse set of recombinant strains. *Plant Disease* 96, 1451–1458.

Glais, L., Tribodet, M. and Kerlan, C. (2005) Specific detection of the PVY[N]-W variant of *Potato virus Y*. *Journal of Virological Methods* 125, 131–136.

Gray, S., De Boer, S., Lorenzen, J., Karasev, A., Whitworth, J., *et al.* (2010) *Potato virus Y*: an evolving concern for potato crops in the United States and Canada. *Plant Disease* 94, 1384–1397.

Gugerli, P. and Fries, P. (1983) Characterization of monoclonal antibodies to *Potato virus Y* and their use for virus detection. *Journal of General Virology* 64, 2471–2477.

Harrington, R., Katis, N. and Gibson, R.W. (1986) Field assessment of the relative importance of different aphid species in the transmission of *Potato virus Y*. *Potato Research* 29, 67–76.

Hataya, T., Inoue, A.K. and Shikata, E. (1994) A PCR-microplate hybridization method for plant virus detection. *Journal of Virological Methods* 46, 223–236.

Hu, X., He, C., Xiao, Y., Xiong, X. and Nie, X. (2009a) Molecular characterization and detection of recombinant isolates of *Potato virus Y* from China. *Archives of Virology* 154, 1303–1312.

Hu, X., Meacham, T., Ewing, L., Gray, S.M. and Karasev, A.V. (2009b) A novel recombinant strain of *Potato virus Y* suggests a new viral genetic determinant of vein necrosis in tobacco. *Virus Research* 143, 68–76.

Jeffries, C.J. (1998) Potato. *FAO/IPGRI Technical Guidelines for the Safe Movement of Germplasm* 19, 62–63.

Jones, R.A.C. (1990) Strain group specific and virus specific hypersensitive reactions to infection with potyviruses in potato cultivars. *Annals of Applied Biology* 117, 93–105.

Kamangar, S.B., Smagghe, G., Maes, M. and De Jonghe, K. (2014) *Potato virus Y* (PVY) strains in Belgian seed potatoes and first molecular detection of the N-Wi strain. *Journal of Plant Diseases and Protection*, 121, 10–19.

Karasev, A.V. and Gray, S.M. (2013a) Genetic diversity of *Potato virus Y* complex. *American Journal of Potato Research* 90, 7–13.

Karasev, A.V. and Gray, S.M. (2013b) Continuous and emerging challenges of *Potato virus Y* in potato. *Annual Review of Phytopathology* 51, 571–586.

Karasev, A.V., Nikolaeva, O.V., Hu, X., Sielaff, Z., Whitworth, J., *et al.* (2010) Serological properties of ordinary and necrotic isolates of *Potato virus Y*: a case study of PVY[N] misidentification. *American Journal of Potato Research* 87, 1–9.

Karasev, A.V., Hu, X., Brown, C.J., Kerlan, C., Nikolaeva, O.V., *et al.* (2011) Genetic diversity of the ordinary strain of *Potato virus Y* (PVY) and origin of recombinant PVY strains. *Phytopathology* 101, 778–785.

Kerlan, C. (2006) *Potato virus Y. AAB descriptions of plant viruses, no 414.* Available at http://www.dpvweb.net/dpv/showdpv.php?dpvno=414 (accessed 15 July 2015).

Kerlan, C., Tribodet, M., Glais, L. and Guillet, M. (1999) Variability of *Potato virus Y* in potato crops in France. *Journal of Phytopathology* 147, 643–651.

Kerlan, C., Nikolaeva, O.V., Hu, X., Meacham, T., Gray, S.M., *et al.* (2011) Identification of the molecular make-up of the *Potato virus Y* strain PVY[Z]: genetic typing of PVY[Z]-NTN. *Phytopathology* 101, 1052–1060.

Lindner, K. (2008) PVY strains in Germany - the period between 1984 and 2006. Report of the 13th European association for Potato research virology section meeting, Coylumbridge, Scotland. *Potato Research* 51, 202.

Lorenzen, J.H., Meacham, T., Berger, P.H., Shiel, P.J., Crosslin, J.M., *et al.* (2006a) Whole genome characterization of *Potato virus Y* isolates collected in the western USA and their comparison to isolates from Europe and Canada. *Archives of Virology* 151, 1055–1074.

Lorenzen, J.H., Piche, L.M., Gudmestad, N.C., Meacham, T. and Shiel, P. (2006b) A multiplex PCR assay to characterize *Potato virus Y* isolates and identify strain mixtures. *Plant Disease* 90, 935–940.

Lorenzen, J., Nolte, P., Martin, D., Pasche, J. and Gudmestad, N. (2008) NE–11 represents a new strain variant class of *Potato virus Y*. *Archives of Virology* 153, 517–525.

Maoka, T., Sugiyama, S., Maruta, Y. and Hataya, T. (2010) Application of cDNA macroarray for simultaneous detection of 12 potato viruses. *Plant Disease* 94, 1248–1254.

McDonald, J.G. and Singh, R.P. (1996a) Host range, symptomatology and serology of isolates of *Potato virus Y* (PVY) that shared properties with both the PVY[N] and PVY[O] strain groups. *American Journal of Potato Research* 73, 309–315.

McDonald, J.G. and Singh, R.P. (1996b) Response of potato cultivars to North American isolates of PVY[NTN]. *American Journal of Potato Research* 73, 317–323.

Mello, A.F.S., Olarte, R.A., Gray, S.M. and Perry, K.L. (2011) Transmission efficiency of *Potato virus Y* strains PVY[O] and PVY[N-Wi] by five aphid species. *Plant Disease* 95, 1279–1283.

Moravec, T., Cerovska, N. and Boonham, N. (2003) The detection of recombinant, tuber necrosing isolates of *Potato virus Y* (PVY^NTN) using a three-primer PCR based in the coat protein gene. *Journal of Virological Methods* 109, 63–68.

Nanayakkara, U.N., Singh, M., Pelletier, Y. and Nie, X. (2012) Investigation of *Potato virus Y* (PVY) strain status and variant population in potatoes in New Brunswick, Canada. *American Journal of Potato Research* 89, 232–239.

Nie, X. and Singh, R.P. (2002a) A new approach for the simultaneous differentiation of biological and geographical strains of *Potato virus Y* by uniplex and multiplex RT-PCR. *Journal of Virological Methods* 104, 41–54.

Nie, X. and Singh, R.P. (2002b) Probable geographical grouping of PVY^N and PVY^NTN based on sequence variation in P1 and 5-UTR of PVY genome and methods for differentiating North American PVY^NTN. *Journal of Virological Methods* 103, 145–156.

Nie, X. and Singh, R.P. (2003) Specific differentiation of recombinant PVY^N:O and PVY^NTN isolates by multiplex RT-PCR. *Journal of Virological Methods* 113, 69–77.

Nikolaeva, O.V., Roop, D.J., Galvino-Costa, S.B.F., dos Reis Figueira, A., Gray, S.M., *et al.* (2012) Epitope mapping for monoclonal antibodies recognizing tuber necrotic isolates of *Potato Virus Y*. *American Journal of Potato Research* 89, 121–128.

Nolte, P., Whitworth, J.L., Thornton, M.K. and McIntosh, C.S. (2004) Effect of seedborne *Potato virus Y* on performance of Russet Burbank, Russet Norkotah, and Shepody potato. *Plant Disease* 88, 248–252.

Ogawa, T., Tomitaka, Y., Nakagawa, A. and Ohshima, K. (2008) Genetic structure of a population of *Potato virus Y* inducing potato tuber necrotic ringspot disease in Japan; comparison with North American and European populations. *Virus Research* 131, 199–212.

Ogawa, T., Nakagawa, A., Hataya, T. and Ohshima, K. (2012) The genetic structure of populations of *Potato virus Y* in Japan; based on the analysis of 20 full genomic sequences. *Journal of Phytopathology* 160, 661–673.

Ohshima, K., Inoue, A.K., Ishikawa, Y., Shikata, E. and Hagita, T. (1990) Production and application of monoclonal antibodies specific to ordinary strain and necrotic strain of *Potato virus Y*. *Annals of the Phytopathological Society of Japan* 56, 508–514.

Ohshima, K., Sako, K., Hiraishi, C., Nakagawa, A., Matsuo, K., Ogawa, T., Shikata, E. and Sako, N. (2000) Potato tuber necrotic ringspot disease occurring in Japan: Its association with *Potato virus Y* necrotic strain. *Plant Disease* 84, 1109–1115.

Piche, L.M., Singh, R.P., Nie, X. and Gudmestad, N.C. (2004) Diversity among PVY field isolates obtained from potatoes grown in the United States. *Phytopathology* 94, 1368–1375.

Quenouille, J., Vassilakos, N. and Moury, B. (2013) *Potato virus Y*: a major crop pathogen that has provided major insights into the evolution of viral pathogenicity. *Molecular Plant Pathology* 14, 439–452.

Rigotti, S. and Gugerli, P. (2007) Rapid identification of *Potato virus Y* strains by one-step triplex RT-PCR. *Journal of Virological Methods* 140, 90–94.

Rigotti, S., Balmelli, C. and Gugerli, P. (2011) Census report of the *Potato virus Y* (PVY) population in Swiss seed potato production in 2003 and 2008. *Potato Research* 54, 105–117.

Schubert, J., Fomitcheva, V. and Sztangret-Wisniewski, J. (2007) Differentiation of *Potato virus Y* using improved sets of diagnostic PCR-primers. *Journal of Virological Methods* 140, 66–74.

Sigvald, R. (1984) The relative efficiency of some aphid species as vectors of potato virus Y^O (PVYO). *Potato Research* 27, 285–90.

Singh, R.P., Valkonen, J.P.T., Gray, S.M., Boonham, N., Jones, R.A.C., *et al.* (2008) The naming of *Potato virus Y* strains infecting potato. *Archives of Virology* 153, 1–13.

Srinivasan, R., Hall, D.G., Servantes, F.A., Alvarez, J.M. and Whithworth, J.L. (2012) Strain specificity and simultaneous transmission of closely related strains of a potyvirus by *Myzus persicae*. *Journal of Economic Entomology* 105, 783–791.

Tian, Y.P., Liu, J.L., Zhang, C.L., Liu, Y.Y., Wang, B., *et al.* (2011) Genetic diversity of *Potato virus Y* infecting tobacco crops in China. *Phytopathology* 101, 377–387.

Tsuda, S., Kameya-Iwaki, M., Hanada, K. and Tomaru, K. (1992) A novel detection and identification technique for plant viruses: Rapid Immunofilter Paper Assay (RIPA). *Plant Disease* 76, 466–469.

Valkonen, J.P.T. (2007) Viruses: economical losses and biotechnological potential. In: Vreugdenhil, D., Bradshaw, J., Gebhardt, C., Govers, F., Mackerron, D.K.L., *et al.* (eds) *Potato Biology and Biotechnology Advances and Prospects*. Elsevier, Oxford, United Kingdom, pp. 619–641.

Van der Vlugt, R.A.A., Verbeek, M., Cuperus, C., Piron, P.G.M., de Haan, E., *et al.* (2008) Strains of *Potato virus Y* in Dutch seed potato culture. Report of the 13th European association for Potato research virology section meeting, Coylumbridge, Scotland. *Potato Research* 51, 191–192.

Verbeek, M., Piron, P.G.M., Dullemans, A.M., Cuperus, C. and van der Vlugt, R.A.A. (2010) Determination of aphid transmission efficiencies for N, NTN and Wilga strains of *Potato virus Y*. *Annals of Applied Biology* 156, 39–49.

Visser, J.C. and Bellstedt, D.U. (2009) An assessment of molecular variability and recombination patterns in South African isolates of *Potato virus Y*. *Archives of Virology* 154, 1891–1900.

Wang, B., Jia, J.L., Wang, X.Q., Wang, Z.Y., Yang, B.H., *et al.* (2012) Molecular characterization of two recombinant *Potato virus Y* isolates from China. *Archives of Virology* 157, 401–403.

Yin, Z., Chrzanowska, M., Michalak, K., Zagórska, H. and Zimnoch–Guzowska, E. (2012) Recombinants of PVY strains predominate among isolates from potato crop in Poland. *Journal of Plant Protection Research* 52, 214–219.

8 Soybean Mosaic

Masarapu Hema,[1] Basavaprabhu L. Patil,[2] V. Celia Chalam[3] and P. Lava Kumar[4]*

[1]Department of Virology, Sri Venkateswara University, Tirupati, India;
[2]National Research Centre on Plant Biotechnology, IARI
(ICAR-NRCPB), Pusa Campus, New Delhi, India; [3]National Bureau of
Plant Genetic Resources (ICAR-NBPGR), Pusa, New Delhi, India;
[4]International Institute of Tropical Agriculture (IITA), Ibadan, Nigeria

8.1 Introduction

Soybean (*Glycine max* (L.) Merr.) is an important annual grain legume widely cultivated between 55°N and 55°S of the equator during warm moist periods for food, cooking oil, animal feed, biofuel and several other culinary and industrial uses (Graham and Vance, 2003; Pimentel and Patzek, 2008). Soybean seed contains more than 40% protein enriched with essential amino acids, about 20% oil, lecithin and vitamins A and D (Sakai and Kogiso, 2008). The crop was first domesticated in China around the 11th century BC. However, its cultivation outside the Asian continent was not recorded until the 18th century AD; first in Europe, followed by the USA (in 1785) and Brazil (in 1882) in the Americas, and Malawi (in 1907) in Africa (Boerma and Specht, 2004). Slow establishment of the crop outside of Asia was attributed to the absence of soybean-specific rhizobia in soils (Boerma and Specht, 2004). Soybean cultivation has since rapidly expanded in the 20th century following the development of improved high-yielding cultivars, some of which have the natural ability to form nodules with local rhizobia and fix atmospheric nitrogen. Artificial inoculation of soils with suitable strains of rhizobia that can nodulate soybean have increased domestic and export market demands and further propelled crop expansion around the world (Boerma and Specht, 2004). Presently, soybean tops grain legume production with an area of 111.7 million ha contributing 276.4 million tonnes of grain annually in over 93 countries (FAOSTAT, 2014). Crop production has expanded by about 31% during the last decade, with highest increases in production in Australia, followed by Europe, Africa, the Americas and Asia (Table 8.1). Soybean plays a significant role in world agriculture and in income and food security of smallholder farmers in developing countries.

Diseases and pests are the major problems associated with soybean production in several countries (Wrather *et al.*, 1997; Hill, 2003). The estimated loss of soybean due to diseases and nematodes in the USA during 2010 was 13 million tonnes valued at US$4.8 billion (Wrather, 2011); US$35 million of these losses were attributed to virus diseases (Hill and Whitham, 2014). Yield losses due to diseases in other countries are also perceived to be significant; however, accurate data are not available. Soybean is host to about 70 viruses, 27 of which are considered serious threats to the soybean industry worldwide (Tolin and Lacy, 2004; Hema *et al.*, 2014; Hill and Whitham, 2014). Some of the more frequently occurring viruses in soybean include *Bean pod mottle virus* (BPMV; genus *Comovirus*), *Cowpea mild mottle virus* (genus

*E-mail: l.kumar@cgiar.org

Table 8.1. Area, production and productivity of soybean in 2013, and percent change in area, production and productivity from 2003 to 2013.

	Area		Production		Yield	
	ha (×100,000)	% Change	t (×100,000)	% Change	t/ha	% Change
World	1112.7	24.8	2764.1	31	2.48	8.24
Africa	18.0	40.5	22.5	52.7	1.25	20.6
Americas	856.3	25.5	2408.3	32.7	2.81	9.57
Asia	206.3	14.7	272.9	6.4	1.32	−9.84
Europe	31.8	61.6	59.4	68.8	1.87	18.66
Oceania	0.411	84.8	0.918	90	2.23	33.92

% change = Per cent increase or decrease compared to 2003 data.
Data source: FAO soybean production statistics for 2013 (FAOSTAT, 2014)

Carlavirus), *Soybean dwarf virus* (genus *Luteovirus*), *Soybean mosaic virus* (SMV, genus *Potyvirus*), *Tobacco ringspot virus* (genus *Nepovirus*), *Tobacco streak virus* (genus *Ilarvirus*), a few begomoviruses (e.g. *Mungbean yellow mosaic virus, Bean golden mosaic virus*) and tospoviruses (*Tomato spotted wilt virus* and *Groundnut bud necrosis virus*) (Hema *et al.*, 2014; Hill and Whitham, 2014). SMV, which causes mosaic disease wherever the crop is grown, is the most important of all the viruses infecting soybean (Cui *et al.*, 2011; Hill and Whitham, 2014). Mosaic disease was first reported in Connecticut in the USA in 1915 (Clinton, 1915) and the causal virus, SMV, was later described in 1921 by Gardner and Kendrick. Severe outbreaks of SMV have been reported worldwide (Hill and Whitham, 2014). This chapter summarizes the characteristics of SMV and the integrated approaches used to manage the disease induced in the various crops.

8.2 *Soybean mosaic virus* Effects in Soybean

Symptoms and yield losses due to SMV vary depending on the soybean genotype, SMV strain, plant age and environmental conditions (Zhou *et al.*, 1995). The most common symptoms are yellow mosaic on trifoliate leaves, rugose leaf lamina, puckering along the veins, wavy leaf margins and downward rolling of the lamina (Fig. 8.1b). Symptoms are most conspicuous in young leaves. Some cultivars develop necrotic lesions on primary leaves, which eventually coalesce into veinal necrosis, leaf yellowing and result in abscission. Different strains often produce different symptoms on the same cultivar (Cui *et al.*, 2011). Some strains of SMV can induce severe stunting, systemic necrosis, leaf yellowing, petiole and stem necrosis, and defoliation followed by death of the infected plants. Temperature has a major effect on the severity of symptom expression as well as the progression of the disease (Li *et al.*, 2009). Symptoms are generally severe at 18°C, mild at 24–25°C and largely masked at 30°C (Hill, 1999; ICTVdB, 2006).

SMV infection affects all the agronomic characters of the crop such as grain weight per plant, leaf area per plant, nodule weight, nodule number, nitrogen fixation, dry weight of the shoot, dry weight of the root, mottled seed rate, plant height, and seed weight, particularly in cases of early infection (Tu *et al.*, 1970; Hill *et al.*, 1987; Zhi *et al.*, 1996). In general, early infection with SMV results in stunting, reduction in pod set, increased mottling of seed coat (Fig. 8.1c) and reduced seed size and weight, whereas late infections (after flowering) do not significantly affect the seed quality or the seed yield (Ren *et al.*, 1997). Pods developed on infected plants are usually dwarfed and flattened, lack hairs and show reduced seed size and oil content (Hill and Whitham, 2014). Cheema *et al.* (2003) also reported on changes in the composition of seed protein and oil in seeds from infected plants compared to those of seeds from healthy plants. Seeds of some cultivars show

(a)

Fig. 8.1. (a) Genome organization of *Soybean mosaic virus* (SMV).The length of each proteolytically derived protein is indicated and given on its top is the number of amino acids; (b) SMV infected soybean plant and (c) seed mottling due to SMV infection.

brown to dark mottling (Bottenberg and Irwin, 1992; Gergerich, 1999; Koning *et al.*, 2003). There is no correlation, however, between the mottled seed and seed transmission (Pacumbaba, 1995). SMV induces cylindrical inclusion bodies typical of *potyvirus* infection such as pin wheels, bundles, scrolls and laminated aggregates in infected cells (Hill, 2003).

SMV is reported to reduce yields by 8–50% under natural field conditions (Arif *et al.*, 2002; Hill, 2003; Cui *et al.*, 2011) and up to 100% in cases of severe outbreaks (Liao *et al.*, 2002). Co-infection of SMV with other viruses such as BPMV, *Cowpea mosaic virus* (genus *Comovirus*), *Alfalfa mosaic virus* (genus *Alfamovirus*) and *Tobacco ringspot virus* (Anjos *et al.*, 1992; Gergerich, 1999; Wang, 2009; Hwang *et al.*, 2011) cause much more severe damage than infection by the individual viruses (Chen *et al.*, 2004; Hill *et al.*, 2007; Malapi-Nelson *et al.*, 2009; Wang, 2009). Reduction in yield may be as high as 66–86% in susceptible cultivars with mixed

infections compared with reductions of 8–25% in cultivars inoculated with SMV and 10% in those inoculated with BPMV. The strain G5H reduced seed yields by more than 50% in both greenhouse and field experiments (Kim *et al.*, 1996). SMV infection has also been shown to predispose some cultivars to pod and stem blight, stem canker and *Phomopsis longicolla* (currently *Diaporthe longicolla*) seed decay that result in poor seed quality, low vigour and loss of seed viability (Koning *et al.*, 2001).

8.3   Transmission

SMV spread in the field is mainly assisted by aphid vectors (Hemiptera: Aphididae). About 32 species of aphids from 15 genera transmit SMV in a non-persistent manner (Steinlage *et al.*, 2002; Hill and Whitham, 2014). The most common vectors are *Acyrthosiphon pisum*, *Aphis fabae*, *A. craccivora*, *A. glycines*, *A. gossypii*, *Myzus persicae*, *Rhopalosiphum maidis* and *R. padi* (Gunasinghe *et al.*, 1986). SMV isolates may show some vector specificity; for example, the SMV-O isolate is transmitted by *M. persicae*, but not by *R. maidis*. The virus is transmissible by sap inoculation and grafting, but not by dodder of the genus *Cuscuta*. Vertical transmission of SMV through seeds of infected cultivars has been reported at a rate of up to 75% (Naik and Murthy, 1997; Gergerich, 1999; Chalam *et al.*, 2004). The extent of seed transmission depends on the virus strain, host genotype and plant stage at infection. The incidence of seed transmission is higher in plants infected before the onset of flowering. The virus in seeds remains infective for a long period of time and can be recovered from seeds that are no longer viable and cannot germinate. In dormant seeds, SMV is found only in the embryo.

Plants grown from infected seed play an important role in the epidemiology of SMV. Such plants are the primary sources of inoculum for secondary spread of the virus within the field by aphid vectors (Nutter *et al.*, 1998). Spread within and among fields is mostly aggregated from a point source, and secondary spread by aphids occurs at a moderately fast rate. Disease incidence varies from 13.3–60% (Fiedorow, 1993; Laguna *et al.*, 2002; Golnaraghi *et al.*, 2004). Seed transmission is linked to two major proteins involved in RNA silencing; Dicer-like 3 and RNA-dependent RNA polymerase 6 (Domier *et al.*, 2011).

8.4 Host Range

SMV can infect several plant species belonging to the families Amaranthaceae (including those species of the former family Chenopodiaceae), Passifloraceae, Schropulariaceae and Solanaceae, but mostly Fabaceae. Soybean is the most economically important host of SMV. Other plant species that are systemically infected by SMV are: *Astragalus monspessulanus, Amaranthus* spp., *Canavalia ensiformis, Cassia occidentalis* (now *Senna ocidentalis*), *Crotalaria spectabilis, Cyamopsis tetragonoloba, Dolichos falcatus* (accepted name *D. trilobus*), *Lespedeza stipulacea* (accepted name *Kummerowia stipulacea*), *L. striata* (accepted name *K. striata*), *Lupinus albus, L. luteus, Macroptilium lathyroides, Mucuna deeringiana* (a synonym of *M. pruriens* var. *utilis*), *Phaseolus lunatus, P. nigricans, P. vulgaris, Physalis longifolia, P. virginiana, Sesbania exaltata, Trigonella caerulea* and *T. foenum-graecum*. SMV causes latent infections in *Hippocrepis multisiliquosa, Lotus tetragonolobus, Lupinus angustifolius, Solanum carolinense,* some cultivars of *P. vulgaris* and *Scorpiurus sulcata*.

SMV causes local lesions on *Chenopodium album, C. quinoa, C. tetragonoloba, Dolichos biflorus* (accepted name *Vigna unguiculata*), *Indigofera hirsuta, Lablab purpureus, Lourea vespertilionis* (accepted name *Christia vespertilionis*), *M. lathyroides, P. lunatus* and some cultivars of *P. vulgaris*. Some isolates of SMV cause both local lesions and systemic symptoms in *C. tetragonoloba* and *M. lathyroides*. Several strains of the virus are recognized on the basis of the reactions on a differential set of soybean cultivars (Bowers and Goodman, 1991).

8.5 Taxonomy and Genome Organization

SMV is classified as a member of the genus *Potyvirus* of the family *Potyviridae* (King *et al.*, 2012). It has been grouped into the '*Bean common mosaic virus* subgroup' of potyviruses based on phylogenetic analysis (Yang *et al.*, 2011). The particles of SMV are non-enveloped, filamentous rods of *circa* 750 nm in length and 15–18 nm in diameter, with the capsid proteins bound to the viral genome in helical symmetry. The genome of SMV is a positive sense, single-stranded monopartite RNA of about 9600 nucleotides. The SMV RNA genome has a poly-A tail at its 3′ end and genome-linked viral protein (VPg) at its 5′ end. The genome encodes a single polyprotein *circa* 350 kDa that is cleaved into multiple proteins by proteases encoded by its own genome. The eleven mature proteins encoded from the 5′ to the 3′ end of the SMV genome are: P1, serine proteinase, the first protein; HC-Pro, helper component cyst-eine proteinase; P3, third protein; P3N-PIPO, where the PIPO ORF is embedded within the P3 cistron and expressed as a fusion product with the N-terminal portion of P3 protein (ca. 25 kDa) (Hillung *et al.*, 2013); 6K1 and 6K2, two 6 kDa proteins; CI, cylindrical in-clusion protein; VPg, viral protein genome-linked; NIa-Pro, main viral proteinase; NIb, replicase; and CP, coat protein (Urcuqui-Inchima *et al.*, 2001, Valli *et al.*, 2007; King *et al.*, 2012) (Fig. 8.1a). Additionally, there is also a short open reading frame embedded within the P3 cistron of the polyprotein of SMV, but translated in the +2 reading frame, which is referred as PIPO. The SMV genome has untranslated regions (UTR) on either end of its RNA genome, which are termed as 5′ UTR and 3′ UTR, and they function as the regulatory elements for translation of the polyprotein. Due to the absence of a 5′ cap-like structure, its genome translates by a cap-independent mechanism.

As with other potyviruses, NIa, a serine protease, is one of the three endopeptidases encoded by SMV that processes the polypro-tein, and NIb is the RNA-dependent RNA polymerase essential for replication of the viral genome (Domier *et al.*, 1987). The pro-teins P1 and HC-Pro are implicated in the suppression of gene silencing, while HC-Pro is also associated with the movement of virus and aphid transmission (Valli *et al.*, 2007; Giner *et al.*, 2008; Seo *et al.*, 2010). Trans-mission of potyviruses by aphids is mostly dependent upon the interaction between HC-Pro and the CP, and is reflected in several conserved domains of these proteins across various potyviruses (Flasinski and Cassidy, 1998; Seo *et al.*, 2010). The KLSK amino acid motif located in the HC-Pro of SMV is implicated in binding to the stylets of aphids (*A. glycines*) (Flasinski and Cassidy, 1998; Seo *et al.*, 2010). The DAG domain of CP is conserved among most potyviruses, including SMV, and facilitates its binding with HC-Pro during aphid transmission (Seo *et al.*, 2010). Studies by Jossey *et al.* (2013) suggest that interactions between CP and HC-Pro may also be essential for SMV to be seed transmitted, and may affect RNA silencing in a host-specific manner. In addition, Seo

et al. (2011) showed that alternations in amino acids in HC-Pro alter symptom expression in resistant cultivars carrying the *Rsv1* gene. The CP of SMV is required for virion assem-bly and viral cell–cell and long-distance movement in plants (Callaway *et al.*, 2001). The C-terminal domain of SMV CP helps in self-interaction, while the charged amino acids on the exposed surface of the C ter-minus help in the inter subunit interactions of CP and thus virus cell-to-cell and long-distance movement and also virion assem-bly (Seo *et al.*, 2013). The CI of potyviruses assists in cell-to-cell movement and it local-izes in special organized structures, notably pinwheel inclusion bodies, which attach to the plasmodesmata immediately after virus infection, align with plasmodesmatal open-ings, and associate with CP (Carrington and Whitham, 1998). It has also been shown that the pinwheel-forming areas are present near the plasmodesmata of SMV infected soy-bean leaf cells (Hunst and Tolin, 1983).

8.6   *Soybean mosaic virus* Diversity

Worldwide, there are reports of numerous strains and isolates of SMV, and in the USA alone at least 98 isolates of SMV have been reported (Cho and Goodman, 1979; Hill and Whitham, 2014). Based on the differential reactions to SMV isolates on the two suscep-tible soybean cultivars 'Clark' and 'Rampage', and the six resistant cultivars 'Buffalo', 'Davis', 'Kwangyo', 'Marshall', 'Ogden' and 'York', the SMV isolates are classified into G1 to G7, with possibly two additional strains G7a and C14 (Cho and Goodman, 1979; Hill and Whitham, 2014). In Japan, five SMV strains (A–E) have been identified (Takahashi *et al.*, 1980). Similarly, a necrotic strain, SMV-N, and a number of G2 isolates are reported in Canada (Tu and Buzzell, 1987; Gagarinova *et al.*, 2008a; Viel *et al.*, 2009). However, the SMV-N strain reported from Canada shares a high degree of sequence similarity with the G2 strain group and hence it is considered as an isolate of the G2 group (Gagarinova *et al.*, 2008b). Comparison of strains from

Japan and the USA suggest that the A and B strains from Japan are similar to the G3 and G2 strains of the USA, respectively (Kanematsu *et al.*, 1998). Analysis of SMV isolates from China resulted in the establishment of 21 distinct strains (Li *et al.*, 2010), in addition to the naturally occurring recombinant strains of SMV (Yang *et al.*, 2014). Reports of differentiation of SMV isolates using soybean cultivars have also been documented from Brazil (Anjos *et al.*, 1985; Almeida *et al.*, 1995). In South Korea, occurrence of all the SMV G strains (G1–G7), SMV-N, G5H, G7a and G7H have been reported (Seo *et al.*, 2009; Hill and Whitham, 2014). However, the dominant SMV strains infecting soybean in South Korea varied at different time points (Cui *et al.*, 2011). In the early 1980s, the G5 strain caused about 80% damage, whereas in the late 1980s, G5H was the dominant strain responsible for over 65% of losses caused by SMV (Cho *et al.*, 1983; Kim *et al.*, 2003). In recent times, G7H has been the most prevalent strain, accounting for about 50% of SMV incidence. In addition, new strains that can break *Rsv*, including CN18, have been identified in South Korea (Kim *et al.*, 2003; Choi *et al.*, 2005; Seo *et al.*, 2009). It is likely that additional strains exist, particularly in countries where SMV infection is prevalent, but strain cataloguing has not been carried out.

The role of mutations in the generation of variability, eventually resulting in novel SMV isolates, some of which are capable of breaking host resistance, have been reported by various groups (Hajimorad *et al.*, 2003; Choi *et al.*, 2005; Gagarinova *et al.*, 2008a,b). Recombination analysis of several different strains and isolates of SMV reveal about 17 different recombination sites across the SMV genome (Gagarinova *et al.*, 2008a). Different isolates/strains of SMV can simultaneously infect the same cells, thus facilitating recombination and the formation of new isolates or strains.

8.7 Diagnostic Methods

Although SMV causes distinctive symptoms in soybean, accurate confirmation of virus infection using diagnostic assays is necessary as a number of other viruses infecting soybean can sometimes also induce symptoms that are more or less similar to SMV. Additionally, diagnostic testing will reveal any latent infections and infections of seed and seedlings and assist in containing the spread of SMV across generations. Conventional diagnostic methods include the use of indicator plants; observations of differential visual symptoms on indicator plants are also commonly used to assess the host range allowing for differentiation between SMV strains (Cui *et al.*, 2011). The most commonly used indicator host species and the associated diagnostic symptoms are: *C. album* or *C. quinoa* producing chlorotic local lesion symptoms; *L. purpureus* showing necrotic local lesions; *M. lathyroides* showing systemic mosaic; *P. vulgaris* displaying systemic mosaic with some strains in some cultivars, but often latent or no infection; and *P. vulgaris* cv. 'Top Crop' produces necrotic local lesions in detached leaves at 30°C (ICTVdB, 2006). SMV causes systemic infection only in *P. vulgaris* cv. 'Double White Princess', unlike the *Bean common mosaic virus* (genus *Potyvirus*) and *Bean yellow mosaic* (genus *Potyvirus*) that systemically infect several *P. vulgaris* cultivars (ICTVdB, 2006). Although inoculation to indicator plants does not require expensive instrumentation and complex protocols, this method is not very sensitive and demands considerable amounts of time and space.

A number of laboratory assays have been developed for the diagnosis of SMV infections (Hill and Whitham, 2014). Several different serological tests are routinely employed for diagnosis of SMV and its strains, which include the direct double-sandwich ELISA, indirect ELISA, tissue-print immunoassay, dot immunobinding assay, immunosorbent electron microscopy, immune-fluorescence and Western blotting (Hema *et al.*, 2014). Additionally, PCR-based techniques such as reverse transcription PCR and real-time PCR are widely employed for the detection of SMV. Of these methods, ELISA-based methods are commonly used for SMV detection. However, SMV polyclonal serum has been shown to cross react with several potyviruses such as *Watermelon mosaic virus, Bean common mosaic virus, Bean yellow mosaic virus, Clover yellow vein virus* and *Lettuce mosaic virus* (Hill, 1999; Gergerich, 1999). Therefore, reconfirmation of ELISA-based SMV diagnosis by reverse transcription PCR assay is necessary or alternatively ELISA can be performed using monoclonal antibodies known to be highly specific to SMV. Although SMV is evenly distributed in all parts of the infected plant, leaf tissues are the preferred source for virus detection by ELISA or reverse transcription PCR.

8.8   Management Strategies

Seed transmission and viruliferous aphids migrating from infected sources (volunteer plants or weed hosts) serve as primary sources of infection to a new soybean crop. Subsequent spread and overall disease incidence is influenced by the density of aphid vector populations which facilitate further spread within and between fields during a cropping season. Polycyclic diseases of this kind can be controlled through the use of virus-free seed, vector control and phytosanitary programmes (Dugje *et al.*, 2009).

A recent study by Zhou *et al.* (2014) demonstrated that SMV incidence can be reduced by the addition of K^+-based fertilizers, indicating the prospect of SMV management through soil nutrient management. Aphid control alone is ineffective in managing SMV (Pedersen *et al.*, 2007). Cultural methods based on an understanding of the disease epidemiology, including an understanding of interactions between the host, virus and vector and habitat, is gaining prominence as eco-friendly integrated pest management approaches to control vector-borne diseases such as SMV (Makkouk *et al.*, 2014). However, cultural control is often difficult to implement under smallholder, mixed farming agricultural systems.

SMV is of quarantine significance in the international exchange of soybean because the virus is seed transmitted and known to consist of different strains. Quarantine control to prevent the spread of infected seeds

is an important requirement to prevent the introduction of SMV or its strains in regions where they are not known to occur. The procedure for detecting and eliminating SMV from soybean seed lots typically involves the removal of mottled seeds, growing seedlings from healthy seeds and testing for SMV by employing a combination of virus detection techniques (e.g. infectivity tests on diagnostic hosts, immunosorbent electron microscopy, ELISA and RT-PCR). A combination of these methods can minimize the risk of virus spread through seeds. Seeds harvested only from virus-free plants should be released from quarantine.

Introgression of host plant resistance in cultivated varieties is one of the most important strategies for the management of any disease, including viral diseases. Three major resistance loci, designated as *Rsv1*, *Rsv3* and *Rsv4*, have been identified and mapped for resistance to SMV, and mostly encode for nucleotide-binding site leucine-rich repeat proteins (Saghai-Maroof *et al.*, 2010). These resistance loci have been shown to condition resistance to different SMV strains. For instance, the *Rsv1* locus identified in soybean line PI 96983 confers resistance to the SMV strains G1, G2 and G3, but not against the strains G5, G6 and G7; the *Rsv3* locus confers resistance against SMV strains G5 to G7; and the *Rsv4* locus identified in soybean line PI 486355 was reported to produce seedling resistance to most SMV isolates, but systemic symptoms can appear as plants mature (Chen *et al.*, 1993; Zhou *et al.*, 2014). Variants to major resistance loci have been identified. There are nine different variants of the *Rsv1* locus in different soybean lines, which confer differential reactions to the SMV strains G1–G7. These are: *Rsv1* from PI 96983; *Rsv1-y* from 'York'; *Rsv1-m* from 'Marshall'; *Rsv1-k* from 'Kwanggyo'; *Rsv1-t* from 'Ogden'; *Rsv1-r* from 'Raiden'; *Rsv1-h* from 'Suweon 97'; *Rsv1-s* from 'LR1'; and *Rsv1-n* from PI 507389 (Hill and Whitham, 2014). Similarly, five alleles have been identified at the *Rsv3* locus and two at the *Rsv4* locus (Hill and Whitham, 2014). Due to a high degree of variability among SMV isolates, the use of multiple sources of resistance has been recommended for effective virus control (Saghai-Maroof *et al.*, 2008, 2010; Shi *et al.*, 2009; Shakiba *et al.*, 2012).

The R-gene mediated defence response against SMV has been studied with major emphasis on the *Rsv1* locus. *Rsv1* is effective against most of the SMV strains including the N isolate (SMV-N) of SMV-G2, against which extreme resistance is obtained; however, the G7 strain of SMV can overcome *Rsv1*-mediated resistance (Hill and Whitham, 2014). The HC-Pro and P3 proteins of SMV have been identified as the elicitors of *Rsv1* mediated resistance, and mutations in these genes can help SMV to overcome *Rsv1*-based resistance (Hajimorad *et al.*, 2005, 2008, 2011; Chowda-Reddy *et al.*, 2011; Wen *et al.*, 2011, 2013; Zhang *et al.*, 2012). However, in contrast to *Rsv1*, the *Rsv3* resistance locus provides resistance against the G5, G6 and G7 strains of SMV, and this resistance does not induce the typical hypersensitive reaction. The elicitors of *Rsv3*-based resistance are mapped to the CI protein (Zhang *et al.*, 2009; Chowda-Reddy *et al.*, 2011). The *Rsv4*-based resistance is effective against G1–G7 strains of SMV, and it is not associated with extreme resistance or hypersensitive reactions, but is characterized by reduced replication and movement of the virus, which is thus termed as non-necrotic late susceptibility (Gunduz *et al.*, 2004). The SMV protein P3 is identified as the elicitor for the *Rsv4* locus (Chowda-Reddy *et al.*, 2011).

Broad-based resistance against various SMV strains is essential for effective disease management. However, this requires pyramiding of various *Rsv* loci, which is complicated considering the involvement of multiple alleles and loci. As an alternative to conventional resistance, pathogen-derived resistance is being exploited to overcome some of these challenges and develop broad-based resistant cultivars (Cillo and Palukaitis, 2014). The CP-mediated transgenic resistance strategy, which was successful in the case of several plant RNA viruses like the *Papaya ringspot virus* (genus *Potyvirus*) (Fitchen and Beachy, 1993; Fuchs and Gonsalves, 2007) has also been employed for SMV (Wang *et al.*, 2001; Furutani *et al.*, 2007). These CP-mediated transgenic soybean lines have been subjected to

field trials and have shown resistance against multiple strains of SMV (Wang *et al.*, 2001; Steinlage *et al.*, 2002). RNA interference-based resistance, which successfully controls infections against several plant viruses (Sudarshana *et al.*, 2007; Cillo and Palukaitis, 2014), has also been employed against SMV in soybean. There is a recent report of a transgenic soybean developed for multiple virus resistance by using RNA interference-technology against *Alfalfa mosaic virus*, BPMV and SMV (Zhang *et al.*, 2011). Zhou *et al.* (2014) employed a novel strategy to induce SMV resistance by over expressing the soybean *GmAKT2* gene responsible for K$^+$ channelling in soybean as an alternative to pathogen gene-mediated resistance. Increase in K$^+$ channelling has been shown to reduce the incidence of SMV (Zhou *et al.*, 2014). Despite significant advances, SMV resistant transgenic soybean has not been deployed for commercial cultivation.

8.9 Concluding Remarks

Since 2003, world soybean cultivation is expanding on an annual average growth of 3% in area and 4% in production (FAOSTAT, 2014). World demand for soybean has been able to absorb ever increasing production, which is still lower than consumption demand which is growing at an average rate of 4.8% per year since 1970 (Flaskerud, 2003). Much of the production gains in soybean have been achieved through increases in production area as yields are stagnant at 2 t/ha on average in the last decade (see Table 8.1; FAOSTAT, 2014). Prevention of on-farm production losses due to cosmopolitan biotic constraints such as SMV is imperative to improving productivity.

Current SMV management is achieved mainly through the use of resistant or tolerant cultivars. However, the occurrence of resistant breaking strains and rapid evolution of new strains that can overcome available resistance is a risk to this strategy. In a number of countries, especially in sub-Saharan Africa, SMV strains have not been characterized.

Efforts should be made to gain this knowledge not only to deploy suitable resistant cultivars, but also to identify the occurrence of any new strains against which available resistance may not be effective. Soybean germplasm needs to be explored for new sources of resistance for known SMV strains as well as other strains prevailing in various countries. Genetic engineering offers immense potential for the development of broad resistance to combat several SMV strains. Transgenic SMV resistance can be combined with resistance to other soybean infecting viruses, for instance, BPMV (Reddy *et al.*, 2001) and *Soybean dwarf virus* (Tougou *et al.*, 2006, 2007), in order to develop multiple virus resistant cultivars. In addition to augmenting host resistance, there is a need for greater emphasis on monitoring soybean seed stocks exchanged between countries to prevent the risk of SMV spread through seed to new regions where the virus or specific strains are not known to occur. For instance in India, SMV was intercepted in soybean germplasm received from Australia, Nigeria, Sri Lanka, Taiwan, Thailand and the USA during quarantine processing (Chalam *et al.*, 2014). This example underscores the significance of seed transmission in long-distance spread of SMV and the importance of quarantine monitoring of soybean seeds received from countries where the virus is established. Similar emphasis is required to prevent the spread of aphid species into new niches. For instance, the recent introduction of *A. glycines*, an efficient vector of SMV, in North America has been a major concern for soybean cultivation in that region (Hill and Whitham, 2014).

Overall, the dynamics of the SMV pathosystem is influenced by evolving SMV strains, vector activity and changes in soybean production systems; all of which are posing new challenges to effective disease management. Integrated approaches combining host plant resistance, cultural practices and quarantine monitoring are critical to effective management.

References

Almeida, A.M.R., Almeida, L.A., Oliveira, M.C.N., Paiva, F.A. and Moreira, W.A. (1995) Strains of *Soybean mosaic virus* identified in Brazil and their influence on seed mottling and rate of seed transmission. *Fitopatologia Brasileira* 20, 227–232.

Anjos, J.R.M., Lin, M.T. and Kitajima, E.W. (1985) Characterization of an isolate of *Soybean mosaic virus*. *Fitopatologia Brasileira* 10, 143–157.

Anjos, J.R., Jarlfors, U. and Ghabiral, S.A. (1992) Soybean mosaic Potyvirus enhances the titer of two comoviruses in dually infected soybean plants. *Phytopathology* 82, 1022–1027.

Arif, M., Stephen, M. and Hassan, S. (2002) Effect of Soybean mosaic Potyvirus on growth and yield components of commercial soybean varieties. *Pakistan Journal of Plant Pathology* 1, 54–57.

Boerma, H.R. and Specht, J.E. (2004) *Soybeans: Improvement, Production, and Uses*, 3rd edn. Agronomy Series No. 16. American Society of Agronomy, Crop Science Society of America & Soil Science Society of America Publishers, Madison, Wisconsin.

Bottenberg, H. and Irwin, M.E. (1992) Using mixed cropping to limit seed mottling induced by *Soybean mosaic virus*. *Plant Disease* 76, 304–306.

Bowers, G.R. and Goodman, R.M. (1991) Strain specificity of *Soybean mosaic virus* seed transmission in soybean. *Crop Science* 31, 1171–1174.

Carrington, J.C. and Whitham, S.A. (1998) Viral invasion and host defence: strategies and counter-strategies. *Current Opinion in Plant Biology* 1, 336–341.

Callaway, A., Giesman-Cookmeyer, D., Gillock, E.T., Sit, T.L. and Lommel, S.A. (2001) The multifunctional capsid proteins of plant RNA viruses. *Annual Review of Phytopathology* 39, 419–460.

Chalam, V.C., Khetarpal, R.K., Mishra, A., Jain, A. and Gupta, G.K. (2004) Seed transmission of *Soybean mosaic virus* in soybean cultivars. *Journal of Mycology and Plant Pathology* 34, 86–87.

Chalam, V.C., Parakh, D.B., Maurya, A.K., Singh, S. and Khetarpal, R.K. (2014) Biosecuring India from seed-transmitted viruses: the case of quarantine monitoring of legume germplasm imported during 2001–2010. *Indian Journal of Plant Protection* 42, 270–279.

Cheema, S.S., Thiara, S.K. and Kang, S.S. (2003) Biochemical changes induced by Soybean yellow mosaic virus in two soybean varieties. *Plant Disease Research* 18, 159–161.

Chen, J., Zheng, H.Y., Lin, L., Adams, M.J., Antoniw, J.F., *et al.* (2004) A virus related to *Soybean mosaic virus* from *Pinellia ternata* in China and its comparison with local soybean SMV isolates. *Archives of Virology* 149, 349–363.

Chen, P., Buss, G.R. and Tolin, S.A. (1993) Resistance to *Soybean mosaic virus* conferred by two independent dominant genes in PI 486355. *Heredity* 84, 25–28.

Cho, E.-K. and Goodman, R.M. (1979) Strains of *Soybean mosaic virus*: classification based on virulence in resistant cultivars. *Phytopathology* 69, 467–470.

Cho, E.-K., Choi, S.H. and Cho, W.T. (1983) Newly recognized *Soybean mosaic virus* mutants and sources of resistance in soybeans. *Research Reports, Office of Rural Development, Korea (SPMU)* 25, 18–22.

Choi, B.K., Koo, J.M., Ahn, H.J., Yum, H.J., Choi, C.W., *et al.* (2005) Emergence of *Rsv*-resistance breaking *Soybean mosaic virus* isolates from Korean soybean cultivars. *Virus Research* 112, 42–51.

Chowda-Reddy, R.V., Sun, H., Hill, J.H., Poysa, V. and Wang, A. (2011) Simultaneous mutations in multi-viral proteins are required for *Soybean mosaic virus* to gain virulence on soybean genotypes carrying different *R* genes. *PLoS One* 6, e28342.

Cillo, F. and Palukaitis, P. (2014) Transgenic resistance. *Advances in Virus Research* 90, 35–146.

Clinton, G.P. (1915) Notes on plant diseases of Connecticut. *Annual Report of the Connecticut Agricultural Experiment Station for* 1915, pp. 421–451.

Cui, X., Chen, X. and Wang, A. (2011) Detection, understanding and control of *Soybean mosaic virus*. In: Sudaric, A. (ed.) *Soybean – Molecular Aspects of Breeding.* INTECH, Rijeka, Croatia, pp. 335–354.

Domier, L.L., Shaw, J.G. and Rhoads, R.E. (1987) Potyviral proteins share amino acid sequence homology with picorna-, como-, and caulimoviral proteins. *Virology* 158, 20–27.

Domier, L.L., Hobbs, H.A., McCoppin, N.K., Bowen, C.R., Steinlage, T.A., *et al.* (2011) Multiple loci condition seed transmission of *Soybean mosaic virus* (SMV) and SMV-induced seed coat mottling in soybean. *Phytopathology* 101, 750–756.

Dugje, I.Y., Omoigui, L.O., Ekeleme, F., Bandyopadhyay, R., Kumar, P.L., *et al.* (2009) *Farmers' Guide to Soybean Production in Northern Nigeria.* IITA, Ibadan, Nigeria, 21 pp.

FAOSTAT (2014) Soybean Production Statistics. http://faostat3.fao.org/download/Q/QC/E (accessed December 22, 2014).

Fiedorow, Z. (1993) Incidence and harmfulness of soybean virus (SMV). *Roczniki Nauk Rolniczych* 23, 13–20.

Fitchen, J.H. and Beachy, R.N. (1993) Genetically engineered protection against viruses in transgenic plants. *Annual Review of Microbiology* 47, 739–763.

Flasinski, S. and Cassidy, B.G. (1998) Potyvirus aphid transmission requires helper component and homologous coat protein for maximal efficiency. *Archives of Virology* 143, 2159–2172.

Flaskerud, G. (2003) Brazil's soybean production and impact. North Dakota State University Extension Service, Fargo, North Dakota. Available at http://www.ag.ndsu.nodak.edu.

Fuchs, M. and Gonsalves, D. (2007) Safety of virus-resistant transgenic plants two decades after their introduction: lessons from realistic field risk assessment studies. *Annual Review of Phytopathology* 45, 173–202.

Furutani, N., Yamagishi, N., Hidaka, S., Shizukawa, Y., Kanematsu, S., *et al.* (2007) *Soybean mosaic virus* resistance in transgenic soybean caused by posttranscriptional gene silencing. *Breeding Science* 57, 123–128.

Gagarinova, A.G., Babu, M., Strömvik, M.V. and Wang, A. (2008a) Recombination analysis of *Soybean mosaic virus* sequences reveals evidence of RNA recombination between distinct pathotypes. *Virology Journal* 5, 143.

Gagarinova, A.G., Babu, M., Poysa, V., Hill, J.H. and Wang, A. (2008b) Identification and molecular characterization of two naturally occurring *Soybean mosaic virus* isolates that are closely related but differ in their ability to overcome *Rsv4* resistance. *Virus Research* 138, 50–56.

Gardener, M.W. and Kendrick, J.B. (1921) *Soybean mosaic virus. Journal of Agricultural Research* 22, 111–114.

Gergerich, R.C. (1999) Comovirus: *Bean pod mottle virus*. In: Hartman, G.L., Sinclair, J.B. and Rupe, J.C. (eds) *Compendium of Soybean Diseases*, 4th edn. APS Press, St. Paul, Minnesota, pp. 61–62.

Giner, A., García-Chapa, M., Lakatos, L., Burgyan, J. and López-Moya, J.J. (2008) Involvement of P1 and HCPro proteins of sweet potato mild mottle ipomovirus (SPMMV) in suppression of gene silencing. In: *Book of Abstracts: International Conference Genetic Control of Plant Pathogenic Viruses and Their Vectors: Towards New Resistance Strategies, ResistVir*, Puerto de Santa Maria, Spain, p. 90.

Golnaraghi, A.R., Shahraeen, N., Pourrahim, R., Farzadfar, S. and Ghasemi, A. (2004) Occurrence and relative incidence of viruses infecting soybeans in Iran. *Plant Disease* 88, 1069–1074.

Graham, P.H. and Vance, C.P. (2003) Legumes: importance and constraints to greater use. *Plant Physiology* 131, 872–877.

Gunasinghe, U.B., Irwin, M.E. and Bernard, R.L. (1986) Effect of a soybean genotype resistant to *Soybean mosaic virus* on transmission-related behaviour of aphid vectors. *Plant Disease* 70, 872–874.

Gunduz, I., Buss, G.R., Chen, P. and Tolin, S.A. (2004) Genetic and phenotypic analysis of *Soybean mosaic virus* resistance in PI 88788 soybean. *Phytopathology* 94, 687–692.

Hajimorad, M.R., Eggenberger, A.L. and Hill, J.H. (2003) Evolution of *Soybean mosaic virus*-G7 molecularly cloned genome in *Rsv1*-genotype soybean results in emergence of a mutant capable of evading *Rsv1*-mediated recognition. *Virology* 314, 497–509.

Hajimorad, M.R., Eggenberger, A.L. and Hill, J.H. (2005) Loss and gain of elicitor function of *Soybean mosaic virus* G7 provoking *Rsv1*-mediated lethal systemic hypersensitive response maps to P3. *Journal of Virology* 79, 1215–1222.

Hajimorad, M.R., Eggenberger, A.L. and Hill, J.H. (2008) Adaptation of *Soybean mosaic virus* avirulent chimeras containing P3 sequences from virulent strains to *Rsv1*-genotype soybeans is mediated by mutations in HC-Pro. *Molecular Plant-Microbe Interactions* 21, 937–946.

Hajimorad, M.R., Wen, R.-H., Eggenberger, A.L., Hill, J.H. and Saghai Maroof, M.A. (2011) Experimental adaptation of an RNA virus mimics natural evolution. *Journal of Virology* 85, 2557–2564.

Hema, M., Sreenivasulu, P., Patil, B.L., Kumar, P.L. and Reddy, D.V.R. (2014) Tropical food legumes: virus diseases of economic importance and their control. *Advances in Virus Research* 90, 431–505.

Hill, J.H. (1999) Soybean mosaic. In: Hartman, G.L., Sinclair, J.B. and Rupe, J.C. (eds) *Compendium of Soybean Diseases*, 4th edn. APS Press, St. Paul, Minnesota, pp. 70–71.

Hill, J.H. (2003) Soybean. In: Loebenstein, G. and Thottappilly, G. (eds) *Virus and Virus-Like Diseases of Major Crops in Developing Countries*. Kluwer Academic, Dordrecht, The Netherlands, pp. 377–396.

Hill, J.H. and Whitham, S.A. (2014) Control of virus diseases in soybeans. *Advances in Virus Research* 90, 355–390.

Hill, J.H., Bailey, T.B., Benner, H.I., Tachibana, H. and Durand, D.P. (1987) *Soybean mosaic virus*: effects of primary disease incidence on yield and seed quality. *Plant Disease* 71, 237–239.

Hill, J.H., Koval, N.C., Gaska, J.M. and Grav, C.R. (2007) Identification of field tolerance to bean pod mottle and *Soybean mosaic virus* in soybean. *Crop Science* 47, 212–218.

Hillung, J., Elena, S.F. and Cuevas, J.M. (2013) Intra-specific variability and biological relevance of P3N-PIPO protein length in potyviruses. *BMC Evolutionary Biology* 13, 249.

Hunst, P.L. and Tolin, S.A. (1983) Ultrastructural cytology of soybean infected with mild and severe strains of *Soybean mosaic virus*. *Phytopathology* 73, 615–619.

Hwang, T.Y., Jeong, S.C., Kim, O., Park, H.-M., Lee, S.-K., *et al.* (2011) Intra-host competition and interactions between *Soybean mosaic virus* (SMV) strains in mixed-infected soybean. *Australian Journal of Crop Science* 5, 1379–1387.

ICTVdB (2006) 00.057.0.01.061. *Soybean mosaic virus*. In: Büchen-Osmond, C. (ed.) *ICTVdB – The Universal Virus Database, Version 4*. Columbia University, New York.

Jossey, S., Hobbs, H.A. and Domier, L.L. (2013) Role of *Soybean mosaic virus*-encoded proteins in seed and aphid transmission in soybean. *Phytopathology* 103, 941–948.

Kanematsu, S., Eggenberger, A.L. and Hill, J.H. (1998) Comparison of *Soybean mosaic virus* strains G2 and G3 with Japanese strains A and B. *Annals of the Phytopathological Society of Japan* 64, 607.

Kim, Y., Noh, J., Kim, M., Im, D., Lee, B., *et al.* (1996) Effects of SMV-G5H strain on plant growth and seed chemical composition of soybeans cv. Danyeobkong. *Korean Journal of Crop Science* 41, 340–347.

Kim, Y.H., Kim, O.S., Lee, B.C., Moon, J.K., Lee, S.C., *et al.* (2003) G7H, a new *Soybean mosaic virus* strain: its virulence and nucleotide sequence of CI gene. *Plant Disease* 87, 1372–1375.

King, A.M.Q., Adams, M.J., Lefkowitz, E.J. and Carstens, E.B. (2012) *Virus Taxonomy: Ninth Report of the International Committee on Taxonomy of Viruses*. Elsevier Academic Press, New York.

Koning, G., TeKrony, D.M., Pfeifferm, T.W. and Ghabrial, S.A. (2001) Infection of soybean with *Soybean mosaic virus* increases susceptibility to *Phomopsis* spp., seed infection. *Crop Science* 41, 1850–1856.

Koning, G., TeKrony, D.M. and Ghabrial, S.A. (2003) Soybean seedcoat mottling: association with *Soybean mosaic virus* and *Phomopsis* spp. seed infection. *Plant Disease* 87, 413–417.

Laguna, I.G., Rodríguez-Pardina, P.E., Truol, G., Ploper, L.D., Arneodo, J., *et al.* (2002) Incidencia de las enfermedades de etiología viral en el cultivo de la soja en Argentina en la campaña 2001–2002. *Revista Avance Agroindustrial* 23, 41–45.

Li, D., Chen, P., Shi, A., Shakiba, E., Gergerich, R., *et al.* (2009) Temperature affects expression of symptoms induced by *Soybean mosaic virus* in homozygous and heterozygous plants. *Journal of Heredity* 100, 348–354.

Li, K., Yang, Q.H., Yang, H., Zhi, J. and Gai, J.Y. (2010) Identification and distribution of *Soybean mosaic virus* strains in southern China. *Phytopathology* 94, 351–357.

Liao, L., Chen, P., Buss, G.R., Yang, Q. and Tolin, S.A. (2002) Inheritance and allelism of resistance to *Soybean mosaic virus* in Zao18 soybean from China. *Journal of Heredity* 93, 447–452.

Makkouk, K., Kumari, S.G., van Leur, J.A.G. and Jones, R.A.C. (2014) Control of plant virus diseases of cool season legumes. *Advances in Virus Research* 90, 207–253.

Malapi-Nelson, M., Wen, R.-H., Ownley, B.H. and Hajimorad, M.R. (2009) Co-infection of soybean with *Soybean mosaic virus* and *Alfalfa mosaic virus* results in disease synergism and alteration in accumulation level of both viruses. *Plant Disease* 93, 1259–1264.

Naik, R.G. and Murthy, K.K.V. (1997) Transmission of *Soybean mosaic virus* through sap, seed and aphids. *Karnataka Journal of Agricultural Science* 10, 565–568.

Nutter, F.W. Jr, Schultz, P.M. and Hill, J.H. (1998) Quantification of within-field spread of *Soybean mosaic virus* in soybean using strain-specific monoclonal antibodies. *Phytopathology* 88, 895–901.

Pacumbaba, R.P. (1995) Seed transmission of *Soybean mosaic virus* in mottled and non-mottled soybean seeds. *Plant Disease* 79, 193–195.

Pedersen, P., Grau, C., Cullen, E., Koval, N. and Hill, J.H. (2007) Potential for integrated management of soybean virus disease. *Plant Disease* 91, 1255–1259.

Pimentel, D. and Patzek, T.W. (2008) Ethanol production using corn, switchgrass and wood; biodiesel production using soybean. In: Pimentel, D. (ed.) *Biofuels, Solar and Wind as Renewable Energy Systems: Benefits and Risks*. Springer, Dordrecht, The Netherlands, pp. 375–396.

Reddy, M.S.S., Ghabrial, S.A., Redmond, C.T., Dinkins, R.D. and Collins, G.B. (2001) Resistance to *Bean pod mottle virus* in transgenic soybean lines expressing the capsid polyprotein. *Phytopathology* 91, 831–838.

Ren, Q., Pfeiffer, T.W. and Ghabrial, S.A. (1997) *Soybean mosaic virus* incidence level and infection time: interaction effects on soybean. *Crop Science* 37, 1706–1711.

Saghai-Maroof, M.A., Jeong, S.C., Gunduz, I., Tucker, D.M., Buss, G.R., *et al.* (2008) Pyramiding of *Soybean mosaic virus* resistance genes by marker-assisted selection. *Crop Science* 48, 517–526.

Saghai-Maroof, M.A., Tucker, D.M., Skoneczka, J.A, Bowman, B.C., Tripathy, S., *et al.* (2010) Fine mapping and candidate gene discovery of the *Soybean mosaic virus* resistance gene, *Rsv4*. *The Plant Genome* 3, 14–22.

Sakai, T. and Kogiso, M. (2008) Soy isoflavones and immunity. *Journal of Medical Investigation* 55, 167–173.

Seo, J.-K., Lee, H.-G., Choi, H.-S., Lee, S.-H. and Kim, K.-H. (2009) Infectious *in vivo* transcripts from a full-length clone of *Soybean mosaic virus* strain G5H. *Plant Pathology Journal* 25, 54–61.

Seo, J.-K., Kang, S.H., Seo, B.Y., Jung, J.K. and Kim, K.H. (2010) Mutational analysis of interaction between coat protein and helper component-proteinase of *Soybean mosaic virus* involved in aphid transmission. *Molecular Plant Pathology* 11, 265–276.

Seo, J.-K., Sohn, S.H. and Kim, K.H. (2011) A single amino acid change in HC-Pro of *Soybean mosaic virus* alters symptom expression in a soybean cultivar carrying *Rsv1* and *Rsv3*. *Archives of Virology* 156, 135–141.

Seo, J.-K, Vo Phan, M.S., Kang, S.H., Choi, H.S. and Kim, K.H. (2013) The charged residues in the surface-exposed C-terminus of the *Soybean mosaic virus* coat protein are critical for cell-to-cell movement. *Virology* 446, 95–101.

Shakiba, E., Chen, P., Gergerich, R., Li, S., Dombek, D., *et al.* (2012) Reactions of commercial soybean cultivars from the mid south to *Soybean mosaic virus*. *Crop Science* 52, 1990–1997.

Shi, A., Chen, P., Li, D., Zheng, C., Zhang, B., *et al.* (2009) Pyramiding multiple genes for resistance to *Soybean mosaic virus* in soybean using molecular markers. *Molecular Breeding* 23, 113–124.

Steinlage, T.A., Hill, J.H. and Nutter, F.W. Jr (2002) Temporal and spatial spread of *Soybean mosaic virus* (SMV) in soybeans transformed with the coat protein gene of SMV. *Phytopathology* 92, 478–486.

Sudarshana, M.R., Roy, G. and Falk, B.W. (2007) Methods for engineering resistance to plant viruses. *Methods in Molecular Biology* 354, 183–195.

Takahashi, K., Tanaka, T., Wataru, I. and Tsuda, T. (1980) Studies on virus diseases and causal viruses of soybean in Japan. *Bulletin of Tohoku National Agricultural Experimental Station* 62, 1–130.

Tolin, S.A. and Lacy, G.H. (2004) Viral, bacterial, and phytoplasma diseases of soybean. In: Boerma, H.R. and Specht, J.E. (eds) *Soybeans; Improvement, Production and Uses*, 3rd edn. ASA/CSSA/SSSA, Madison, Wisconsin, pp. 765–819.

Tougou, M., Furutani, N., Yamagishi, N., Shizukawa, Y., Takahata, Y., *et al.* (2006) Development of resistant transgenic soybeans with inverted repeat-coat protein genes of *Soybean dwarf virus*. *Plant Cell Reports* 25, 1213–1218.

Tougou, M., Yamagishi, N., Furutani, N., Shizukawa, Y., Takahata, Y., *et al.* (2007) *Soybean dwarf virus*-resistant transgenic soybeans with the sense coat protein gene. *Plant Cell Reports* 26, 1967–1975.

Tu, J.C. and Buzzell, R.I. (1987) Stem-tip necrosis: a hypersensitive, temperature dependent, dominant gene reaction of soybean to infection by *Soybean mosaic virus*. *Canadian Journal of Plant Science* 67, 661–665.

Tu, J.C., Ford, R.E. and Quiniones, S.S. (1970) Effects of *Soybean mosaic virus* and/or *Bean pod mottle virus* infection on soybean nodulation. *Phytopathology* 60, 518–523.

Urcuqui-Inchima, S., Haenni, A.L. and Bernardi, F. (2001) Potyvirus proteins: a wealth of functions. *Virus Research* 74, 157–175.

Valli, A., López-Moya, J.J. and García, J.A. (2007) Recombination and gene duplication in the evolutionary diversification of P1 proteins in the family *Potyviridae*. *Journal of General Virology* 88, 1016–1028.

Viel, C., Ide, C., Cui, X., Wang, A., Farsi, M., *et al.* (2009) Isolation, partial sequencing, and phylogenetic analyses of *Soybean mosaic virus* (SMV) in Ontario and Quebec. *Canadian Journal of Plant Pathology* 31, 108–113.

Wang, A. (2009) *Soybean mosaic virus*: research progress and future perspectives. *Proceedings of World Soybean Research Conference VIII* (www.wsrc2009cn), Beijing, China.

Wang, X., Eggenberger, A.L., Nutter, F.W. Jr and Hill, J.H. (2001) Pathogen derived transgenic resistance to *Soybean mosaic virus* in soybean. *Molecular Breeding* 8, 119–127.

Wen, R.-H., Saghai-Maroof, M.A. and Hajimorad, M.R. (2011) Amino acid changes in P3, and not the overlapping PIPO-encoded protein, determine virulence of *Soybean mosaic virus* on functionally immune *Rsv1*-genotype soybean. *Molecular Plant Pathology* 12, 799–807.

Wen, R.-H., Khatabi, B., Ashfield, T., Saghai-Maroof, M.A. and Hajimorad, M.R. (2013) The HC-Pro and P3 cistrons of an avirulent *Soybean mosaic virus* are recognized by different resistance genes at the complex *Rsv1* locus. *Molecular Plant-Microbe Interactions* 26, 203–215.

Wrather, J.A. (2011) Diseases put a lid on soybean yields in the USA-research in needed. http://agebb. missouri.edu/commag/news/archives/v20n3/news7.htm (accessed 25 November 2014).

Wrather, J.A., Anderson, T.R., Arsyad, D.M., Gai, J., Ploper, L.D., *et al.* (1997) Soybean disease loss estimates for the top 10 soybean producing countries in 1994. *Plant Disease* 81, 107–110.

Yang, Y., Gong, J., Li, H., Li, C., Wang, D., *et al.* (2011) Identification of a novel *Soybean mosaic virus* isolate in China that contains a unique 5′ terminus sharing high sequence homology with *Bean common mosaic virus*. *Virus Research* 157, 13–18.

Yang, Y., Lin, J., Zheng, G., Zhang, M. and Zhi, H. (2014) Recombinant *Soybean mosaic virus* is prevalent in Chinese soybean fields. *Archives of Virology* 159, 1793–1796.

Zhang, C., Hajimorad, M.R., Eggenberger, A.L., Tsang, S., Whitham, S.A., *et al.* (2009) Cytoplasmic inclusion cistron of *Soybean mosaic virus* serves as a virulence determinant on *Rsv3*-genotpe soybean and a symptom determinant. *Virology* 391, 240–248.

Zhang, C., Grosic, S., Whitham, S.A. and Hill, J.H. (2012) The requirement of multiple defense genes in soybean *Rsv1*-mediated extreme resistance to *Soybean mosaic virus*. *Molecular Plant-Microbe Interactions* 25, 1307–1313.

Zhang, X., Sato, S., Ye, X., Dorrance, A.E., Morris, T.J., *et al.* (2011) Robust RNAi-based resistance to mixed infection of three viruses in soybean plants expressing separate short hairpins from a single transgene. *Phytopathology* 101, 1264–1269.

Zhi, H., Hu, Y. and Liu, T. (1996) Effects of SMV on the growth of soybean varieties. *Oil Crops of China* 18, 44–47.

Zhou, L., He, H., Liu, R., Han, Q., Shou, H., *et al.* (2014) Overexpression of *GmAKT2* potassium channel enhances resistance to *Soybean mosaic virus*. *BMC Plant Biology* 14, 154.

Zhou, Y., Hou, Q. and Yang, D. (1995) Identification of soybean viruses from ten provinces of China. *Soybean Science* 14, 246–250.

9

Yam Mosaic

Angela O. Eni*

Department of Biological Sciences, Covenant University, Ota, Nigeria

9.1 Introduction

Yam mosaic virus (YMV), genus *Potyvirus*, infects and causes mild to severe leaf symptoms both in domesticated edible yam species and their wild relatives (Thouvenel and Fauquet, 1979; Goudou-Urbino *et al.*, 1996a) in all locations where yams are grown (Africa, the Caribbean, Latin America and the South Pacific) (Goudou-Urbino *et al.*, 1996b; Hughes *et al.*, 1997; Eni *et al.*, 2008, 2010; Odedara *et al.*, 2011). Several other potyviruses described in various yam-growing countries in the 1970s and 1980s including Dioscorea green-banding mosaic virus reported in Togo (Reckhaus and Nienhaus, 1981), yam virus in Nigeria (Terry, 1976), and Dioscorea trifida virus reported in the Caribbean and in South America (Migliori and Cadilhac, 1976), are synonymous with YMV and were all found to be related to YMV both serologically and in host range (Porth *et al.*, 1987; Goudou-Urbino *et al.*, 1996a). *Japanese yam mosaic virus* (JYMV), another *Potyvirus* isolated from *D. japonica* in Japan in 1974 was reported as a strain of YMV (Okuyama and Saka, 1978); however, comparative genomic studies revealed considerable differences between JYMV and YMV, indicating that JYMV is a different *Potyvirus* species (Aleman *et al.*, 1996; Fuji and Nakamae, 1999). YMV is the most fully characterized of the several viruses that infect yam.

9.2 Disease and Symptoms

Yam mosaic disease, caused by *Yam mosaic virus*, is currently the most economically important virus disease of yam (*Dioscorea* spp.) in major yam-producing areas in the world. The disease results in varying shades of leaf discoloration and malformation symptoms including green vein banding, systemic mosaic, chlorosis, mottling, shoe stringing, curling, and stunted growth (Fig. 9.1a,b). The symptoms induced on an infected plant are usually variable even in leaves of the same plant, and several of these symptoms may simultaneously be present in a single infected plant. Symptom severity assessment for YMV infection is usually done on a scale of 1–5, where 1 indicates no symptoms, 2 indicates moderate or mild symptoms, 3 indicates severe symptoms, 4 indicates very severe symptoms, and 5 indicates distortion malformation of leaf or stem.

9.3 Distribution

Yam mosaic virus was first isolated from *D. cayenensis* in Côte d'Ivoire in 1979 and has subsequently been isolated from a number of other yam species. Although YMV is reportedly present in all yam-growing areas of the world, studies on the country-wide and regional distribution of YMV have specifically been documented in Nigeria (Odedara *et al.*, 2011; Asala *et al.*, 2012), Ghana (Olatunde, 1999; Oppong *et al.*, 2007; Eni *et al.*, 2010), Côte d'Ivoire (Séka *et al.*, 2009; Toualy *et al.*, 2014), Benin (Eni *et al.*, 2008), Togo (Eni *et al.*, 2010), Cameroon (Njukeng *et al.*, 2014) and Burkina Faso (Goudou-Urbino *et al.*, 1996b). Such extensive field studies are concentrated

*E-mail: angela.eni@covenantuniversity.edu.ng

in West Africa because over 90% of global yam production takes place in this region annually (FAO, 2013). In addition, YMV isolates from Guadeloupe, French Guiana,

Fig. 9.1. Green vein (a) banding and (b) shoe stringing symptoms on *Yam mosaic virus*-infected yam.

Puerto Rico (Bousalem *et al.*, 2000b), Brazil and New Caledonia (Goudou-Urbino *et al.*, 1996a) have also been analyzed.

Incidence of YMV in the various distribution studies across West Africa ranged from 26.2% in five major yam-producing states and the Federal Capital Territory (FCT) in the Guinea savanna zone of Nigeria (Asala *et al.*, 2012), to 52% in the north-west and south-west regions of Cameroon (Njukeng *et al.*, 2014). In 2005, YMV infection was detected in all the major yam-producing zones sampled in Ghana, Togo and Benin and was present in almost all the farms sampled (Fig. 9.2). The incidence of YMV in the zones ranged from 35.8% (Guinea savanna) to 46.5% (forest-savanna transition) in Ghana, 22.1% (Ouest Atacora) to 48.6% (Cotonnière du Centre Bénin) in Benin, and finally from 29.4% (Savane Cotière) to 34.2% (Savane Derivée Humide) in Togo (Fig. 9.3).

9.4 Economic Impact

Millions of people in tropical and subtropical regions of the world depend on yam tubers for food. Yam productivity, in particular yam tuber size, depends on the ability of

Fig. 9.2. Distribution of *Yam mosaic virus* in yam farms in (a) Ghana, (b) Togo and (c) Benin in 2005 (countries drawn to different scales).

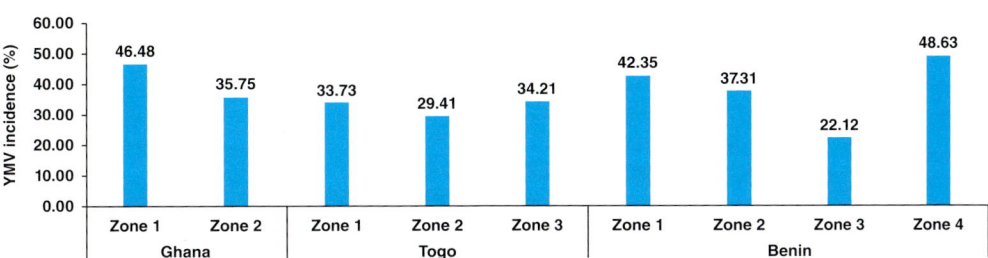

Fig. 9.3. Incidence of *Yam mosaic virus* (YMV) in major yam-producing agroecological zones in Ghana, Benin and Togo in 2005. Ghana: zone 1, Forest-Savanna Transition, zone 2, Zone Guinea savanna; Benin: zone 1, Zone Cotonnière du Nord Bénin, zone 2: Zone Vivrière du Sud Borgou, zone 3: Zone Ouest Atacora, zone 4: Zone Cotonnière du Centre Bénin; Togo: zone 1, Forêt Decidue et emi decidue de montagne, zone 2, Savane Cotière, zone 3, Savane Derivée Humide.

healthy yam leaves to efficiently trap, convert and sink the sun's light energy into chemical energy during photosynthesis. The chemical energy is then stored in tubers for food. The leaf symptoms associated with yam mosaic disease reduce the photosynthetic efficiency of infected plants, thus reducing tuber yield and quality (Amusa *et al.*, 2003). Tuber yield losses resulting from YMV infection depends on the growth stage at which a plant becomes infected, and infection at an early growth stage usually results in greater yield losses. Data on yam tuber yield loss due to single infection by YMV are few. Séka *et al.* (2014) observed between 24.8% and 48.0% tuber yield loss as a result of single YMV infection of various local and improved varieties of yam. The least average tuber yield loss was observed in 'TDr 89/02665', whereas the highest tuber yield loss was observed in a local variety, 'Bètè-bètè'. Similarly, single YMV infection resulted in 65.3% and 52.62% average tuber yield losses per plot in 'TDr 93-31' and 'TDr 95-127', respectively, when experimentally inoculated with YMV (Adeniji *et al.*, 2012). The high tuber yield losses observed in both studies highlights the detrimental effects of YMV infection on the food and income security of the millions of people who depend on yam for food, and on resource-poor farmers in the region who also depend on the crop for income. Furthermore, the presence of YMV in yam tubers hinders the international trading of yam tubers and the international movement of yam germplasm

required for research and improvement purposes (Brunt *et al.*, 1989).

9.5   Causative Virus

Yam mosaic virus belongs to the genus *Potyvirus*, family *Potyviridae* (Thouvenel & Fauquet, 1979). Like other potyviruses, YMV has a linear, monopartite, single-stranded positive-sense RNA genome of about 9608 nucleotides encased in a flexuous or filamentous particle that is 785 nm long (Aleman *et al.*, 1996). The particle is encapsidated by approximately 2000 copies of a 34 kDa coat protein. The YMV genome has a 3′ poly(A) tract and a viral genome-linked protein (VPg) that is covalently attached to the 5′ end. The genome contains a single open reading frame that is translated into a 3103 amino acid polyprotein believed to be proteolytically processed into nine or more smaller proteins like in other potyviruses (Riechmann *et al.*, 1992; Aleman *et al.*, 1996). Genetic variability among YMV isolates is reported to be high and has been attributed to recombination events and the differential accumulation of mutations (Aleman-Verdaguer *et al.*, 1997; Bousalem *et al.*, 2000b). Discrimination of YMV isolates on the basis of sequence diversity of the P1, HC, P3, NIb and CP genomic regions conformed to previous serological grouping observed for the same isolates (Goudou-Urbino *et al.*, 1996a; Aleman-Verdaguer *et al.*, 1997), although the number of isolates (6) used in these studies were few.

Relative to eight other potyviruses, YMV shows the most variability in the CP and P1 regions compared to the other regions analyzed; the N terminus of the CP was found to be the most variable. The YMV P3 protein was relatively conserved among the YMV isolates studied though the P3 region was found to be variable between *Potyvirus* species (Aleman-Verdaguer *et al.*, 1997; Bousalem *et al.*, 2000b; Ayisah and Gumedzoe, 2012). Variability among YMV isolates in these studies was not consistently correlated to origin of the isolate either geographic or host species.

In earlier studies it was suggested that YMV consists of two serotypes based on the reaction pattern of 69 YMV isolates to four monoclonal antibodies prepared against the first Côte d'Ivoire isolate (Goudou-Urbino *et al.*, 1996a). The study suggested that at least three different epitopes may be involved in the observed reaction patterns. This invariably implied that monoclonal antibodies prepared against any YMV isolate must be adequately characterized using all the diverse YMV serotypes that exist. This is to ensure accuracy in reporting and to prevent the certification of possibly infected materials that may escape detection.

Besides partial sequences of diverse genes, the complete genome sequence of YMV deduced from the Côte d'Ivoire isolate is publicly available at the GenBank (NC_004752; Aleman *et al.*, 1996).

9.6   Host Range

Following its isolation and characterization from *D. cayenensis* in Côte d'Ivoire, YMV was mechanically transmitted to several *Dioscorea* species and has subsequently been isolated from *D. liebrechtsiana, D. rotundata, D. alata, D. preussii, D. esculenta, D. trifida, D. abyssinica, D. praehensilis*, and *D. mangenotiana* (Thouvenel & Fauquet, 1979; Goudou-Urbino *et al.*, 1996a). The virus is mechanically transmissible to *Nicotiana benthamiana, N. megalosiphon* and *Chenopodium amaranticolor*. More recently, natural infection of YMV has been observed

in several weed plants collected from yam fields in Nigeria, including *Abelmoschus esculentus, Amaranthus spinosus, Physalis angulata, Phyllanthus amarus, Ludwigia abyssinica, Galinsoga quadriradiata, Justicia flara, Euphorbia heterophylla, Melanpodium divaricatum, Sacciolepis africana, Crotalaria retusa, Puerraria phaseoloides, Platostoma africanum, Conyza sumatrensis*, and *Chroniolea oduratiu* (Asala *et al.*, 2014). The diversity of families and genera of plants able to host the virus is a fact that must be taken into consideration when trying to elucidate the epidemiology of the disease and management strategies.

9.7   Transmission

Yam mosaic virus is naturally transmitted through infected planting materials, insect vectors and experimentally by mechanical inoculation (Thouvenel and Dumont, 1988). Yam is vegetatively propagated using pieces of purchased tuber or stored tuber from the previous year's harvest. Therefore, one of the most significant means of YMV transmission is through the use of infected vegetative planting material. Usually, larger tubers that are likely produced from healthy mother plants are sold due to their higher market value whereas smaller yam tubers, which may have been produced from an infected plant, are used for planting. Such uninformed selection of planting materials generally results in virus transmission, and early infection of a plant will ultimately result in greater tuber yield losses since the culpable leaf symptoms are most likely to appear at the onset of plant growth.

A wide range of aphid vectors including *Aphis gossypii, A. craccivora, A. fabae, A. citricidus, Myzus persicae* and *Rhopalosiphum maidis* transmit YMV in a non-persistent manner. However, the relative importance of each of these vectors is unknown. Some of these vectors, particularly *A. craccivora*, are useful for vector transmission studies. Experimental mechanical sap transmission of YMV from yam to yam and to other indicator plants has been demonstrated. Transmission efficiency was found to be higher by mechanical sap transmission than by experimental vector transmission (Odu *et al.*, 2006). Transmission efficiency ranged from 0–100% and from 0–66.6% for mechanical sap transmission and vector transmission, respectively, with the 29 susceptible genotypes of *D. alata* assessed.

9.8 Diagnostic Methods

Rapid and specific diagnosis of YMV is crucial for yam mosaic disease management. The occurrence of symptomless YMV infected plants (Odu *et al.*, 2004, 2006) and variability of symptoms associated with YMV infection makes symptomatology unsuitable for specific diagnosis of YMV (Bock; 1982; Rossel and Thottappilly, 1985; Brunt *et al.*, 1990). Advances in immunology and molecular biology have played significant roles in the development of rapid, specific and sensitive assays for the detection of YMV. Thus, several serological and nucleic acid-based assays have been described for the specific diagnosis of YMV, including triple antibody sandwich (TAS)-ELISA, dot-blot immunoassay, direct tissue blotting immunoassay, immunosorbent electron microscopy and immunocapture reverse transcription PCR (IC-RT-PCR) (Mumford and Seal, 1997; Bousalem *et al.*, 2000a; Njukeng *et al.*, 2002). Of these methods, TAS-ELISA and IC-RT-PCR are more commonly used for specific diagnosis of YMV, with TAS-ELISA being the assay of choice for most researchers and quarantine agencies in sub-Saharan Africa due to the expertise, equipment and reagent demand associated with the nucleic acid-based assay, IC-RT-PCR (Olatunde, 1999; Odu *et al.*, 2004; Oppong *et al.*, 2007).

Monoclonal and polyclonal antibodies useful for the detection of YMV are available in the antibody bank of the International Institute of Tropical Agriculture, Nigeria, and at the Leibniz-Institut DSMZ - Deutsche Sammlung von Mikroorganismen und Zellkulturen GmbH (DSMZ, Germany). In the TAS-ELISA, the well of microtitre plates are coated with either diluted YMV polyclonal or monoclonal antibodies for efficient trapping of the virus, particularly in samples with low virus titre. Following virus trapping from the test sample, a primary detecting monoclonal or polyclonal antibody is added to the wells and subsequently detected by a secondary anti-IgG conjugated antibody which are often commercially available.

The most commonly used primer pair for PCR amplification of YMV is the pair designed by Mumford and Seal (1997) to amplify partial coat protein CP and 3' end of the UTR regions of YMV (YMV forward primer, 5'-ATCCGGGATGTGGACAATGA-3' and YMV reverse primer, 5'-TGGTCCTCCGCCACAT-CAAA-3'). The test format is either RT-PCR (Toualy *et al.*, 2014) or IC-RT-PCR as described by Mumford and Seal (1997).

Simultaneous testing of field leaf samples collected from Ghana and Togo in 2005 revealed that as much as 79 YMV positive samples detected by IC-RT-PCR were negative for YMV when tested by TAS-ELISA (Eni *et al.*, 2010). Since the same polyclonal antibody was used for virus trapping in both assays, while monoclonal antibody and PCR primers were used for detection in TAS-ELISA and IC-RT-PCR, respectively, it was postulated that the monoclonal antibody may be the culprit and may be non-reactive to some YMV serotypes (Goudou-Urbino *et al.*, 1996a). This worrisome observation resulted in the re-evaluation of YMV detection methods. In the re-assessment, three sap dilutions of 18 previously confirmed YMV positive yam leaves were retested by TAS-ELISA and IC-RT-PCR as in the previous study (Eni *et al.*, 2010). A third ELISA format, the protein-A sandwich ELISA (PAS-ELISA), was also used. In PAS-ELISA, the ELISA plates are pre-coated with protein-A which has great affinity to the F_c region of antibodies. The use of protein-A ensures that the trapping antibody orientation is such that the antibody F_{ab} regions are adequately positioned to interact with and trap virus particles that may be present in the test samples (Edwards and Cooper, 1985; Naidu and Hughes, 2001). Furthermore, the same polyclonal antibody was used both as trapping and detecting antibody thereby bypassing the use of the monoclonal antibody.

The frequency of YMV detection was highest with IC-RT-PCR at all sap dilutions, followed by PAS-ELISA and then by TAS-ELISA (Fig. 9.4). Although clearly more sensitive than the ELISA assays, IC-RT-PCR failed to detect YMV in some infected leaves. This may be due to deterioration of the viral RNA since the test samples used in the study were calcium chloride dried samples that had been in storage for 3 years. The negative IC-RT-PCR results may also be due to interference of the PCR by polyphenols and glutinous polysaccharides contained in yam leaves (Wilson, 1997). Both at 1:10 and 1:50 dilutions, IC-RT-PCR and PAS-ELISA detected YMV in 12 (66.7%) and 11 (61.1%) of the samples, respectively. At 1:100 sap dilutions, YMV detection by IC-RT-PCR increased to 77.8%, whereas detection by PAS-ELISA reduced to 50%. YMV detection by TAS-ELISA was low at all sap dilutions with 38.9% detection at 1:10 and 16.7% detection at 1:50 and 1:100 sap dilutions (Eni *et al.*, 2012).

The greater detection sensitivity observed in PAS-ELISA may be attributed to the use of protein-A in PAS-ELISA, which ensured that a greater proportion of the antibodies were appropriately aligned for virus trapping, therefore increasing the sensitivity of the PAS-ELISA over the TAS-ELISA where the trapping antibody orientation is not purposely enhanced. Secondly, the use of a monoclonal detecting antibody may also be responsible for the lower sensitivity of TAS-ELISA since the monoclonal antibody may be YMV serotype specific. Goudou-Urbino *et al.* (1996a) defined two YMV serogroups using monoclonal antibodies response patterns. These findings suggest the need to re-characterize previously categorized yam varieties/genotype/accessions, especially where the TAS-ELISA format with a monoclonal detecting antibody was used in categorizing these materials as resistant, susceptible or tolerant.

Finally, RT-LAMP has also been successfully employed for the detection of the RNA genome molecule of a relative of YMV (*Japanese yam mosaic virus*) from leaves, propagules and roots of Japanese yam as reported elsewhere (Fukuta *et al.*, 2003).

9.9 Management Strategies

The principle of exclusion remains the best approach for the management of YMV as with all other plant viruses. Where absolute

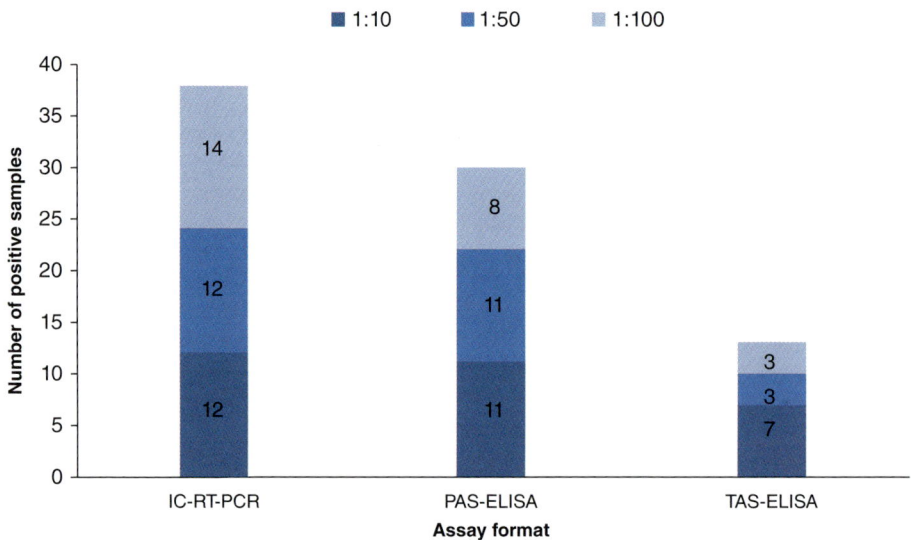

Fig. 9.4. Sensitivity of IC-RT-PCR, PAS-ELISA and TAS-ELISA for the detection of *Yam mosaic virus* in three sap dilutions of infected leaves.

exclusion is not achievable, measures to delay the onset of infection must be employed to reduce the eventual effect of virus infection on tuber yield and quality. Unfortunately, both exclusion and delay of YMV infection have been difficult to achieve in Africa mainly because YMV is both vector-transmitted and also transmitted vegetatively through infected planting materials. The use of meristem-tip culture either alone or in combination with chemotherapy and/or thermotherapy can be used for the generation of YMV-free tissue culture plantlets for further multiplication as clean planting materials. This has been used for other vegetatively propagated crops (Faccioli and Marani, 1998; Mink *et al.*, 1998). In the case of *D. opposita*, cryotherapy of shoot tips allowed Shin *et al.* (2013) to obtain YMV-free plantlets. The use of these control measures are, however, not applicable on a large scale given the facilities and funds required for these techniques.

Integrated pest management approaches have often resulted in the most sustainable virus control efforts for several crops; however, proper consideration of the specific contributions of the various interplaying factors involved in the yam mosaic disease cycle will ultimately result in a more effective YMV control strategy. Implementation of quarantine measures, vector control, roguing, farm hygiene for eradication of possible sources of inoculum, use of virus-free planting material, use of resistant or tolerant yam varieties/genotypes, and making appropriate changes in cropping practices, can be collectively employed in the management of yam mosaic disease.

Unrestricted movement of planting materials across the porous land borders in West Africa has contributed to the widespread distribution of YMV in the region (Oppong *et al.*, 2007; Eni *et al.*, 2008, 2010; Seka *et al.*, 2009; Odedara *et al.*, 2011; Asala *et al.*, 2012; Toualy *et al.*, 2014). Implementation and enforcement of existing quarantine laws, particularly at the land borders across West African countries should reduce the spread of YMV.

Vector transmission of YMV by several species of aphids is another critical factor to consider. Aphid transmission is particularly important because aphids facilitate virus spread both within a field and over long distances. The non-persistent transmission of YMV by aphids makes the use of insecticides inefficient, as most insecticides are systemic and may not act fast enough to prevent transmission of stylet-borne viruses which are acquired and transmitted within seconds or minutes. The exclusion of migrant aphids by the use of non-susceptible barrier plants such as maize may also minimize cross transmission of YMV from infected farms. This is particularly important during the early growth stages of the yam plant, since delayed infection would reduce the overall effect of the virus on tuber yield. The use of various types of mulches for the effective control of non-persistent stylet-borne plant viruses have also been reported (Cradock *et al.*, 2001); such mulches may be useful for the control of YMV.

Farmer education is imperative. Farmers must be educated on the need to consciously clean out debris from the previous year's crops from their farm lands prior to new planting. This will prevent the regrowth of left over tubers from infected plants that may serve as sources of YMV inoculum, and prevent the infection of the current plants at an early growth stage. Weeding and removal of wild species of yam that may be an alternate host for YMV (Asala *et al.*, 2014) will also contribute to preventing YMV infection; in addition, weeding will also ensure availability of necessary nutrients for the plants by preventing competition for nutrients. These farm hygiene practices will also contribute to the control of the aphid vectors of the virus (Evans, 1954).

The use of certified virus-free yam planting materials contributes to the management of YMV. The success of this approach is, however, dependent on the absence of initial sources of YMV inoculum and aphid vectors around the farm. The use of resistant/tolerant yam varieties has been considered to be the most effective and environmentally friendly means of YMV control as part of integrated pest management. The progress in breeding yam for YMV resistance is very slow due to the myriad of challenges associated with yam breeding. The long breeding cycle, sparse flowering, poor synchronization of male and female flowering phases and inefficient pollination mechanisms are some of the major challenges to yam breeding. These factors, as well as poor understanding of the yam genome and the dormancy phenomenon, all complicate yam breeding efforts irrespective of the primary breeding objective. Several advances in molecular biology and biotechnology have provided some solutions to these challenges. Studies have been undertaken with the view of identifying sources of YMV resistance, and some yam genotypes/accessions have been identified as resistant/tolerant to YMV (Mignouna *et al.*, 2001, 2002; Odu *et al.*, 2006).

The successful management of yam mosaic disease, and possibly replacement of YMV infected planting stock, will result in greater yam yield, which is a desired outcome for the teeming population of Africa, where food security issues and high food prices remain a huge concern.

9.10   Future Perspectives

The extensive spread of YMV within the West African sub-region remains a challenge to global yam production, since most of the world's yams are produced in the region. However, the huge volume of knowledge acquired from years of intensive research on YMV must be holistically appraised along with more recently acquired information and breakthroughs in research, to combat yam mosaic disease and meet the ever increasing demand for yam tubers. Although the multiplicity of yam viruses remains a challenge, currently available specific YMV molecular diagnostics must be employed and where the need arises, re-characterization of previously characterized germplasm must be undertaken to ensure accuracy. Furthermore, available molecular data, both of the virus and of known resistant genes/markers, must be employed for the development of specific and sensitive rapid YMV diagnostics while exploring the recent advances and improvements on disease diagnostics. Finally, the interplay between the various biotic and abiotic factors involved in the yam mosaic virus dynamics should be further investigated. The recent detection of YMV in several weed samples in Nigeria suggests that alternative/reservoir hosts of importance to the YMV disease dynamics, may exist that are currently unknown. Knowledge and importance of vectors must also be updated.

References

Adeniji, M.O., Shoyinka, S.A., Ikotun, T., Asiedu, R., Hughes, J.d'A., *et al.* (2012) Yield loss in guinea yam (*Dioscorea rotundata* Poir.) due to infection by *Yam mosaic virus* (YMV) genus *Potyvirus*. *Ife Journal of Science* 14, 237–244.

Aleman, M.E., Marcos, J.F., Brugidou, C., Beachy, R.N. and Fauquet, C. (1996) The complete nucleotide sequence of *Yam mosaic virus* (Ivory Coast isolate) genomic RNA. *Archives of Virology* 141, 1259–1278.

Aleman-Verdaguer, M.E., Goudou-Urbino, C., Dubern, J., Beachy, R.N. and Fauquet, C. (1997) Analysis of the sequence diversity of the P1, HC, P3, NIb and CP genomic regions of several yam mosaic Potyvirus isolates: implications for the intraspecies molecular diversity of potyviruses. *Journal of General Virology* 78, 1253–1264.

Amusa, N.A., Adegbita, A.A., Muhammed, S. and Daiyewu, R. (2003) Yam diseases and its management in Nigeria. *African Journal of Biotechnology* 2, 497–502.

Asala, S., Matthew, D., Alegbejo, M.D., Kashina, B.D., Banwo, O.O., *et al.* (2012) Distribution and incidence of viruses infecting yam (*Dioscorea* spp.) in Nigeria. *Global Journal of Bio-Science and Biotechnology* 1, 163–167.

Asala, S., Alegbejo, M.D., Kashina, B.D., Banwo, O.O. and Shinggu, C.P. (2014) Viruses in weeds in *Dioscorea* yam fields in Nigeria. *African Crop Science Journal* 22, 109–115.

Ayisah, K.D. and Gumedzoe, Y.M.D. (2012) Genetic diversity among *Yam mosaic virus* (YMV) isolates infecting yam of the complex *Dioscorea cayenensis-rotundata* in Togo. *International Journal of Biological and Chemical Sciences* 6, 1090–1101.

Bock, K.R. (1982) The identification and partial characterization of plant viruses in the tropics. *Tropical Pest Management* 28, 399–411.

Bousalem, M., Dallot, S. and Guyader, S. (2000a) The use of phylogenic data to develop molecular tools for the detection and genotyping of *Yam mosaic virus*: potential application in molecular epidemiology. *Journal of Virological Methods* 90, 25–36.

Bousalem, M., Douzery, E.J. and Fargette, D. (2000b) High genetic diversity, distant phylogenetic relationships and intraspecies recombination events among natural populations of *Yam mosaic virus*: a contribution to understanding *Potyvirus* evolution. *Journal of General Virology* 81, 243–255.

Brunt, A.A., Crabtree, K., Dallwitz, M.J., Gibbs, A.J. and Watson, L. (1990) *Viruses of tropical plants*. CAB International, Wallingford, UK.

Brunt, A.A., Jackson, G.V.H. and Frison, E.A. (1989) Technical guidelines for the safe movement of germplasm. FAO, Rome.

Cradock, K.R., Da Graça, J.V. and Laing, M.D. (2001) Control of aphid virus-vectors in *Cucurbita pepo* L. in KwaZulu-Natal, South Africa. *Subtropical Plant Science Journal* 53, 49–54.

Edwards, M.L. and Cooper, J.I. (1985) Plant virus detection using a new form of indirect ELISA. *Journal of Virological Methods* 11, 309–319.

Eni, A.O., Hughes, J.d'A. and Rey M.E.C. (2008) Survey of the incidence and distribution of five viruses infecting yam in the major yam producing zones in Benin. *Annals of Applied Biology* 153, 223–232.

Eni, A.O., Hughes, J.d'A., Asiedu, R. and Rey, M.E.C. (2010) Survey of the incidence and distribution of viruses infecting yam (*Dioscorea* spp.) in Ghana and Togo. *Annals of Applied Biology* 156, 243–251.

Eni, A.O., Hughes, J.d'A., Asiedu, R. and Rey, M.E.C. (2012) Re-evaluation of *Yam mosaic virus* (YMV) detection methods. *Academic Journal of Plant Sciences* 5, 18–22.

Evans, A.C. (1954) Groundnut rosette disease in Tanganyika. I. Field studies. *Annals of Applied Biology* 41, 189–206.

Faccioli, G. and Marani, F. (1998) Virus elimination by meristem tip culture and tip micrografting. In: Hadidi, A., Khetarpal, R.K. and Koganezawa, H. (eds) *Plant Virus Disease Control*. APS Press, St. Paul, Minnesota, pp. 346–380.

FAO (2013) Food and Agricultural Organisation of the United Nations. FAO Statistics 2008. FAO, Rome. http://faostat.fao.org/.

Fuji, S. and Nakamae, H. (1999) Complete nucleotide sequence of the genomic RNA of a *Japanese yam mosaic virus*, a new Potyvirus in Japan. *Archives of Virology* 144, 231–240.

Fukuta, S., Iida, T., Mizukami, Y., Ishida, A., Ueda, J., *et al.* (2003) Detection of *Japanese yam mosaic virus* by RT-LAMP. *Archives of Virology* 148, 1713–1720.

Goudou-Urbino, C., Givord, L., Quiot, J.B., Boeglin, M., Konate, G., *et al.* (1996a) Differentiation of yam virus isolates using symptomatology, Western blot assay and monoclonal antibodies. *Journal of Phytopathology* 144, 235–240.

Goudou-Urbino, C., Konate, G., Quiot, J.B. and Dubern, J. (1996b) Aetiology and ecology of a yam mosaic disease in Burkina Faso. *Tropical Sciences* 36, 34–40.

Hughes, J.d'A., Dongo, L.N. and Atiri, G.I. (1997) Viruses infecting cultivated yams (*Dioscorea alata* and *D. rotundata*) in Nigeria. *Phytopathology* 87, S45.

Migliori, A. and Cadilhac, B. (1976) Contribution to the study of a virus disease of yam: *Dioscorea trifida* in Guadeloupe. *Annals of Phytopathology* 8, 73–78.

Mignouna, H.D., Njukeng, P., Abang, M.M. and Asiedu, R. (2001) Inheritance of resistance to *Yam mosaic Potyvirus* in white yam (*Dioscorea rotundata*). *Theoretical and Applied Genetics* 103, 1196–2000.

Mignouna, H.D., Abang, M.M., Onasanya, A., Agindotan, B. and Asiedu, R. (2002) Identification and potential use of RAPD markers linked to *Yam mosaic virus* resistance in white yam (*Dioscorea rotundata* Poir.). *Annals of Applied Biology* 140, 163–169.

Mink, G.I., Wample, R. and Howell, W.E. (1998) Heat treatment of perennial plants to eliminate phytoplasmas, viruses and viroids while maintaining plant survival. In: Hadidi A., Khetarpal, R.K. and Koganezawa, H. (eds) *Plant Virus Disease Control*. APS Press, St. Paul, Minnesota, pp. 332–345.

Mumford, R.A. and Seal, S.E. (1997) Rapid single-tube immunocapture RT-PCR for the detection of two yam potyviruses. *Journal of Virology Methods* 69, 73–79.

Naidu, R.A. and Hughes, J.d'A. (2001) Methods for the detection of plant virus diseases. In: Hughes, J.d'A. and Odu, B.O. (eds) *Plant virology in Sub-Saharan Africa*. Proceedings of a Conference Organized by IITA, International Institute of Tropical Agriculture, Ibadan, Nigeria, pp. 233–260.

Njukeng, A.P., Atiri, G.I., Hughes, J.d'A., Agindotan, B.O., Mignouna, H.D., *et al.* (2002) A sensitive TAS-ELISA for the detection of some West African isolates of *Yam mosaic virus* in *Dioscorea* spp. *Tropical Science* 42, 65–74.

Njukeng, A.P., Azeteh, I.N. and Mbong, G.A. (2014) Survey of the incidence and distribution of two viruses infecting yam (*Dioscorea* spp.) in two agro-ecological zones of Cameroon. *International Journal of Current Microbiology and Applied Sciences* 3, 1153–1166.

Odedara, O.O., Ayo-John, E.I., Gbuyiro, M.M., Falade, F.O. and Agbebi, S.E. (2011) Serological detection of yam viruses in farmers' fields in Ogun state, Nigeria. *Archives of Phytopathology and Plant Protection* 45, 840–845.

Odu, B.O., Asiedu, R., Hughes, J.d'A. Shoyinka, S.A. and Oladiran, O.A. (2004) Identification of resistance to *Yam mosaic virus* (YMV), genus *Potyvirus* in white Guinea yam (*Dioscorea rotundata* Poir.). *Field Crops Research* 89, 97–105.

Odu, B.O., Asiedu, R., Shoyinka, S.A. and Hughes, J.d'A. (2006) Screening of water yam (*Dioscorea alata* L.) genotypes for reactions to viruses in Nigeria. *Journal of Phytopathology* 154, 716–724.

Okuyama, S. and Saka, H. (1978) *Yam mosaic virus*. Science Report of the Faculty of Agriculture, Ibaraki University 26, Ami, Ibaraka, Japan, pp. 29–34.

Olatunde, O.J. (1999) *Viruses of yam in Ghana*. MSc. thesis, University of Greenwich, London.

Oppong, A., Lamptey, J.N.L., Ofori, F.A., Anno-Nyako, F.O., Offei, S.K., *et al.* (2007) Serological detection of *Dioscorea alata Potyvirus* on white yams (*Dioscorea rotundata*) in Ghana. *Journal of Plant Sciences* 2, 630–634.

Porth, A., Lesemann, D.E. and Vetten, H.J. (1987) Characterization of *Potyvirus* isolates from West Africa yams (*Dioscorea* spp.). *Journal of Phytopathology* 120, 166–183.

Reckhaus, P. and Nienhaus, F. (1981) Etiology of a virus disease of a white yam (*Dioscorea rotundata*) in Togo. *Journal of Plant Disease and Protection* 88, 492–509.

Riechmann, J.L., Lain, S. and García, J.A. (1992) Highlights and prospects of Potyvirus molecular biology. *Journal of General Virology* 73, 1–16.

Rossel, H.W. and Thottappilly, G. (1985) *Virus Diseases of Important Food Crops in Tropical Africa*. International Institute of Tropical Agriculture, Ibadan, Nigeria.

Séka, K., Diallo, A.H. Kouassi, N.K. and Ake, S. (2009) Effect of *Yam mosaic virus* (YMV) and *Cucumber mosaic virus* (CMV) on *Dioscorea* spp. cultivars grown in Bouaké and Toumodi in Côte d'Ivoire. *International Journal of Biological and Chemical Sciences* 3, 694–703.

Séka, K., Etchian, A.O., Assiri, P.K., Toualy, M.N.Y., Diallo, H.A., *et al.* (2014) Yield loss caused by *Yam mosaic virus* (YMV) and *Cucumber mosaic virus* (CMV) on the varieties of *Dioscorea* spp. *International Journal of Agronomy and Agricultural Research* 5, 64–71.

Shin, J.H., Kang, D.K. and Sohn, J.K. (2013) Production of *Yam mosaic virus* (YMV)-free *Dioscorea opposita* plants by cryotherapy of shoot-tips. *Cryo Letters* 34, 149–157.

Terry, E.R. (1976) Incidence, symptomatology and transmission of a yam virus in Nigeria. In: Cock, J., McIntyre, R. and Graham, M. (eds) *Proceedings of the 4th Symposium of the International Society for Tropical Root Crops*. Cali, Colombia, pp. 170–173.

Thouvenel, J.C. and Dumont, R. (1988) An epidemiological approach to the study of *Yam mosaic virus*es in the Ivory Coast. *Proceedings of the International Society for Tropical Root Crops*. 30 Oct–5 Nov 1988, Bangkok, Thailand, pp. 643–649.

Thouvenel, J.C. and Fauquet, C. (1979) Yam mosaic, a Potyvirus infecting *Dioscorea cayenensis* in the Ivory Coast. *Annals of Applied Biology* 93, 279–283.

Toualy, M.N.Y., Diallo, H.A., Akinbade, S.A., Séka, K. and Lava-Kumar, P. (2014) Distribution, incidence and severity of viral diseases of yam (*Dioscorea* spp.) in Côte d'Ivoire. *African Journal of Biotechnology* 13, 465–470.

Wilson, I.G. (1997) Inhibition and facilitation of nucleic acid amplification. *Applied and Environmental Microbiology* 63, 3741–3751.

10 Sugarcane Mosaic

Laura Silva-Rosales,[1]* Ricardo I. Alcalá-Briseño[2] and Fulgencio Espejel[1]

[1]*Plant-Virus Interaction Laboratory, Department of Genetic Engineering at Cinvestav-Unidad Irapuato, Guanajuato, Mexico; [2]Department of Plant Pathology, University of Florida, Gainesville, Florida, USA*

Monocot species, in particular grasses, are cultivated over large areas worldwide for human and animal consumption and lately for biomass energy production. However, viruses like *Sugarcane mosaic virus* (SCMV), alone or in conjunction with other viruses or microorganisms, have emerged in some regions as devastating problems for their cultivation. Here we present the taxonomy, distribution, diversity and economic importance of this virus that infects maize and sugarcane as well as provide some insights into its evolution. Efforts to obtain resistance through classical breeding and transgenic approaches are also described.

10.1 Structure, Taxonomy and Diversity

SCMV, a member of the genus *Potyvirus* in the *Potyviridae* family of plant viruses, belongs to the replication group IV. As such, its genome consists of a single-stranded (+) RNA molecule. Its length of 9.6 kb is encapsidated by approximately 2,000 monomers of the coat protein (CP) forming flexuous filaments of about 750 nm in length (Riechmann *et al.*, 1992; Adams *et al.*, 2005). Instead of a canonical cap (5′m7G) at the 5′ untranslatable region, there is a covalently linked viral protein (VPg) and a poly(A) tract at the 3′ untranslatable region. The genome codes for ten proteins and a small frameshift-derived peptide (PIPO) (Adams *et al.*, 2005; Chung *et al.*, 2008). After

entering the plant cell, virions release their RNAs into the cytoplasm where they function as mRNAs and yield a single large polyprotein upon translation. This product is subsequently cleaved by three viral proteinases: Protein 1 (P1), helper component (HC-Pro) and nuclear inclusion a-endopeptidase (NIa-Pro). The rest of the viral proteins are involved in replication and movement (Urcuqui-Inchima *et al.*, 2001; Adams *et al.*, 2005).

In the early 1960s, *Maize dwarf mosaic virus* (MDMV) was identified in Ohio, and further classified into several strains: A, B, C, D, E and F (Williams and Alexander, 1965; Adams and Antoniw, 2006). At that time, plant virologists separated virus species based on host range. As such, it was determined that MDMV-B was not able to infect Johnson grass but could infect sugarcane, whereas MDMV-A was incapable of infecting sugarcane but easily infected Johnson grass. Discrimination between MDMV and SCMV did not occur until much later through the pioneering efforts of Shukla and co-workers in 1989. They examined SCMV and MDMV isolates infecting sugarcane and maize, respectively, using cross-absorbed antisera against MDMV-A and MDMV-B. According to the cross reactions against 17 strains of MDMV/SCMV, they grouped the strains into: (i) the *Johnsongrass mosaic virus* group containing two strains, MDMV-O and SCMV-JG; (ii) the *Sorghum mosaic virus* (SrMV) group with the three strains SCMV-H, SCMV-I and

*E-mail: lsilva@ira.cinvestav.mx

SCMV-M; (iii) the MDMV group with four strains (MDMV-A, MDMV-D, MDMV-E and MDMV-F); and (iv) the SCMV group containing eight strains (SCMV-A, SCMV-B, SCMV-D, SCMV-E, MDMV-B, SCMV-SC, SCMV-BC and SCMV-Sabi) (Shukla *et al.*, 1989). Once nucleic acid sequences were available, pairwise similarity values and phylogenetics were used for the delimitation of the virus species and genera. Based on the species demarcation criteria in the genus *Potyvirus* (Adams *et al.*, 2005), the SCMV group currently consists of the following species: *Cocksfoot streak virus*, *Johnsongrass mosaic virus*, *Pennisetum mosaic virus*, SrMV, MDMV, *Zea mosaic virus* and SCMV (Chen *et al.*, 2002; Gibbs and Ohshima, 2010). Of note, this is the only group of viruses within the genus *Potyvirus* that is able to infect members of the family Poaceae, and it is also one of the oldest members of the genus, which presumably emerged around 7250 years ago in Northern Africa and South-East Asia (Gibbs and Ohshima, 2010).

10.2   Genetics, Strains and Phylogenetics

There are some 866 sequences related to SCMV in the database of the National Center for Biotechnology Information. The isolates were obtained from different hosts in 26 countries. Analysis of CP nucleotide gene sequences has primarily been used to decipher correlations between host and geographic origin. In 2002, Chen's group analyzed the *Potyvirus* phylogeny by infecting diverse grasses and reported a correlation between the CP gene sequence and the host from which the isolate was derived: two well-defined clades, one from sugarcane and one from maize, were revealed (Chen *et al.*, 2002). Alegria *et al.* (2003), on the other hand, phylogenetically grouped SCMV isolates from the USA, Germany, China and Africa in two major monophyletic groups, sugarcane (SCE) and maize (MZ), plus 13 minor additional clades. In 2005, the SCMV group I was re-analyzed into three clades: (i) the group I of MZ that was split into two subgroups with some geographic correlation; (ii) the Amero-European group IA (isolates from Germany, Spain and Mexico); and (iii) the Asian group IB (isolates from China). Group II included sugarcane isolates from Australia, China and the USA, whereas Group III included novel isolates from Thailand, a unique SCMV-MDB isolate from the USA plus mixed isolates from maize and sugarcane (Gemechu *et al.*, 2005). Subsequent analysis in 2008 of SCMV isolates infecting sugarcane, both noble (*Saccharum officinarum*) and hybrid cultivars (*Saccharum* interspecific hybrids), reported on five groups and two unique strains. The SCE group consisted of isolates from hybrid cultivars of sugarcane, noble sugarcane, weeds and maize from China, the USA, Australia, Africa, Pakistan, India, Brazil and Iran. The MZ group was formed with isolates from maize and sorghum from China, the USA, Spain, Mexico and Argentina. A third group, designated SCE/MZ, was comprised exclusively of an isolate from Thailand. Additionally, a new group with isolates from noble sugarcane emerged which are geographically separated into Southern and Eastern China isolates. Finally, a fifth and tentative group was formed by Brazilian isolates. Two unique strains, SCMV-MDB (probably now SCVM-Ohio) from maize isolated in the USA and SCMV-Abaca from *Musa textilis* isolated in the Philippines (Xu *et al.*, 2008), were reported. The most recent phylogenetic study of the SCMV clade, that rendered five groups (A to E), was performed in 2011 based on the analysis of the CP gene sequence and, for the first time, complete genome sequences (Gao *et al.*, 2011). Group A (MZ) consisted of maize isolates from China, Germany, Argentina, Mexico and Spain; group B (noble SCE) with isolates from noble sugarcane; group C (SCE/MZ) with a combination of isolates from maize and sugarcane from Thailand, Vietnam and one isolate from China; group D (SCE) with isolates from maize within the USA/Mexico, and sugarcane from Iran, USA, Australia, South Africa and China; and the new group E that clustered together the most divergent sugarcane isolates of Vietnam and China. Furthermore, 14 whole genomes were analyzed, 9 from maize and 5 from sugarcane; a more consistent clustering was observed. That is, maize isolates (group I) divided into two subgroups, IA with maize isolates of China, and IB with maize isolates of Europe and America; the isolates from sugarcane (group II) had only four Chinese sequences; Group III emerged with the unique sequences of an Australian virus isolated from sugarcane.

Our unpublished data (Alcalá-Briseño *et al.*, unpublished data) of full-length CP gene sequences were analyzed by maximum likelihood and using PhyML (Guindon *et al.*, 2010) with an estimation model calculated by jModelTest2 (Darriba *et al.*, 2012) (GTR+ G+I) and showed three SCMV lineages (I to III in Fig. 10.1). Lineage III is represented by the ancestral group of sugarcane, composed of isolates of Vietnam and the province of Yunnan, China; so far, the closest isolates to Java island which is described as the origin of the mosaic disease in sugarcane (Brandes, 1919). Lineage II is the sugarcane group formed exclusively of sugarcane isolates from China, the USA, India, Argentina, Iran, Cameroon and Australia. The lineage I is formed mainly by maize isolates making four maize groups MZa–MZd plus the hybrid group H, with interspecific isolates from sugarcane, *Maranta arundinacea,* sugarcane and maize. In the same cluster, the group MZa clusters isolates from the Americas and Europe, whereas the group MZb, has only maize isolates from China. The next clade is represented by a divergent group of maize isolates (MZc) from North America, Rwanda and Brazil. Finally, the group MZd contains a mixture of sugarcane and maize isolates from South-East Asia and isolates infecting *Musa* spp. and *Setaria* spp. from Thailand, Vietnam and China. Overall, this grouping could possibly reflect the evolutionary history of SCMV from the ancestral isolates. The latter isolates that exclusively infect sugarcane are grouped in lineage III, presumably isolates that maintained the sugarcane host (lineage II) emerged from this group and later spread to other hosts (lineage I, H) which eventually included maize (MZa, MZb, MZc and MZd).

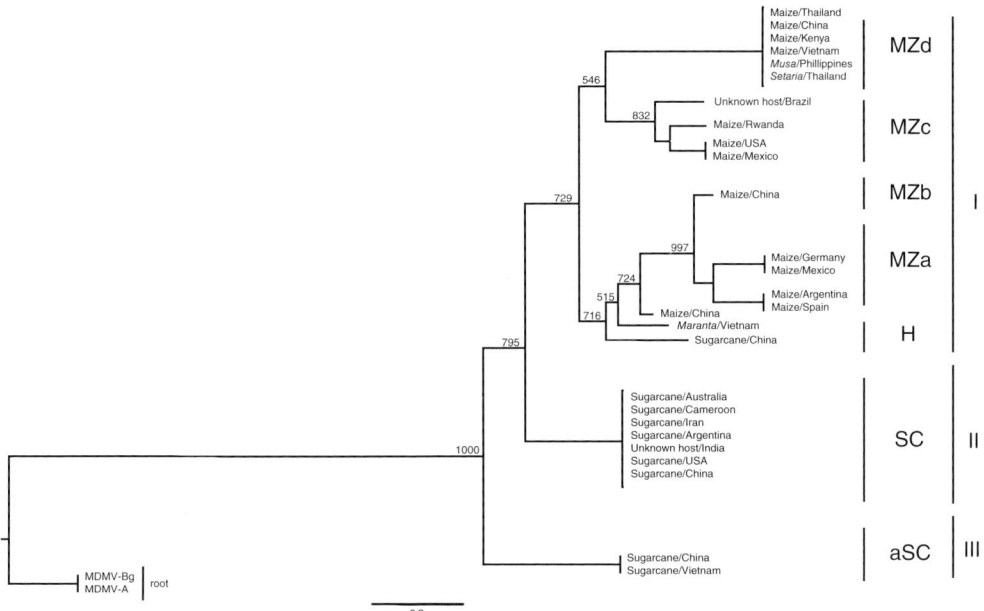

Fig. 10.1. Phylogenetic reconstruction using 102 coat protein sequences. Bootstrap support of 1000 replications was evaluated and values under 50% were collapsed. MDMV sequences were used as the outgroup. Names were condensed for visualization purposes. aSC, ancestral sugarcane; SC, sugarcane, MZ, maize.

When 22 complete genome sequences were used to reconstruct the phylogenetic history of SCMV, by means of maximum likelihood and Bayesian estimations, two large clades based on sugarcane and maize hosts were observed. Recombination and phylogenetic analysis of SCMV genomes indicated widespread events in the five continents (manuscript in preparation). The diversity of SCMV over time presented the same scenario; that is, the presence of two major groups, sugarcane and maize.

10.3   Disease Symptoms, Host Range and Transmission

Mosaic patterns, streaks and stunting have been described as common symptoms of the SCMV disease in different hosts like sorghum, maize (Fig. 10.2) and sugarcane, among other grasses. The viral symptomatology was described by Brandes in 1919 in sugarcane as 'irregular light-coloured streaks or spots on the leaves', with a great diversity of patterns and strong differences between the affected and unaffected areas. Emphasis was put on the position, coloration and streaking pattern depending on the host (sugarcane, maize, sorghum, rice, millet, crab-grass, foxtail or *Panicum*). Virus-infected maize plants were described as short plants showing 'mosaic or mottle at the base of young leaves which remain quite diffuse, but often light

areas coalesce into narrow continuous or broken streaks along the veins', whereas in younger leaves 'uniformly yellow (mosaic) or chlorotic areas may be more pronounced at the sides and tips or as streaks between the margins and midribs' (Brandes, 1919; Williams and Alexander, 1965).

Natural and experimental hosts of SCMV have been known for some time; a useful resource is available at http://ictvdb.bio-mirror.cn/ICTVdB/00.057.0.01.062.htm (Adams and Antoniw, 2006). SCMV produces mosaic and/or ringspots on natural hosts like *Saccharum* spp., *Sorghum bicolor, Panicum* spp., *Eleusine* spp., *Setaria* spp. and *Zea mays*; stunting has also been reported. Mosaics and/or necrosis are induced on *Sorghum halepense* by particular strains. Under experimental conditions, susceptibility to SCMV is found in several families, but the most susceptible species belong to the families Poaceae and Fabaceae (subfamily Papilionoideae).

As with other potyviruses (Blackman and Eastop, 2000; Gibbs *et al.*, 2008), SCMV is transmitted by a number of aphid species. Studies from the 1970s established that the isolate SCMV-MDB (possibly now SCMV-Oh) is not only transmitted by *M. persicae*, but also by *Schizaphis graminum, M. euphorbiae, Aphis fabae, A. pisum* and *Rhopalosiphum padi* (Louie and Knoke, 1975). Insects belonging to these taxa possess 'piercing–sucking' mouthparts for the efficient extraction of plant sap, which also facilitate the

(a)   (b)   (c)

Infected plant with SCMV-VER1   Healthy plant

Fig. 10.2. Symptoms of *Sugarcane mosaic virus* (SCMV) VER1 infection in maize plants. (a) Plant infected with the SCMV-VER 1 isolate showing typical symptoms of severe mosaic 15 days post inoculation. (b) Details of streaks in a leaf exhibiting severe mosaic. (c) Healthy or mock-inoculated plant.

uptake and inoculation of viruses to plants in a non-persistent manner (Backus, 1985). Details on the mode of transmission of SCMV have been obtained mainly from transmission studies with other potyviruses, notably *Potato virus Y* (Froissart *et al.*, 2002) and *Zucchini yellow mosaic virus* (ZYMV).The HC-Pro of ZYMV contains a PTK motif that binds to the DAG motif of the CP at their exposed regions. ZYMV isolates without the PTK motif are not aphid-transmitted. Another motif in this protein (KITC), binds to the aphid stylet, forming a functional bridge between this insect structure and the virion CP (Peng *et al.*, 1998). The KITC and PTK motifs of SCMV are at amino acid positions 287 and 1634; suggesting a similar mechanism for aphid transmission as with ZYMV.

10.4 Distribution

SCMV has been reported in at least 25 countries. The first account of SCMV as the causal agent of an anomaly of sugarcane was in Java in 1882. Its characterization as a viral entity in sugarcane and other grasses came later in 1919 in Puerto Rico. Brandes (1919) reported that the source of infection originated in Java and noted rapid spread to Argentina, Brazil, Peru and the USA (Brandes, 1919; Abbot, 1929; Koike and Gillaspie, 1989). The virus has also been reported in maize in sub-Saharan Africa since the 1930s (Cronjé, 2001). In 1963, an outbreak in maize was described in the USA. The disease was named maize dwarf mosaic and it was attributed to infections by a number of different virus isolates (Louie and Knoke, 1975). Evidence of SCMV infections were later presented during the 1960s in sugarcane fields located in India, Thailand and Taiwan (Abbot and Stokes, 1966; Sharma *et al.*, 2002) and in China in maize (Chen *et al.*, 2002). The virus has since been detected in maize, sorghum and sugarcane in Kenya and Australia (Teakle and Grylls, 1973; Louie and Darrah, 1980), Morocco (Fischer and Lockhart, 1974), Italy (Tosic *et al.*, 1977), Cameroon, Pakistan and Iran (Gillaspie *et al.*, 1978), Egypt, Japan and Colombia (Gillaspie and

Mock, 1979), tropical Africa (Thottappilly *et al.*, 1993), Germany (Oertel *et al.*, 1997), Mexico (Delgadillo, 1987; Espejel *et al.*, 2006) and Spain (Achon *et al.*, 2007).

SCMV co-infections occur with *Maize chlorotic mottle virus* (family *Tombusviridae*, genus *Machlomovirus*). The two diseases are mutually reinforcing and give rise to the devastating maize lethal necrosis (Uyemoto *et al.*, 1980; Wangai *et al.*, 2012). On occasion other cereal potyviruses are involved (CABI, 2014). In Mexico, SCMV is associated in co-infections with MDMV, *Maize chlorotic mottle virus* and *Maize white line mosaic virus* (family *Tombusviridae*, genus *Aureusvirus*), and in some instances a spiroplasma (unpublished data). A study in Southern China revealed a high incidence of mixed infections of SCMV and SrMV in hybrid sugarcane and in noble sugarcane (Xu *et al.*, 2008). Other studies on mixed infections of SCMV with *Maize rayado fino virus* (family *Tymoviridae*, genus *Marafivirus*) have been reported from Brazil in maize fields surrounded by sugarcane fields (Gonçalves *et al.*, 2007). In Turkey, Ilbaği *et al.* (2006) described the presence of mixed infections in maize with *Barley yellow dwarf virus*-PAV and MDMV, or SCMV and JGMV (one sample contained the four viruses). In Pakistan, Yasmin *et al.* (2011) found SCMV together with *Maize streak virus*, MDMV and *Sugarcane bacilliform virus* in a sugarcane field and weed species in close proximity. Yahaya *et al.* (2014) documented co-infections of SCMV and SrMV in sorghum. Finally, recent outbreaks of maize lethal necrosis disease have been reported due to the mixed infections with *Maize chlorotic mottle virus* and SCMV in sub-Saharan Africa (Wangai *et al.*, 2012).

10.5 Economic Impact

It is well documented that SCMV and MDMV are two of the most important viruses infecting maize. These viruses cause serious yield losses (Louie *et al.*, 1991; Fuchs and Grüntzig, 1995; Lapierre and Signoret, 2004; Ali and Yan, 2012), ranging for example in China between 20% and 80% in maize production (Chen *et al.*, 2002; Jiang and Zhou, 2002). Prevalence of the viruses and their impact, however, differ between countries and according to the type of germplasm cultivated as well as the climate and soil conditions. In Europe, there are more reports of the disease in Spain and Germany than in other countries. Since 2007, a 10% increase was noted in the occurrence of SCMV in Spain (Achon *et al.*, 2007), and in Germany more cases of detection have been documented since the mid-1990s (Oertel *et al.*, 1997). While SCMV is responsible for major losses in different parts of Europe and Asia, the virus does not seem to have the same striking effect in the Americas, except for a few sporadic outbreaks (Gilbert *et al.*, 2005). Similarly, mosaic disease has been under control with sporadic outbreaks of economic significance in Kenya and the sub-Saharan region (Cronjé, 2001; Wangai *et al.*, 2012).

10.6   Host Resistance and Plant–Virus Interactions

Cultivation of resistant varieties is the most important and effective method of limiting yield losses associated with SCMV. Typically, breeding resistant varieties involves screening of germplasm along with a dissection of the genetic basis of resistance (Xia *et al.*, 1999). The first studies on SCMV resistance were reported in the USA and Europe mainly with the inbred lines Pa405 and FAP1360A. The maize line Pa405 is highly resistant to SCMV, MDMV and *Wheat streak mosaic virus*. All studies have since confirmed that at least two dominant genes (*Scmv1* and *Scmv2*) are required for resistance to SCMV and other viruses in different maize genotypes (Table 10.1). These genes are located on chromosomes 6 and 3, respectively. The identification of the number of resistance genes contributing to SCMV resistance involved mapping of specific regions of the maize genome using different molecular approaches and markers such as amplified fragment length polymorphism (AFLP), restriction fragment length polymorphism (RFLP)

and simple sequence repeat. High-resolution mapping of chromosome regions harbouring the SCMV resistance genes in BC populations of the cross FAP1360A × F7 (Duβle *et al.*, 2003) revealed the presence of two closely linked genes in the *Scmv1* region (*Scmv1a* and *Scmv1b*). Yuan *et al.* (2003), using quantitative trait locus analysis of the cross F7 × FAP1360A, identified two QTLs in the *Scmv1* region, *Scmv1a* and *Scmv1b*. Conversely, in a South American maize genotype, two QTLs clustered on chromosome 3 (*Scm2a* and *Scm2b*), and one QTL (*Scm1*) on chromosome 6 in F_2 plants derived from a cross between L520 (resistant) × L19 (De Souza *et al.*, 2008). In Chinese 'Siyi' germplasm, two dominant complementary genes were found in F_2 and BC populations. Both genes were mapped on chromosome 3 and chromosome 6 (Wu *et al.*, 2002, 2007). Findings similar to those with the American inbred line Pa405 were obtained with a Chinese inbred line, Huangzao4. Zhang *et al.* (2003) later detected five QTLs and Liu *et al.* (2010) identified one major QTL (*Qscmv6*) on chromosome 6 in a F_9 immortal recombinant inbred line from a cross between Huangzao4 and Mo17.

McMullen and Louie (1989) confirmed one major resistance gene (*Mdm1*) for MDMV. This gene co-segregates with the nucleolus organizer region and was linked to the RFLP marker *umc85* on chromosome 6 (Simcox *et al.*, 1995). Three major resistance genes were confirmed in F_2 plants (sweetcorn × Pa405) challenged with a mixture of MDMV-A and MDMV-B (SCMV) (Mikel *et al.*, 1984). Other studies with 'Pa405' also demonstrated the same resistance regions (McMullen *et al.*, 1994; Jones *et al.*, 2011). The dominant *Wsm1* gene confers similar levels of SCMV resistance. *Mdm1*, *Scmv1* and *Wsm1* all co-segregate with the marker(s) *umc85* and/or *bnl6.29*. It is not yet known whether these three markers are very closely linked genes or whether a single gene confers resistance to all viruses.

More recently, Zambrano *et al.* (2014) determined the genetic resistance to six different viruses (including SCMV) in a maize resistant inbred line Oh1VI (Zambrano *et al.*, 2013). A recombinant inbred line population

Table 10.1. Resistant maize genotypes, location of resistance genes and markers against SCMV and other viruses.

Resistant material	Gene	Chromosome and (bin)[a]	Linkage markers	Virus	Mapping population	References
Pa405	Mdm1	6 (6.00)	umc85	MDMV	BC	McMullen and Louie, 1989
	Mdm1	6 (6.00)	umc85/bn16.9	MDMV	BC1	Simcox et al., 1995
	Wsm1	6(6.00/01)	umc85/npi235	WSMV	F2	McMullen et al., 1994
	Wsm2	3(3.04/05)	umc102/bn16.06	WSMV	F2	McMullen et al., 1994
	Wsm3	10	umc163/umc44	WSMV	F2	McMullen et al., 1994
	Wsm1	6	umc85/umc1887	SCMV, MDMV, WSMV	NIL	Jones et al., 2011
FAP1360A	Scmv2	3(3.04/05)	phi029/phi053	SCMV	BC	Melchinger et al., 1998
	Scmv1a	6(6.00/01)	phi126/phi077	SCMV	BC	Dußle et al., 2003; Yuan et al., 2003
	Scmv1b	6(6.00/01)	phi126/phi077	SCMV	BC	Dußle et al., 2003; Yuan et al., 2003
	5 QTL	1, 3, 5, 6, 10				Xia et al., 1999
	Scm1	6 (6.01)	bn16.29	SCMV	BC5	Xu et al., 1999
	Scm2	3 (3.04	umc92/umc102	SCMV	BC5	Xu et al., 1999
Siyi	Rscmv1	6(6.00/01)	umc2311/bnlg1371	SCMV	F2, BC	Wu et al., 2007
	Rscmv2	3(3.04/05)	umc1527/phi053	SCMV	F2, BC	Wu et al., 2007
L520	Scm1	6(6.01)	umc1018	SCMV	F2:3, F2	de Souza et al., 2008
	Scm2a	3(3.04)	nc030	SCMV	F2:3, F2	de Souza et al., 2008
	Scm2b	3(3.04)	umc2002	SCMV	F2:3, F2	de Souza et al., 2008
Huangzao4	5 QTL	1, 5 and 10	umc1160, umc1822, bnlg594	SCMV	F2:3	Zhang et al., 2003
		3(3.04/05)	phi029/phi053	SCMV	F2:3	Zhang et al., 2003
		6(6.00/01)	bnlg161/phi423796	SCMV	F2:3	Zhang et al., 2003
	Qscmv6	6(6.00/01)	bnlg1600/phi077	SCMV	RIL	Liu et al., 2009a
Oh1VI	Scmv1	6	PHM15S961-13/PZA03047-12	SCMV, MDMV, WSMV	RIL	Zambrano et al., 2014
	Scmv2	3	PZA00627-1/PHM13420-11	MCDV, MFSV, MMV	RIL	Zambrano et al., 2014

[a]A bin is the statistically defined interval between two fixed core markers. Rscmv1/Scmv1/Qscmv1/Qscmv6/Wsm1, and Rscmv2/Scmv2/Wsm2 allude to the same corresponding regions, but with different genes nomenclature. BC, backcross; MDMV, Maize dwarf mosaic virus; NIL, nearly isogenic line; RIL, recombinant inbred line; SCMV, Sugarcane mosaic virus; WSMV, Wheat streak mosaic virus; MCDV, Maize chlorotic dwarf virus; MFSV, Maize five streak virus; MMV, Maize mosaic virus.

was derived from an Oh1VI × Oh28. Two QTLs for resistance were identified, and the largest cluster (for all six viruses) was located on the short arm of chromosome 6 (*Scmv1*). Resistance QTLs for five viruses (including SCMV) mapped to the same, or nearby, regions of chromosome 3 (*Scmv2*), but it is not known whether these QTL clusters contain single or multiple genes. Dominant, recessive or additive gene effects could explain the segregating resistance for all and each of the six viruses. To determine whether translation factors (Diaz-Pendon *et al.*, 2004) were responsible for the phenotype associated with the QTLs, the positions of their coding sequences were mapped and identified in the B73 v.2 genome. Two eIF4e genes were found on chromosome 3 within the interval defining the virus resistance QTL, but not contained in the *Scmv2/Rscmv2* region that was identified in the fine-mapping studies. However, the possibility that eIF4E plays a role in virus resistance in maize has not been ruled out. In the maize *Scmv2/Rscmv2* region, candidate genes were located based on their possible roles in resistance and the response to various stimuli: a Rho GTPase activating protein and an auxin binding protein-1 gene (Ding *et al.*, 2012), in addition to heat shock protein 70, general vesicular transport factor p115, Rho GTPase activating protein, and syntaxin/t-SNARE genes were identified (Ingvardsen *et al.*, 2010). No genes homologous to the NBS-LRR containing class of *R* genes were found in either genomic region.

A recent study allowed for the identification of the position of homologous candidate genes on the *Scmv1* region of the B73 genome (Tao *et al.*, 2013). In this study, combined linkage and association mapping analysis suggested two predicted genes, the cycloartenol synthase (*CAS1*-like homolog) and the thioredoxin h (*Zmtrx-h*) genes as the most likely candidates for *Scmv1*, with potential roles in disease defence responses. CAS1 is involved in the biosynthesis of secondary metabolites, such as campesterol and stigmasterol (Babiychuk *et al.*, 2008; Griebel and Zeier, 2010) and contains a terpene synthase motif. Thioredoxin is a regulator of cellular redox

status, and h-type thioredoxin functions in defence responses to viruses (Sun *et al.*, 2010). However, the Zmtrx-h protein in the *Scmv1* region lacks the sequence for the conserved WC(G/P)PC redox motif, making it unlikely to have any effect on the cellular redox status. Hence, additional investigations are required to uncover the mechanism by which these candidate genes confer resistance to SCMV.

Given that both candidate genes have putative roles in basal resistance, *Scmv1* likely confers broad-spectrum resistance. According to a model proposed by Kou and Wang (2010), a single gene conferring broad-spectrum resistance may function in overlapping pathways that confer race-specific resistance, or at sites of cross-talk between different defence pathways. The existence of two closely linked resistance genes (*Scmv1a* and *Scmv2*) within the *Scmv1* region would support the idea that the *CAS-1* and *Zmtrx-h* genes lie in this region. However, this idea needs further investigation. All in all, mapping studies, even those derived from the analysis of different mapping populations, consistently identified the *Scmv1* and *Scmv2* genomic regions as being critical in conferring resistance against SCMV.

10.7   Diagnosis and Management

As with most other viruses, SCMV was described in the early 1970s after the purification of its viral particles, as well as by the host reactions it elicited on differential plants. Later in the 1980s and presently, serological detection methods such as ELISA and ELISA variants are commonly used in the diagnosis of the virus. Also, routinely used is the amplification of the CP gene from its corresponding cDNA obtained with the aid of reverse transcriptases, followed by amplification mediated by DNA polymerase. This conventional reverse transcription-PCR detection method has been further developed into one step (Xie *et al.*, 2009) or multiplex (Shuba-Reddy *et al.*, 2011) protocols. Analyses of RFLPs are used to specifically discriminate between strains of SCMV and

SrMV (Yang and Mirkov, 1997). More recently, metagenomic strategies are being applied to virus detection and characterization (Adams et al., 2014).

10.8 Management Strategies

Generally, controlling weed host plants, avoiding contaminated host plants near and within the crop and reducing aphid populations are the best ways to manage SCMV. Tissue culture and thermotherapy can be used to obtain SCMV-free plants (El-Nasr et al., 1989). Tolerant hybrids are recommended for certain regions based on resistance testing. For example, maize H614C, H611(R) C5, H612C, H5020 or EAH6302, are recommended for areas of East Africa where the virus is prevalent (Louie and Darrah, 1980). Other groups have focused on the generation of transgenic plants using mostly the CP gene (Table 10.2). For instance, in Florida virus-resistant transgenic sugarcane lines harbouring the CP of SCMV have been evaluated in the field for their resistance to the virus, SCMV incidence, their sucrose content per hectare, yield traits and economic index (Gilbert et al., 2005). In China, there are examples of transgenic maize carrying the NIb cistron that shows resistance to SCMV. High levels of resistance (83%) were obtained via RNA interference-mediated resistance derived from the generation of double-stranded RNA (Bai et al., 2008).

Another example that used the CP cistron showed varying levels of resistance to SCMV (Liu et al., 2009b). Transgenic plants harbouring MDMV-B CP were resistant not only to MDMV, but also to Maize chlorotic mosaic virus (Murry et al., 1993). Ingelbrecht et al. (1999) examined the SrMV CP cistron as a transgene source and obtained resistance against SrMV-SCI and -SCM strains, but plants were susceptible to SCMV, which shares only 75% similarity at the nucleotide level with the source of the viral transgene. This type of resistance appears to have induced posttranscriptional gene silencing (PTGS). More recently, other approaches to resistance have been based on the use of double-stranded RNA molecules derived from the CP cistron using an in planta transformation strategy (Gan et al., 2014). Whether transgenic plants could be an alternative strategy for managing SCMV in the field remains to be tested, especially in comparison with resistant lines obtained by conventional breeding assisted by molecular markers, and developed from the regions of viral resistance detected on chromosomes 3 and 6.

10.9 Concluding Remarks

SCMV, the longest-known damaging virus disease of cereal crops, is currently widely distributed in tropical and subtropical areas in single or mixed infections. Although the

Table 10.2. Summary of the transgenic approaches applied to SCMV control.

Host plant	Transgene	Molecular strategy	Resistance against	Transgenic lines	References
Maize	CP	RNA	MDMV-A/MDMV-B (SCMV)	R901084-1	Murry et al., 1993
Sugarcane	CP of SrMV-SCH	RNA	SrMV-SCI, SrMV-SCM	Group-lines11, 16, 17	Ingelbrecht et al., 1999
Sugarcane	CP	RNA	SCMV-E	CP 80-127, CP 80-1198	Gilbert et al., 2005
Maize	NIb	RNAi	SCMV	L8-10, L8-11	Bai et al., 2008
Maize18-599 (red)	CP	RNA	SCMV-MDB	CP7-3	Liu et al., 2009b
Maize	CP	RNAi	SCMV	8112	Gan et al., 2014

CP, coat protein; MDMV, *Maize dwarf mosaic virus*; RNAi, interference RNA; SCMV, *Sugarcane mosaic virus*; SrMV, *Sorghum mosaic virus*.

molecular characterization of SCMV strains is known in many countries, there are perhaps more subgroups to be described that are capable of infecting plants belonging to the family Poaceae. It is anticipated that these will become evident with increased application of the new wide-sequencing techniques. The development of resistance against this virus needs more research, particularly on the *Scmv1/Rscmv1* and *Scmv2/ Rscmv2* regions of chromosomes 6 and 3, respectively. Further research into innovative diagnostic techniques as well as new recommendations for managing the disease is needed to minimize the impacts of SCMV on maize and sugarcane production.

References

Abbot, E.V. (1929) Mosaic of sugarcane in Peru. *Science, New Series* 69, 381.

Abbot, E.V. and Stokes, I.E. (1966) A world survey of sugar cane mosaic virus strains. *Sugar y Azucar* 61, 27–29.

Achon, M.A., Serrano, L., Alonso-Dueñas, N. and Porta, C. (2007) Complete genome sequences of *Maize dwarf mosaic* and *Sugarcane mosaic virus* isolates coinfecting maize in Spain. *Archives of Virology* 15, 2073–2078.

Adams, I.P., Harju, V.A., Hodges, T., Hany, U., Skelton, A., *et al.* (2014) First report of maize lethal necrosis disease in Rwanda. *New Disease Reports* 29, 22.

Adams, M.J. and Antoniw, J.F. (2006) DPVweb: a comprehensive database of plant and fungal virus genes and genomes. *Nucleic Acids Research* 34, D382–D385.

Adams, M.J., Antoniw, J.F. and Beaudoin, F. (2005) Overview and analysis of the polyprotein cleavage sites in the family *Potyviridae*. *Molecular Plant Pathology* 6, 471–487.

Alegria, O.M., Royer, M., Bousalem, M., Chatenet, M., Peterschmitt, M., *et al.* (2003) Genetic diversity in the coat protein coding region of eighty-six *Sugarcane mosaic virus* isolates from eight countries, particularly from Cameroon and Congo. *Archives of Virology* 148, 357–372.

Ali, F. and Yan, J. (2012) Disease resistance in maize and the role of molecular breeding in defending against global threat. *Journal of Integrative Plant Biology* 54, 134–151.

Babiychuk, E., Bouvier-Nave, P., Compagnon, V., Suzuki, M., Muranaka, T., *et al.* (2008) Allelic mutant series reveal distinct functions for *Arabidopsis* cycloartenol synthase 1 in cell viability and plastid biogenesis. *Proceedings of the National Academy of Sciences USA* 105, 3163–3168.

Backus, E.A. (1985) Anatomical and sensory mechanisms of leafhopper and planthopper feeding behavior. In: Nault, L.R. and Rodriguez, R.J. (eds) *The Leafhoppers and Planthoppers*. John Wiley & Sons, Inc., New York, pp. 83–103.

Bai, Y., Yang, H., Qu, L., Zheng, J., Zhang, J., *et al.* (2008) Inverted-repeat transgenic maize plants resistant to *Sugarcane mosaic virus*. *Frontiers of Agriculture in China* 2, 125–130.

Blackman, R.L. and Eastop, V.F. (2000) *Aphids on the World's Crops: An Identification and Information Guide*. John Wiley & Sons, Inc., New York.

Brandes, E.W. (1919) The mosaic disease of sugarcane and other grasses. *Technical Bulletin United States-Department of Agriculture No.* 829.

CABI (2014) Maize lethal necrosis disease datasheet. Invasive Species Compendium. http://www.cabi.org/isc/datasheet/119663 (accessed 15 September 2014).

Chen, J., Chen, J. and Adams, M.J. (2002) Characterization of potyviruses from sugarcane and maize in China. *Archives of Virology* 147, 1237–1246.

Chung, B.Y., Miller, W.A., Atkins, J.F. and Firth, A.E. (2008) An overlapping essential gene in the *Potyviridae*. *Proceedings of the National Academy of Sciences USA* 105, 5897–5902.

Cronjé, C.P.R. (2001) Sugarcane viruses in Sub-Saharan Africa. In: Hughes, J., and Odu, B.O. (eds) *Plant Virology in Sub-Saharan Africa: Proceedings of a Conference Organized by IITA*. Ibadan, Nigeria, pp. 492–497.

Darriba, D., Taboada, G.L., Doallo, R. and Posada, D. (2012) jModelTest 2: more models, new heuristics and parallel computing. *Nature Methods* 9, 772.

de Souza, I.R.P., Schuelter, A.R., Guimarães, C.T., Schuster, I., de Oliveira, E., *et al.* (2008) Mapping QTL contributing to SCMV resistance in tropical maize. *Hereditas* 145, 167–173.

Delgadillo, F. (1987) Identificación de la enfermedad de la necrosis letal del maíz en el estado de Guanajuato. *Revista Mexicana de Fitopatología* 5, 21–26.

Diaz-Pendon, J.A., Truniger,V., Nieto, C., Garcia-Mas, J., Bendahmane, A., *et al*. (2004) Advances in understanding recessive resistance to plant viruses. *Molecular Plant Pathology* 5, 223–233.

Ding, J.Q., Li, H.M., Wang, Y.X., Zhao, R.B., Zhang, X.C., *et al*. (2012) Fine mapping of *Rscmv2*, a major gene for resistance to *Sugarcane mosaic virus* in maize. *Molecular Breeding* 30, 1593–1600.

Duβle, C.M., Quint, M., Melchinger, A.E., Xu, M.L. and Lübberstedt, T. (2003) Saturation of two chromosome regions conferring resistance to SCMV with SSR and AFLP markers by targeted BSA. *Theoretical and Applied Genetics* 106, 485–493.

El-Nasr, A.M.A., Fahmy, F.G. and Fushdi, M.H. (1989) Elimination of sugarcane mosaic disease by tissue culture and hot water treatment. *Asian Journal of Agricultural Sciences* 20, 277–292.

Espejel, F., Jeffers, D., Noa-Carrazana, J.C., Ruiz-Castro, S. and Silva-Rosales, L. (2006) Coat protein gene sequence of a Mexican isolate of *Sugarcane mosaic virus* and its infectivity in maize and sugarcane plants. *Archives of Virology* 151, 409–412.

Fischer, H.U. and Lockhart, B.E. (1974) Identity of a strain of *Sugarcane mosaic virus* occurring in Morocco. *Plant Disease Reporter* 58, 1121–1123.

Froissart, R., Michalakis, Y. and Blanc, S. (2002) Helper component transcomplementation in the vector transmission of plant viruses. *Phytopathology* 92, 576–579.

Fuchs, E. and Grüntzig, M. (1995) Influence of *Sugarcane mosaic virus* (SCMV) and *Maize dwarf mosaic virus* (MDMV) on the growth and yield of two maize varieties. *Journal of Plant Disease and Protection* 102, 44–50.

Gan, D., Ding, F., Zhuang, D., Jiang, H., Jiang, T., *et al*. (2014) Application of RNA interference methodology to investigate and develop SCMV resistance in maize. *Journal of Genetics* 93, 305–311.

Gao, B., Cui, X.W., Li, X.D., Zhang, C.Q. and Miao, H.Q. (2011) Complete genomic sequence analysis of a highly virulent isolate revealed a novel strain of *Sugarcane mosaic virus*. *Virus Genes* 43, 390–397.

Gemechu, A.L., Chiemsombat, P., Attathom, S., Reanwarakorn, K. and Lersrutaiyotin, R. (2005) Cloning and sequence analysis of coat protein gene for characterization of *Sugarcane mosaic virus* isolated from sugarcane and maize in Thailand. *Archives of Virology* 151, 167–172.

Gibbs, A. and Ohshima, K. (2010) Potyviruses and the digital revolution. *Annual Review of Phytopathology* 48, 205–223.

Gibbs, A.J., Ohshima, K., Phillips, M.J. and Gibbs, M.J. (2008) The prehistory of potyviruses: their initial radiation was during the dawn of agriculture. *PloS One* 3, e2523.

Gilbert, R.A., Gallo-Meagher, M., Comstock, J.C., Miller, J.D. and Abouzid, A. (2005) Agronomic evaluation of sugarcane lines transformed for resistance to *Sugarcane mosaic virus* strain E. *Crop Science* 45, 2060–2067.

Gillaspie, A.G. Jr and Mock, R.G. (1979) Recent survey of *Sugarcane mosaic virus* strains from Colombia, Egypt and Japan. *Sugarcane Pathologists' Newsletter* 22, 21–23.

Gillaspie, A.G. Jr, Mock, R.G. and Smith, F.F. (1978) Identification of *Sugarcane mosaic virus* and characterization of strains of the virus from Pakistan, Iran, and Cameroon. *Proceedings of the International Society of Sugar Cane Technologists* 16, 347–355.

Gonçalves, M.C., Maia, I.G., Galleti, S.R. and Fantin, G.M. (2007) Infecção mista pelo *Sugarcane mosaic virus* e *Maize rayado fino virus* provoca danos na cultura do milho no Estado de São Paulo. *Summa Phytopathologica* 33, 22–26.

Griebel, T. and Zeier, J. (2010) A role for beta-sitosterol to stigmasterol conversion in plant-pathogen interactions. *Plant Journal* 63, 254–268.

Guindon, S., Dufayard, J.F., Lefort, V., Anisimova, M., Hordijk, W. and Gascuel, O. (2010) New algorithms and methods to estimate maximum-likelihood phylogenies: assessing the performance of PhyML 3.0. *Systematic Biology* 59(3), 307–321.

Ilbaği, H., Rabenstein, F., Habekuss, A., Ordon, F. and Çitir, A. (2006)Incidence of virus diseases in maize fields in the Trakya region of Turkey. *Phytoprotection* 87, 115–122.

Ingelbrecht, I.L., Irvine, J.E. and Mirkov, E. (1999) Posttranscriptional gene silencing in transgenic sugarcane. Dissection of homology-dependent virus resistance in a monocot that has a complex polyploid genome. *Plant Physiology* 119, 1187–1197.

Ingvardsen, C.R., Xing, Y.Z., Frei, U.K. and Lübberstedt, T. (2010) Genetic and physical fine mapping of *Scmv2*, a potyvirus resistance gene in maize. *Theoretical and Applied Genetics* 120, 1621–1634.

Jiang, J.X. and Zhou, X.P. (2002) Maize dwarf mosaic disease in different regions of China is caused by *Sugarcane mosaic virus*. *Archives of Virology* 147, 2437–2443.

Jones, M.W., Boyd, E.C. and Redinbaugh, M.G. (2011) Responses of maize (*Zea mays* L.) near isogenic lines carrying *Wsm1*, *Wsm2*, and *Wsm3* to three viruses in the *Potyviridae*. *Theoretical and Applied Genetics* 123, 729–740.

Koike, H. and Gillaspie, A.G. (1989) Mosaic. In: Ricaud, C., Eagan, B.T. and Gillaspie, A.G. (eds) *Mosaic Diseases of Sugarcane-Major Diseases*. Science Publishers, Amsterdam, pp. 301–322.

Kou, Y.J. and Wang, S.P. (2010) Broad-spectrum and durability: understanding of quantitative disease resistance. *Current Opinion in Plant Biology* 13, 181–185.

Lapierre, H. and Signoret, P.A. (2004) *Viruses and Virus Diseases of Poaceae*. INRA Editions, Paris.

Liu, X.H., Tan, Z.B. and Rong, T.Z. (2009a) Molecular mapping of a major QTL conferring resistance to SCMV based on immortal RIL population in maize. *Euphytica* 167, 229–235.

Liu, X., Tan, Z., Li, W., Zhang, H. and He, D. (2009b) Cloning and transformation of SCMV CP gene and regeneration of transgenic maize plants showing resistance to SCMV strain MDB. *African Journal of Biotechnology* 8, 3747–3753.

Liu, X.H., Zheng, Z.P., Tan, Z.B., Li, Z. and He, C. (2010) Genetic analysis of two new quantitative trait loci for ear weight in maize inbred line Huangzao4. *Genetics and Molecular Research* 9, 2140–2147.

Louie, R. and Darrah, L.L. (1980) Disease resistance and yield loss to *Sugarcane mosaic virus* in East African-adapted maize. *Crop Science* 20, 638–640.

Louie, R. and Knoke, J.K. (1975) Strains of *Maize dwarf mosaic virus*. *Plant Disease Reporter* 59, 518–522.

Louie, R., Findley, W.R., Knoke, J.K. and McMullen, M.D. (1991) Genetic basis of resistance in maize to five *Maize dwarf mosaic virus* strains. *Crop Science* 31, 14–18.

McMullen, M.D. and Louie, R. (1989) The linkage of molecular markers to a gene controlling the symptom response in maize to *Maize dwarf mosaic virus*. *Molecular Plant-Microbe Interactions* 2, 309–314.

McMullen, M.D., Louie, R., Simcox, K.D. and Jones, M.W. (1994) Three genetic loci control resistance to *Wheat streak mosaic virus* in the maize inbred Pa405. *Molecular Plant-Microbe Interactions* 7, 708–712.

Melchinger, A.E., Kuntze, L., Gumber, R.K., Lübberstedt, T. and Fuchs, E. (1998) Genetic basis of resistance to *Sugarcane mosaic virus* in European maize germoplasm. *Theoretical and Applied Genetics* 96, 1151–1161.

Mikel, M.A., D'Arcy, C.J., Rhodes, A.M. and Ford, R.E. (1984) Genetics of resistance of two corn inbreds to *Maize dwarf mosaic virus* and transfer of resistance into sweet corn. *Phytopathology* 74, 467–473.

Murry, L.E., Elliott, L.G., Capitant, S.A., West, J.A., Hanson, K.K., *et al.* (1993) Transgenic corn plants expressing MDMV strain B coat protein are resistant to mixed infections of *Maize dwarf mosaic virus* and *Maize chlorotic mottle virus*. *Nature Biotechnology* 11, 1559–1564.

Oertel, U., Schubert, J. and Fuchs, E. (1997) Sequence comparison of the 3′ terminal parts of the RNA of four German isolates of sugarcane mosaic potyvirus (SCMV). *Archives of Virology* 142, 675–687.

Peng, Y.H., Kadoury, D., Gal-On, A., Wang, Y. and Raccah, B. (1998) Mutations in the HC-Pro gene of Zucchini yellow mosaic potyvirus: effects on aphid transmission and binding to purified virions. *Journal of General Virology* 79, 897–904.

Riechmann, J.L., Laín, S. and García, J.A. (1992) Highlights and prospects of potyvirus molecular biology. *Journal of General Virology* 73, 1–16.

Sharma, P.N., Sharma, O.P. and Sharma, S.K. (2002) Virus diseases of cereal crops. In: Gupta, V.K. and Paul, Y.S. (eds) *Diseases of Field Crops*. Indus Publishing Company, New Delhi, pp. 353–371.

Shuba-Reddy, C.V., Sreenivasulu, P. and Sekhar, G. (2011) Duplex-immunocapture-RT-PCR for detection and discrimination of two distinct potyviruses naturally infecting sugarcane (*Saccharum* spp. hybrid). *Indian Journal of Experimental Biology* 49, 68–73.

Shukla, D.D., Tosic, M., Jilka, R.E., Ford, R.W. and Langham, M.A.C. (1989) Taxonomy of potyviruses infecting maize, sorghum, and sugarcane in Australia and the United States as determined by reactivities of polyclonal antibodies directed towards virus-specific N-termini of coat proteins. *Phytopathology* 79, 223–229.

Simcox, K.D., McMullen, M.D. and Louie, R. (1995) Co-segregation of the *Maize dwarf mosaic virus* resistance gene, *Mdm1*, with the nucleolus organizer in maize. *Theoretical and Applied Genetics* 90, 341–346.

Sun, L., Ren, H., Liu, R., Li, B., Wu, T., *et al.* (2010) An h-type thioredoxin functions in tobacco defense responses to two species of viruses and an abiotic oxidative stress. *Molecular Plant-Microbe Interactions* 23, 1470–1485.

Tao, Y.F., Jiang, L., Liu, Q.Q., Zhang, Y., Zhang, R., *et al.* (2013) Combined linkage and association mapping reveals candidates for *Scmv1*, a major locus involved in resistance to *Sugarcane mosaic virus* (SCMV) in maize. *BMC Plant Biology* 13, 162.

Teakle, D.S. and Grylls, N.E. (1973) Four strains of *Sugarcane mosaic virus* infecting cereals and other grasses in Australia. *Australian Journal of Agricultural Research* 24, 465–477.

Thottappilly, G., Bosqe-Pérez, N.A. and Rossel, H.W. (1993) Viruses and virus diseases of maize in tropical Africa. *Plant Pathology* 42, 494–509.

Tosic, M., Benetti, M.P. and Conti, M. (1977) Studies on *Sugarcane mosaic virus* (SCMV) isolates from northern and central Italy. *Annales de Phytopathologie* 9, 387–393.

Urcuqui-Inchima, S., Haenni, A.L. and Bernardi, F. (2001) Potyvirus proteins: a wealth of functions. *Virus Research* 74, 157–175.

Uyemoto, J.K., Bockelman, D.L. and Claflin, L.E. (1980) Severe outbreak of corn lethal necrosis disease in Kansas. *Plant Disease* 64, 99–100.

Wangai, A., Redinbaugh, M.G., Kinyua, Z., Miano, D., Leley, P., *et al.* (2012) First report of *Maize chlorotic mottle virus* and maize (corn) lethal necrosis in Kenya. *Plant Disease* 96, 1582–1583.

Williams, L.E. and Alexander, L.J. (1965) Maize dwarf mosaic, a new corn disease. *Phytopathology* 55, 802–804.

Wu, J.Y., Tang, J.H., Xia, J.L. and Chen, W.C. (2002) Molecular tagging of a new resistance gene to *Maize dwarf mosaic virus* using microsatellite markers. *Acta Botanica Sinica* 44, 177–180.

Wu, J.-Y., Ding, J.-Q., Du, Y.-X., Xu, Y.-B. and Zhang, X.-C. (2007) Genetic analysis and molecular mapping of two dominant complementary genes determining resistance to *Sugarcane mosaic virus* in maize. *Euphytica* 156, 355–364.

Xia, X., Melchinger, A.E., Kuntze, L. and Lübberstedt, T. (1999) Quantitative trait loci mapping of resistance to *Sugarcane mosaic virus* in maize. *Phytopathology* 89, 660–667.

Xie, Y., Wang, M., Xiu, D., Li, R. and Zhou, G. (2009) Simultaneous detection and identification of four sugarcane viruses by one-step RT-PCR. *Journal of Virological Methods* 162, 64–69.

Xu, D.L., Park, J.W., Mirkov, T.E. and Zhou, G.H. (2008) Viruses causing mosaic disease in sugarcane and their genetic diversity in Southern China. *Archives of Virology* 153, 1031–1039.

Xu, M.L., Melchinger, A.E., Xia, X.C. and Lübberstedt, T. (1999) High-resolution mapping of loci conferring resistance to sugarcane mosaic virus in maize using RFLP, SSR and AFLP markers. *Molecular and General Genetics* 261, 574–581.

Yahaya, A., Dangora, D.B., Khan, A.U. and Zangoma, M.A. (2014) Detection of sugarcane mosaic disease in crops and weeds associated with sugarcane fields in Makarfi and Sabon Gari local government areas of Kaduna state, Nigeria. *International Journal of Current Science* 11, E99–E104.

Yang, Z.N. and Mirkov, T.E. (1997) Sugarcane mosaic and Sorghum mosaic virus strains and development of RT-PCR-based RFLPs for strain discrimination. *Phytopathology* 87, 932–939.

Yasmin, T., Iqbal, S., Farooq, A., Zubair, M. and Riaz, A. (2011) Prevalence, distribution and incidence of major sugarcane infecting viruses in NWFP and Punjab. *Pakistan Journal of Phytopathology* 23, 24–30.

Yuan, L., Duβle, C.M., Melchinger, A.E., Utz, H.F. and Lübberstedt, T. (2003) Clustering of QTL conferring SCMV resistance in maize. *Maydica* 48, 55–62.

Zambrano, J.L., Francis, D.M. and Redinbaugh, M.G. (2013) Identification of resistance to *Maize rayado fino virus* in maize inbred lines. *Plant Disease* 97, 1418–1423.

Zambrano, J.L., Jones, M.W., Brenner, E., Francis, D.M., Tomas, A., *et al.* (2014) Genetic analysis of resistance to six virus diseases in a multiple virus-resistant maize inbred line. *Theoretical and Applied Genetics* 127, 867–880.

Zhang, S.H., Li, X.H., Wang, Z.H., George, M.L., Jeffers, D., *et al.* (2003) QTL mapping for resistance to SCMV in Chinese maize germplasm. *Maydica* 48, 307–312.

11 Papaya Ringspot

Gustavo Fermin,[1]* Melaine Randle[2] and Paula Tennant[2,3]

[1]Instituto Jardín Botánico de Mérida, Faculty of Sciences, Universidad de Los Andes, Mérida, Venezuela; [2]Biotechnology Centre, The University of the West Indies, Mona Campus, Jamaica; [3]Department of Life Sciences, The University of the West Indies, Mona Campus, Jamaica

11.1 Introduction: Disease and Symptoms

Papaya ringspot disease caused by *Papaya ringspot virus* (PRSV) is perhaps the most serious disease of papaya (Tennant *et al.*, 2007; Tripathi *et al.*, 2008), a crop that is well adapted to intensive commercial orchards and backyard stands in tropical and subtropical regions (Purcifull *et al.*, 1984). High prevalence of the disease has been noted in the Caribbean islands, the USA (Florida, Texas and Hawaii), South America, the Philippines, Taiwan, Thailand and the southern region of China (Gonsalves, 1998) as far back as the 1930s (Fermin *et al.*, 2010). The aetiological agent exists as two serologically indistinguishable biotypes (Purcifull *et al.*, 1984; Gonsalves, 1994): type P, which is pathogenic to papaya and cucurbits, and type W, previously designated as *Watermelon mosaic virus* 1 (WMV-1), refers to the biotype that only infects cucurbits (Milne *et al.*, 1969; Purcifull and Hiebert, 1979; Yeh *et al.*, 1984).

Several early reports indicate that more than 30 species in the three plant families Caricaceae, Cucurbitaceae and Amaranthaceae can be infected experimentally by PRSV type P isolates, including commercial papaya varieties (Jensen, 1949; Conover, 1964; Wang *et al.*, 1978). Recent studies by Laney *et al.* (2012), however, demonstrate susceptibility of the ornamental legume tree, black locust (*Robinia pseudoacacia* L.), of the family Fabaceae to PRSV as well as a high incidence of virus transmission by seed. Later Noguera and colleagues (personal communication) provide evidence of infection of yet another member of the family Fabaceae, namely, *Vigna vexillata*. The host range, at least for the P biotypes, seems to be wider than initially reported and might be extended as more plants are studied. Cucurbit hosts of economic importance that are impacted by PRSV type W include squash, watermelon, cucumber and cantaloupe. Some 38 species in 11 genera of Cucurbitaceae can be infected by this biotype.

An array of symptoms is induced on infection with PRSV. Papaya seedlings infected with type P isolates, for example, generally show prominent vein clearing and downward cupping of the young leaves in about 1–2 weeks post-infection. After several weeks, the leaves become mottled and distorted (Fig. 11.1a,b), the lobes being markedly reduced in size (Fig. 11.1c; Conover, 1964). Plants of all ages are susceptible to PRSV and generally express mottle patterns on the leaves 2 to 3 weeks after inoculation, followed by severe leaf distortion and reduction similar to damage caused by mites. Some strains such as those found in Taiwan induce symptoms of wilting along with mosaic

*E-mail: fermin@ula.ve

Fig. 11.1. Typical *Papaya ringspot virus* P symptoms in papaya. (a) Leaf chlorosis and mottled lamina of an adult-infected plant. (b) Leaf showing signs of chlorosis, slight deformation and some green silencing areas in the lobes. (c) Heavily deformed leaf of papaya in the later stages of infection. (d) As the disease advances, the complete physiology of the plant is compromised to the point that the plant eventually dies—in many occasions after bearing fruits. (e) Typical ringspot blemishes over the surface of a green papaya fruit from an infected plant.

(Chang, 1979). Spots or streaks with a greasy or water-soaked appearance may also appear on stems (Jensen, 1949). Plants that become infected with the virus shortly after planting do not produce fruit because of flower abortion. Fruits that are produced, however, exhibit water-soaked concentric ring blemishes (Fig. 11.1d,e), the diagnostic symptom after which the name of the virus is derived (Jensen, 1949). Misshapen fruits may sometimes be produced (Purcifull *et al.*, 1984) and brix levels and fruit yield are noticeably lower when compared to those of healthy plants (Gonsalves, 1998). Infected plants eventually die (Fig. 11.1d). Various observations suggest that symptom expression can be temperature sensitive as well as dependent on the stage of plant development at the time of infection, plant vigour and virus strain (Gonsalves, 1994).

Symptoms caused by type P isolates in cucurbit hosts (zucchini, squash, pumpkin and bitter melon) include prominent mosaic patterns and leaf distortion, and in some cases there is the development of mild mottle patterns (Gonsalves and Ishii, 1980; Chin *et al.*, 2007a). Local lesions are induced on *Chenopodium quinoa* and *C. amaranticolor* (Purcifull *et al.*, 1984). Foliar symptoms typically associated with W biotypes are mosaic patterns, dark green blisters and distortion. The Guadeloupe strain causes stripe mosaic and distortion (Purcifull *et al.*, 1984). *Cucumis metuliferus* (horned cucumber) is a diagnostic species in which systemic mottle or mosaic symptoms are induced by both type P and type W isolates (Provvidenti and Robinson, 1977; Provvidenti and Gonsalves, 1982). *Nicotiana benthamiana* Domin (Solanaceae), a diagnostic species for many different viruses, is not susceptible to either P or W biotypes (Purcifull *et al.*, 1984).

11.2 Distribution

Due to a history of misidentification or mis-naming of the aetiological agent of papaya ringspot, it is difficult to accurately state when the virus was described for the first time in many parts of the world (Fermin *et al.*, 2010). Nonetheless, it is more or less widely accepted that the virus can be found wher-ever papayas are grown along with these three conditions: (i) availability of virus in-oculum; (ii) environmental conditions that are conducive for the manifestation of the disease; and (iii) the presence of the vector. One of the most contentious issues regard-ing the epidemiology of ringspot in papaya is the potential transmission of the virus through seeds (Bayot *et al.*, 1990). In the past, it was assumed that the virus could only be transmitted mechanically or by its vectors (aphid species belonging to various genera of the family Aphididae, see below). Recent evidence from new hosts has dem-onstrated, however, that the virus can be seed-transmitted in hosts other than *Carica papaya* (Laney *et al.*, 2012). Seed transmis-sion is epidemiologically important as in-fected seeds serve as the primary source of inoculum. Moreover, movement of germplasm as seeds fosters introductions into new areas, which will invariably affect the genetic struc-ture and dynamics of the virus population. Regardless of the mode of transmission, PRSV has been documented in virtually all tropical areas where the plant is grown for commercial or domestic purposes (CABI, 2014). In the neotropics, the virus has a ubiquitous distribution from northern Mexico to nor-thern Argentina, as well as in all the Caribbean and the southernmost part of Florida. A simi-lar situation is found in Asia, where most of the tropical area of the continent shows suit-able conditions for the development of the disease. Increasingly, there are more new reports of the virus in Africa (Taylor, 2001), including the Canary Islands (Melgarejo-Nárdiz *et al.*, 2010), as well as other potyvi-ruses closely related to PRSV. In Oceania,

too, virus outbreaks have been confirmed in all papaya-growing areas of Australia, Hawaii and other islands in the Pacific Ocean.

11.3 Economic Impact

There are no clear statistics regarding world-wide losses due to PRSV infections in pa-paya. Available data indicate that in Taiwan, for example, PRSV-P destroyed almost all commercial plantations of the southern re-gion of the island with a 60% reduction in yields within a few years after the virus was detected in 1975 (Yeh *et al.*, 1988). Yield reduc-tions of 70% have been reported in Brazil (Rezende and Costa, 1993). In the Philip-pines, a decrease of 85% in production has also been observed (Opina, 1991), whereas recent data from Mexico (Mora-Aguilera *et al.*, 1996) indicate a loss of 90% of plants within a year of establishing orchards. Sur-veys conducted in India and neighbouring countries show that losses can get up to 100% (Kenganal, 2009). Although more data are needed, these examples can be used as a rough guide as to the scope of the losses incurred by this virus disease. The figures point to losses that are costly in terms of food security and income to farm-ers and others whose livelihoods depend on agriculture. Besides losses attributed to dir-ect effects of the disease, there are also the losses associated with interventions or in-vestments by industry, governmental agen-cies and academia to manage the disease or buffer its impact.

11.4 Causative Virus

PRSV consists of 800–900 nm long filament-ous particles that contain a monopartite posi-tive sense genomic RNA. The virus genome contains about 10,326 nucleotides with a viral genome-linked protein (VPg) covalently at-tached to the 5′ end and a poly(A) tail at the 3′ end (Yeh *et al.*, 1992). Like other potyviruses, the order of the genes is 5′ non-translated region, 63 K P1, 52 K helper component

(HC)-proteinase (HC-Pro), 46 K P3, 6 K1, 72 K cylindrical inclusion (CI), 6 K2, 48 K nuclear inclusion a (NIa), 59 K nuclear inclusion b (NIb), 35 K coat protein (CP), and 3' non-coding region (Fig. 11.2). The virus encodes almost all of its proteins in the form of a single polyprotein (*circa* 340–368 kDa), which via a combination of co-translational, post-translational, autoproteolytic and transproteolytic processing is subsequently cleaved into mature proteins by three viral proteases (Yeh and Gonsalves, 1985). Based on work with PRSV and other potyviruses, the roles of the functional proteins are believed to include: CP for genome encapsidation; NIb, NIa, P1, and CI for genome replication (along possibly with P3, 6 K1 and 6 K2); NIa, P1, HC-Pro for polyprotein processing; CP and CI for cell-to-cell movement; CP, HC-Pro and VPg for long distance movement; and HC-Pro and CP for aphid transmission (Shulka *et al.*, 1994; Urcuqui-Inchima *et al.*, 2001). P3, HC-Pro and NIa-Pro have also been implicated in viral pathogenicity and the disruption of host defence responses (Vance *et al.*, 1995; Pruss *et al.*, 1997; Shi *et al.*,

1997; Chiang *et al.*, 2007; Gao *et al.*, 2012a; Sahana *et al.*, 2012). Presumably the P3N-PIPO protein, which plays a central role as a movement protein, is produced from a separate small open reading frame (Chung *et al.*, 2008). Inoculation studies with PRSV mutants derived from type P and type W isolates demonstrated that NIa and NIb are important symptom determinants in papaya (Chen *et al.*, 2001, 2008). Similar studies with type P mutants indicate that HC-Pro is a major determinant for the development of local lesions in *C. quinoa* (Lee *et al.*, 2001; Chiang *et al.*, 2007). More recently, P3N-PIPO has been shown to interact with RuBisCO, thereby potentially contributing to symptom development.

High-throughput biotechnologies have contributed towards a better understanding of many of the interactions between *Potyvirus* proteins and their interplay with host factors (Ivanov *et al.*, 2014). Based on a yeast two-hybrid system, Shen *et al.* (2010b) demonstrated self-interactions with HC-Pro, VPg, NIa-Pro and CP, as well as crossed interactions in the pairs VPg-P1, VPg-P3, VPg-CI,

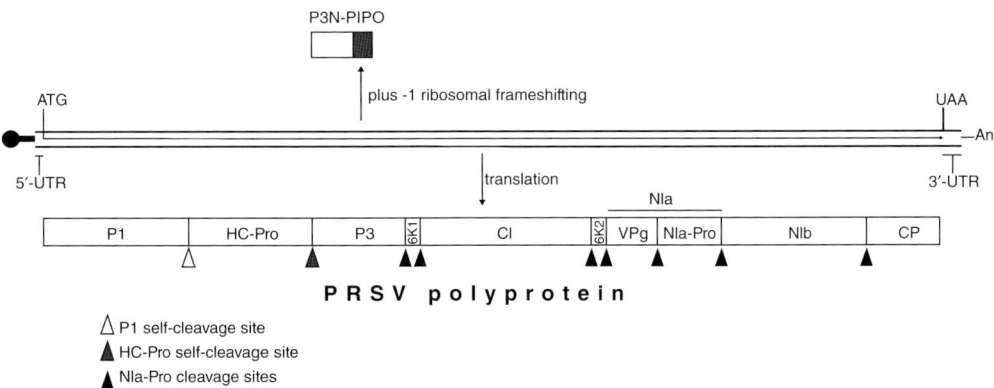

Fig. 11.2. Schematic representation of the PRSV genome based on the sequence of *Papaya ringspot virus*, GenBank NC_001785. The figure is drawn to scale and depicts the covalently linked VPg protein (black circle at the 5' end) to the PRSV RNA genome. Translation (internal arrow of the box representing the genome, from the ATG start codon to the UAA stop codon) generates a long polyprotein that is proteolytically processed by the proteases P1 (white triangle), HC-Pro (grey triangle) and NIa (black triangles). An additional protein, P3N-PIPO (PIPO for Pretty Interesting Potyviridae ORF; Chung *et al.*, 2008), was recently described; exact processing details are still unknown- but apparently involve the creation of a fused N terminal end of P3 to a -1 translation frameshift product derived from a 'hidden' *pipo* cistron. The potential for functional short overlapping ORFs has been largely overlooked in potyviruses. This discovery raises questions about the possibility of even more short overlapping coding sequences.

VPg-CP, NIaPro-CI, and NIb-P3. Using a similar approach, Gao *et al.* (2012b) provided evidence of direct interaction between NIa-Pro and proteins found in papaya, namely, EIF3G (translation initiation factor 3G), FBPA1 (fructose 1,6-bisphosphate-aldolase class 1 protein), FK506BP (fk 506-binding protein), GTPBP (GTP-binding family protein), MSRB1 (methionine sulfoxide reductase B1) and MTL (metallothionein-like protein); NIa-Pro was also found to interact with calreticulin (Shen *et al.*, 2010a). In the potyvirus *Turnip mosaic virus*, P3N-PIPO interacts with the cell protein PCaP1 to facilitate cell-to-cell movement of the CI-virion structure via plasmodesmata (Vijayapalani *et al.*, 2012). A summary of PRSV sequence/product functions and interactions is provided in Table 11.1. Besides this wealth of protein–protein interactions, PRSV modulates the

Table 11.1. Gene and gene products of *Papaya ringspot virus* (*Potyviridae*): functions and virus-host interactions

Sequence/Product	Viral function	Interactions
5′-UTR	IRES-mediated translation initiation; replication, recombination	Covalent link with VPg via a Tyr bridge. Efficiently interacts (or recruits) eIF4G. It may also interact with eIF4A and eIF4B
P1	P1 proteinase (EC 3.4-.-): serine protease of the C terminus while in the PRSV polyprotein; modulation of replication and host defence responses; keeps viral amplification below detrimental levels	Interacts with the PRSV polyprotein: cleavage of the P1/HC-Pro junction; inhibits host defence RNA silencing processes by modulation of HC-Pro activities through self-cleavage. See also VPg
HC-Pro	Helper component-proteinase (EC 3.4.22.45): the N-terminal region is required for aphid transmission and efficient genome amplification, the central region is required for long-distance movement in plants, and the C-terminal domain has cysteine endopeptidase activity on its own C terminus (peptidase family C6)	Self-interaction and homologous stabilizing activity on CP: it helps stabilizing/regulating virus replication, translation and encapsidation; interacts with the host microtubule associated protein HIP2, the calmodulin-related rgsCaM protein, the Ca^{+2} binding protein calreticulin, the ethylene-inducible transcriptional factor RAV2, and eIF4; participates in virus replication. See also VPg
P3	Protein P3: part of VRCs	Localizes only in the cytoplasm; in other potyviruses it can be found in the nucleus or associated with the ER; it associates in VRCs along with CI, NIa and NIb. Interacts with the large subunit of Rubisco. See also VPg
P3N-PIPO	Movement protein; does not derive from the PRSV polyprotein but from a – 1 ribosomal frameshifting at the P3 coding region, and fused to part of the canonical N portion of P3	Involved in cell-to-cell movement in combination with CI and CP in a vesicle-mediated viral-RNA transport system; interacts with PCaP1, and both the small and large subunits of Rubisco, ER and plasmodesmata
6K1	6 kDa protein 1: involved in virus replication. 6 K1 is also essential for the proper proteolytic separation of P3 from CI	PRSV polyprotein and/or NIa; present in VRCs
CI	Cytoplasmic inclusion protein (EC 3.6.4.-): ATP-dependent RNA helicase (DExH/D family); acts on acid anhydrides; involved in virus replication, and cellular and sub-cellular movement	Interacts with the PRSV genomic RNA; it also forms cylindrical inclusions; involved in virus replication, cell-to-cell and systemic movement possibly interacting with P3N-PIPO and CP; interacts with the inhibitor of the cell KK2 kinase, P58[IPK]

Continued

Table 11.1. Continued.

Sequence/ Product	Viral function	Interactions
6K2	6 kDa protein 2: involved in virus replication	Membrane associated peptide; induces ER-derived vesicles formation involved in replication and intracellular movement of viral RNA towards plasma membrane, chloroplast (replication) and plasmodesmata; it moves cell-to-cell. Interacts with the PRSV RNA and NIb; co-localizes with siRNA bodies
NIa	Nuclear inclusion protein A (EC 3.4.22.44). NIa-VPg is a viral genome-linked protein involved in genome stabilization, silencing suppression, and other cellular functions. VPg is uridylylated by the polymerase and is covalently attached to the 5′ end of the genomic RNA. This uridylylated form acts as a nucleotide-peptide primer for the polymerase. NIa-Pro is an endopeptidase (peptidase family 4) involved in the PRSV polyprotein processing (except for those carried out by P1 or HC-Pro)	VPg is covalently linked to the 5′ end of the PRSV genome ('capping'), binding to the anionic cell phospholipid bilayer; regulation of viral proteolytic activity and ATPase activity in the fusion NIa-VPg::NIa-Pro. Other unknown activities by cell-derived phosphorylation. Tip of virions as VPg::CI::HC-Pro, viral replication complex (fused to NIa-Pro), nuclear translocation with CP. It also interacts with itself and the PSRV RNA, P1, P3 and NIb. Other cell protein interactions: eIF4E, eIF4A, PABP, PVIP, fibrillarin, RNA helicase R8. DNase activity in the nucleus- especially when associated with cellular PVIP, and accumulates in the nucleolus
NIb	Nuclear inclusion protein B (EC 2.7.7.48): viral replicase (RNA-directed RNA polymerase); catalyzes RNA-template-directed extension of the 3′ end of an RNA strand one nucleotide at a time	Virus replication. Interacts with the host SCE1 (SUMO conjugating enzyme) in the cytoplasm and the nucleus; in combination with NIa forms crystalline nuclear inclusion bodies. Like NIa, it also interacts with PABP. Regulated by SUMOylation. See also VPg
CP	Capsid protein: virus genome encapsidation, aphid transmissibility, virus translation and replication, cell-to-cell and systemic movement, host-specific pathogenicity determinant (N terminus)	Interacts with itself, PRSV RNA, HC-Pro, aphid stylet proteins; cell interactions include HSP70 and cochaperone CPIP
3′ UTR	Recruitment of proteins and PRSV sequences (3′ portion of the CP coding region) for the initiation of the synthesis of the virus genome RNA (−) strand	Interacts mostly with NIb and PABP

References for the potyvirus functions and interactions summarized are given in the text. CP, coat protein; CPIP, coat protein interacting protein; eIF, eukaryotic (translational) initiation factor; ER, endoplasmic reticulum; HC-Pro, helper component–proteinase; IRES, internal ribosome entry site; PABP, poly(A)-binding protein; PRSV, *Papaya ringspot virus*; PVIP, potyviral VPg-interacting protein; siRNAs, short interference RNA molecules; UTR, untranslated region; VPg, viral genome-linked protein; VRC, viral replication complexes.

activity of the resident cell ubiquitin/26S proteasome (responsible for protein turnover as well as regulation of plant-pathogen interactions) by means of interactions be-tween HC-Pro and the papaya orthologus of *Arabidopsis* PAA (α1 subunit of the core 20S proteasome), as reported by Sahana *et al.* (2012).

11.5   Taxonomy, Strains and Phylogenetics

PRSV belongs to the genus *Potyvirus* of the family *Potyviridae,* which has not yet been assigned to an order. Based on host range, the virus consists of two pathotypes that are capable of infecting members of the Caricaceae and cucurbits (type P), or only cucurbits (type W). While the type W (WMV-1) is still regarded as a synonym of PRSV-W, some authorities consider WMV-1 as a separate entity (CABI, 2014). Both types induce local lesions in some members of the family Amaranthaceae presumably because of a single amino acid change in the NIa-Pro protein (Chen *et al.*, 2008). On the other hand, it is now clear that papaya mosaic disease, which has been attributed to PRSV in the past, is due to infection with the less prevalent *Papaya mosaic virus* (*Potexvirus*, family *Alphaflexiviridae*, order Tymovirales).

PRSV has an apparent Asian origin. This was initially proposed by Bateson and coworkers in 2002 (Bateson *et al.*, 2002) and later corroborated by others (Gibbs *et al.*, 2008; Olarte-Castillo *et al.*, 2011). However, the emergence of the pathotypes P and W is attributed to divergence in local populations. Gibbs *et al.* (2008) showed that the outbreak of PRSV reported in 1993 originated from a population of PRSV-W that was already endemic in Australia. Olarte-Castillo *et al.* (2011), on the other hand, suggest that the P type derived multiple times from W types and that the ancestral origin of the virus could be either of the two, P or W. According to Mangrauthia *et al.* (2008, 2009a), a P pathotype from Thailand is the major parent of most Asian W pathotypes, whereas a P type from India is the major parent of the W pathotypes in the Americas. The hypothesis that the type W is an ancestral virus of PRSV is compelling because cucurbit hosts are affected by potyviruses other than those that affect papaya, the sequence variability among isolates from the Indian subcontinent is the highest, and although emergence of the virus almost coincided with the introduction of papaya in India, the virus most probably existed before its encounter with

new hosts, where its fitness is probably higher. Potyviruses radiated concomitantly with the emergence and advance in agriculture some 6600 years ago (Gibbs *et al.*, 2008), whereas PRSV seems to have emerged as a separate entity about 400 years ago (Olarte-Castillo *et al.*, 2011). It was eventually introduced into the Americas shortly before or early in the 18th century.

PRSV belongs to a cluster of closely related potyviruses that include *Algerian watermelon mosaic virus* (Yakoubi *et al.*, 2008b), *Moroccan watermelon mosaic virus* (Fischer and Lockhart, 1974; Yakoubi *et al.*, 2008a), *Zucchini yellow fleck virus* (Desbiez *et al.*, 2007) and the still unassigned *Zucchini tigré mosaic virus* (Romay *et al.*, 2014). As previously stated by Desbiez *et al.* (2007), some viruses belong to 'clusters' that share similarities in host range, molecular sequences and immunological relatedness, giving support to the hypothesis of a continuum of isolates within this particular group of viruses. Of note, the potyviruses *Moroccan watermelon mosaic virus* (Arocha *et al.*, 2008), *Zucchini yellow mosaic virus* and *Papaya leaf distortion mosaic virus* (for an example in the Philippines, see Sta. Cruz *et al.*, 2009), as well as other unidentified putative members of the group, are also able to infect papaya (Tennant *et al.*, 2007).

To date the complete genome sequence of PRSV has been reported for isolates from Hawaii (type P; Yeh *et al.*, 1992), India (type P, Parameswari *et al.*, 2007; types P and W Mangrauthia *et al.*, 2009a), Brazil (type W; Inoue-Nagata *et al.*, 2007), China (type P; Lu *et al.*, 2008; Zhang *et al.*, 2014), Mexico (type P; Noa-Carrazana *et al.* 2007), South Korea (type W, unpublished but available at the GenBank database), Taiwan (types P and W; unpublished but available at the GenBank database) and Thailand (type P by Charoensilp *et al.* (2003) and type W by Attasart *et al.* (2002)). The most variable product among the putative proteins encoded by the genome seems to be P1, and the most conserved, CI (Charoensilp *et al.*, 2003; Parameswari *et al.*, 2007). Values of similarity among viral proteins from isolates of the Americas and Asia differ greatly. As regards resemblance to other potyviruses, the NIb protein is most conserved (Charoensilp *et al.*, 2003).

The CP gene of PRSV has so far been the sequence that is most frequently used in testing the relatedness hypothesis of the species, and among conspecific isolates of the virus. In Bangladesh, for example, sequence variability at the amino acid level of up to 14% was recorded among P isolates from six papaya-growing areas (Akhter *et al.*, 2013). In analyses of two PRSV outbreaks in the Caribbean (Jamaica) and South America (Venezuela) that were separated by time, Chin *et al.* (2007b) found that variability at the nucleotide level among isolates of the second outbreak was higher than that of isolates from the first outbreak. The rate of change in isolates between both countries also differed. On the other hand, the CP sequence variation in isolates from India, where the virus most probably emerged, shows an 18.4% and 15% divergence at the nucleotide and amino acid levels, respectively. As expected, the Indian isolates grouped together in a well delimited clade in this analysis, separate from isolates from other parts of the world; the clustering pattern of the isolates did not correlate well with their geographical origin (Srinivasulu and Sai-Gopal, 2011). On the contrary, Noa-Carrazana *et al.* (2006) reported that the variation of PRSV CP sequences correlated well with the geographical origin of isolates used in their analysis of PRSV in Mexico (Noa-Carrazana *et al.*, 2006). Various conclusions can be drawn from the analysis of the PRSV CP sequence variability: (i) local geographical clades are almost always well supported but restricted to isolates from the areas the virus samples were collected; and (ii) the relationship between distinct geographical/local clades is less clear when globally analyzed and this is likely due to differing selection pressures. Nonetheless, it appears that the main contributor to the parental strains for the Americas is India, and for Asia, Thailand.

Besides the CP gene, other regions of PRSV (part of the coding region of NIb, coding region of CP, 3′ untranslated region and poly(A) tail) have been studied and similar conclusions drawn. For example, Abdalla and Ali (2012) in their analysis of the 3′ end sequences of PRSV-W isolates from Oklahoma found that differences at local levels fit well with the places where the variants originated and that clear, wider geographical correlations depended upon the sequences examined in the comparison. However, it was also noted that unambiguous global correlations between distribution and variation are not apparent. The short 5′ untranslated region of PRSV, the smallest of all known members of the genus *Potyvirus*, is composed of a highly conserved 5′ most end region and a highly variable stretch of nucleotides upstream of the translation initiation site. The 3′ untranslated region, on the other hand, is highly conserved (94–95%) among isolates from the Americas and from Asia (92–96%) (Parameswari *et al.*, 2007). Mutation, recombination, local and long distance movement and probably seed transmission in hosts other than *C. papaya* seemingly contribute to the dynamics of the PRSV type P population and perhaps changes (and reversion) in the host range of type W isolates (Chin *et al.*, 2007b; Fernández-Rodríguez *et al.*, 2008; Mangrauthia *et al.*, 2008, 2009a).

11.6   Transmitting Vector

Approximately 60 species of aphids (Hemiptera: *Aphididae*) transmit PRSV in a nonpersistent manner (Gonsalves, 1998). This implies quick acquisition of a few seconds to minutes, and a similarly short inoculation period. There is no latent period for PRSV transmission since aphids do not colonize papaya, and transmission occurs as a result of brief exploratory probes of transient aphid vectors. Early studies indicate that type P strains are transmitted by 21 species in 11 genera, including *Myzus persicae* (Jensen, 1949; Conover, 1964; Zettler *et al.*, 1968) and *Aphis gossypii* (Purcifull *et al.*, 1984). More recent studies with these insects and *A. craccivora* suggest that *A. gossypii* is more efficient in transmitting PRSV to more plants in the field than *M. persicae* or *A. craccivora* (Kalleshwaraswamy and Krishna Kumar, 2008). In addition, it was shown that *M. persicae* and *A. gossypii* use significantly less time for the initiation of the first probe on inoculation test plants compared to *A. craccivora*. This could possibly translate to increased and efficient transmission of the virus in the field. Type W strains are transmitted by 24 species of aphids (including *Aphis craccivora*, *Aulacorthum solani*, *M. persicae* and *Macrosiphum euphorbiae*) that belong to 15 different genera (Karl and Schmelzer, 1971). The virus is transmitted by *M. persicae* following acquisition probes of 15–45 seconds (Milne *et al.*, 1969), and by *M. persicae* and *A. fabae* following acquisition probes of 10–60 seconds and inoculation periods of 1 hour (Adlerz, 1974). Leafminer flies (*Liriomyza sativae*) were found to transmit two type W isolates from squash to squash in greenhouse trials, but transmission frequencies were low (Zitter and Tsai, 1977).

Besides aphid number and the innate efficiency of aphid populations to transmit the virus, other vector-mediated components contribute to transmission efficiency. These include the ability to retain the virus following acquisition and the ability to infect a series of susceptible host plants following a single acquisition. Well-documented molecular studies related to the mechanisms of non-persistent transmission for the *Potyvirus* group indicate that the CP and the HC-Pro are essential components for aphid transmissibility (Pirone and Thornbury, 1983; Atreya *et al.*, 1995). Changes in the CP and HC-Pro have resulted in a loss in the ability of aphids to transmit certain potyviruses by constant mechanical probes (Atreya *et al.*, 1990; Husted, 1995). Aphid transmission is facilitated by the DAG sequence (Asp-Ala-Gly) found in the CP (Atreya *et al.*, 1990, 1991), whereas attachment of virus particles to sites in the aphid's food canal is facilitated by the HC-Pro (Gray, 1996; Blanc *et al.*, 1997). The N and C termini of the potyvirus CP are trypsin sensitive and are found on the exterior of the virions. The N terminus has virus-specific sequences such as the conserved DAG motif (Urcuqui-Inchima *et al.*, 2001) in which mutations do not commonly occur and hence, the assumption that it is needed for transmission. Typically, the N terminus is cleaved by trypsin-like enzymes in the aphid saliva that releases the virions from the mouth parts of the aphids for inoculation (Harrison and Robinson, 1988). The HC-Pro may connect the DAG region of the virion to retention

sites in the aphid, or may indirectly mediate interaction between the DAG and the aphid (Pirone, 1991). Further studies showed there could be a reduction or loss in transmissibility if there were changes in the DAG motif or changes in nearby bases (Gal-On *et al.*, 1992; Atreya *et al.*, 1995). Investigations to ascertain which components in HC-Pro are vital for transmission by aphids revealed that the degree of transmission is dependent on the species of aphid (Urcuqui-Inchima *et al.*, 2001). This could be attributed to the origin of HC-Pro and the ability for the retention of the virus particles by HC-Pro on the insect's stylet (Wang *et al.*, 1998; Blanc *et al.*, 2014).

11.7   Diagnostic Methods and Management Strategies

Typically, infection with PRSV is initially diagnosed by the observation of ringspot blemishes on immature papaya fruits or mosaic symptom expression on the foliage (Tennant *et al.*, 2007). This is subsequently followed by molecular diagnostic tests to confirm the presence of the pathogen as leaf discoloration or distortion symptoms can be very similar to those induced by other pests or by nutrient deficiencies (Gonsalves *et al.*, 2010) as well as environmental stresses, for example, temperature. Temperature has a marked effect on PRSV symptom expression in papaya. Symptom expression is not as obvious at low temperatures of 15–20 °C, or higher temperatures at 40 ± 5 °C (Mangrauthia *et al.*, 2009b). The optimum temperature for PRSV symptom development in papaya appears to be 26–31 °C.

Several diagnostic methods, such as ELISA, Western blotting, dot-blot immunobinding assay using PSRSV-specific polyclonal antibodies, reverse transcription-polymerase chain reaction (RT-PCR), and real time (RT) PCR have been applied to the detection of PRSV (Ling *et al.*, 1991; Ruiz-Castro and Silva-Rosales, 1997; Chiang *et al.*, 2001; Noa-Carrazana *et al.*, 2006; Cruz *et al.*, 2009; Shen *et al.*, 2010b; Sreenivasulu and Sai-Gopal, 2010; Usharani *et al.*, 2013). However, ELISA (Tennant *et al.*, 1994; Yeh *et al.*, 1998) is used worldwide as a reliable technique for PRSV detection in papaya, and commercial ELISA detection kits are readily available. In recent years, reverse transcription loop-mediated isothermal amplification (RT-LAMP) has been developed as a novel diagnostic tool for rapid detection of various plant RNA viruses, including PRSV (Shen *et al.*, 2014). Using a number of specifically designed primers, target virus gene sequences are amplified under isothermal conditions within a short period (Notomi *et al.*, 2000). The reactions are easily monitored by real-time measurements of turbidity (Notomi *et al.*, 2000; Parida *et al.*, 2008). RT-LAMP is touted as more cost-effective than RT-PCR and real-time RT-PCR.

Although PRSV is readily diagnosed, there are limited or no effective disease control systems. Disease control is best accomplished by preventative measures through the practice of quarantine, eradication and avoidance, and in some regions disease control involves planting papaya as an annual or biannual crop rather than a semi-perennial crop. Genetic resistance has not been found in papaya, but there is at least one monogenic dominant gene that confers resistance to the virus in wild but related species, such as *Vasconcellea cauliflora* (Jacq.) A. DC., *V. stipulata* (V.M. Badillo) V.M. Badillo, *Vasconcellea goudotiana* Triana & Planch. and *Vasconcellea cundinamarcensis* V.M. Badillo. Interspecific reproductive barriers have, however, limited the introgression of the gene into papaya. Nonetheless, in 2006, reports of promising fertile hybrids between *C. papaya* and *V. quercifolia* were released (Drew *et al.*, 2006). More recently, Azad *et al.* (2012) generated PRSV-resistant papayas by means of interspecific crosses involving *C. papaya* × *V. goudotiana* and *C. papaya* × *V. cauliflora*. It is anticipated that progeny derived from these materials might be used to facilitate the introgression of PRSV-P resistance from *Vasconcellea* species to papaya varieties of economic importance through traditional crosses. More recently, however, it has been shown that *Vasconcellea* is actually distantly related to *C. papaya* (a monospecific genus); more importantly, since the sister clade of *Carica* is the one

composed by *Horovitzia* and *Jarilla* (Carvalho and Renner, 2012), plants belonging to these two genera are the ones where resistance genes must be investigated. Other sources of resistance genes might be potentially found in 'natural', if not feral, populations of dioecious papayas (Brown *et al.*, 2012). A more phylogenetically distant plant, *Cucumis metuliferus* (Cucurbitaceae), also seems to harbour genes involved in resistance against PRSV and may potentially be useful in biotechnological projects aimed at engineering PRSV-resistant papayas (Lin *et al.*, 2013).

Other control measures against PRSV have included the use of tolerant plants, cross protection and cultural practices – albeit with mixed results (for examples in Brazil, Mexico and Venezuela see Vegas *et al.*, 2000; Lima *et al.*, 2001; Rivas-Valencia *et al.*, 2003). Cross protection involves the use of attenuated strains of the virus to protect plants against economic damage caused by infections with severe strains of the same virus (Chiang *et al.*, 2007). Logistic problems and lack of complete protection have contributed to limited use, and later abandonment, of this strategy (Gonsalves *et al.*, 2007). Although better economic returns have been reported with planting tolerant papaya varieties, a steady production of papaya is not always guaranteed. Roguing has also proven to be an inefficient measure of disease control from an economic point of view. In Mexico, for example, cases have been reported in which the removal of infected plants delayed damages caused by the virus, but contributed little to delaying or prohibiting the establishment of the disease (Hernández-Castro *et al.*, 2003). Even less efficient, at least under the experimental conditions used by the authors, is the use of neem (*Azadirachta indica* A. Juss) extracts, combined with the use of barriers comprised of plantings of maize and *Hibiscus sabdariffa* (Hernández-Castro *et al.*, 2004) or *Hibiscus sabdariffa* alone (Rivas-Valencia *et al.*, 2008). More recent studies have reported on the successful use of plant barriers or intercropping in papaya orchards. In Cuba, intercropping of papaya plants with *Zea mays* L. allowed a 25% reduction in virus disease incidence and 17% in disease severity (Cabrera-Mederos *et al.*, 2013). Using a combination of different intercropping strategies involving plantain (*Musa acuminata* × *balbisiana*, AAAB), Castro and colleagues (personal communication) observed an almost disease-free plot of papayas in Venezuela.

The plot was surrounded by heavily infected plants that were subjected to applications of pesticides for the control of aphids (Fig. 11.3).

To date, the most practical measure against papaya ringspot is the use of transgenic papaya plants harbouring the CP gene of

Fig. 11.3. Comparison of intercropping and chemical insect-control management strategies in an experimental papaya orchard south of Lake Maracaibo, Venezuela (2013–2014). Using the early flowering cv. 'Maradol', the experimental plot was divided in two blocks, one received treatments for the chemical control of aphids (CCA), and the other was intercropped with plantains and plant barriers (IPB). (a) A few months old healthy papaya plants of the CCA block (far from the plot, plantains of the IPB block are clearly visible). (b) Two months later, the papayas of the CCA block (right) are still healthy, and adjacent to the barrier plantains of the IPB block (left). (c) After fruit set, the CCA papayas started to show symptoms of ringspot (PRSV infection was confirmed by ELISA and RT-PCR), the disease worsened, and by August 2014 all 400 CCA plants were heavily infected, and thus, destroyed. On the contrary: (d) the IPB block was surrounded (or delimited) by a row of *Ocimum americanum* L., a row of *Hibiscus sabdariffa* L. and two rows of the tetraploid plantain FHIA-20 (*Musa acuminata* X *balbisiana*, AAAB), followed by rows of plantain alone alternating with mixed rows of papaya/plantains (the block ended up having 240 papaya plants). (e) Papaya plants were greenhouse germinated and then transplanted to the block when the plantains were sufficiently large to provide full shade conditions. (f) Papayas in the IPB block set fruits slightly later than the ones in CCA, but as August 2014 only eight plants showed symptoms of infection (further confirmed by ELISA and RT-PCR). The photograph of a heavily PRSV-infested papaya in Fig. 11.1d is from the CCA block. Data kindly provided by Castro and colleagues (2014, unpublished results).

PRSV (Gonsalves, 1998; Tecson-Mendoza et al., 2008; Li et al., 2014). Resistance is attained by gene silencing and the resistant phenotype is highly dependent on gene dosage, plant development and a close nucleotide similarity between the resident transgene and the incoming infecting strain of PRSV (Tennant et al., 2001); that is, resistance is homologous. Most transgenic papaya plants also show limited levels of resistance to different heterologous strains of the virus (Fermin et al., 2004; Kalam-Azad et al., 2014) and, by extension, closely PRSV-related potyviruses (Bau et al., 2008). Despite the many benefits that could derive from the adoption of papaya transgenic plants (i.e. the first transgenic fruit in consumer markets (Stokstad, 2008), they are well characterized to the molecular level (Suzuki et al., 2008), there is solid evidence of the lack of allergenic potential (Fermin et al., 2011) and limited impact on the soil microsphere (Hsieh and Pan, 2006)), much of papaya production worldwide derives from non-transgenic plants (Fermin et al., 2010). There is limited production in Hawaii (Gonsalves, 1998), and the Guandong province and Hainan Island of China (Li et al., 2014), as well as a few unauthorized transgenic plantings in other parts of the world (for examples see Ohmori et al., 2008; Nakamura et al., 2014).

11.8 Concluding Remarks

Papaya is more than a commodity in most regions where the plant is commercially cultivated; it is an important food source because of its year-round availability and nutritional properties. In rural communities of Thailand, for example, papaya is a subsistence crop that is consumed green in large quantities in salads (Sakuanrungsirikul et al., 2005). The lack of access, then, to a rich source of vitamins, or a main staple food, cannot be underestimated. PRSV was recorded in north-east Thailand in the 1970s and has subsequently severely affected most orchards in the country, resulting in a decline in cultivation and shortages of the fruit. The virus has been thoroughly characterized at the biological and molecular levels, diagnostic systems are more robust, and yet appropriate and effective management systems have not been developed in most regions to mitigate the economic and social effects of this devastating disease of papaya.

References

Abdalla, O.A and Ali, A. (2012) Genetic diversity in the 3′-terminal region of *Papaya ringspot virus* (PRSV-W) isolates from watermelon in Oklahoma. *Archives of Virology* 157, 405–412.

Adlerz, W.C. (1974) Spring aphid flights and incidence of watermelon mosaic viruses I and II in Florida. *Phytopathology* 64, 350–353.

Akhter, M.S., Basavaraj, Y.B., Akanda, A.M., Mandal, B. and Jain, R.K. (2013) Genetic diversity based on coat protein of *Papaya ringspot virus* (pathotype P) isolates from Bangladesh. *Indian Journal of Virology* 24, 70–73.

Arocha, Y., Vigheri, N., Nkoy-Florent, B., Bakwanamaha, K., Bolomphety, B., *et al.* (2008) First report of the identification of *Moroccan watermelon mosaic virus* in papaya in Democratic Republic of Congo (DRC). *Plant Pathology* 57, 387.

Atreya, C.D., Raccah, R. and Pirone, T.P. (1990) A point mutation in the coat protein abolishes aphid transmissibility of a *Potyvirus*. *Virology* 178, 16–165.

Atreya, P.L., Atreya, C.D. and Pirone, T.P. (1991) Amino acid substitutions in the coat protein result in loss of insect transmissibility of a plant virus. *Proceedings of the National Academy of Science USA* 88, 7887–7891.

Atreya, P.L., Lopez-Moya, J.J., Chu, M., Atreya, C.D. and Pirone, T.P. (1995) Mutational analysis of the coat protein N-terminal amino acids involved in *Potyvirus* transmission by aphids. *Journal of General Virology* 76, 265–270.

Attasart, P., Charoensilp, G., Kertbundit, S., Panyim, S. and Juricek, M. (2002) Nucleotide sequence of a Thai isolate of *Papaya ringspot virus* type W. *Acta Virologica* 46, 241–246.

Azad, M.A.K., Rabbani, M.G. and Amin, L. (2012) Plant regeneration and somatic embryogenesis from immature embryos derived through interspecific hybridization among different *Carica* species. *International Journal of Molecular Sciences* 13, 17065–17076.

Bateson, M.F., Lines, R.E., Revill, P., Chaleeprom, W., Ha, C.V., *et al.* (2002) On the evolution and molecular epidemiology of the Potyvirus *Papaya ringspot virus*. *Journal of General Virology* 83, 2575–2585.

Bau, H.-J., Kung, Y.-J., Raja, J.A.J., Chan, S.-J., Chen, K.-C., *et al.* (2008) Potential threat of a new pathotype of *Papaya leaf distortion mosaic virus* infecting transgenic papaya resistant to *Papaya ringspot virus*. *Phytopathology* 98, 848–856.

Bayot, R.G., Villegas, V.N., Magdalita, P.M., Jovellana, M.D., Espino, T.M., *et al.* (1990) Seed transmissibility of *Papaya ringspot virus*. *Philippines Journal of Crop Science* 15, 107–111.

Blanc, S., López-Moya, J.J., Wang, R.Y., García-Lampasona, S., Thornbury, D.W., *et al.* (1997) A specific interaction between coat protein and helper component correlates with aphid transmission of a Potyvirus. *Virology* 231, 141–147.

Blanc, S., Drucker, M. and Uzest, M. (2014) Localizing viruses in their insect vectors. *Annual Review of Phytopathology* 52, 403–425.

Brown, J.E., Bauman, J.M., Lawrie, J.F., Rocha, O.J. and Moore, R.C. (2012) The structure of morphological and genetic diversity in natural populations of *Carica papaya* (Caricaceae) in Costa Rica. *Biotropica* 44, 179–188.

CABI (2014) Papaya ringspot virus. Available at www.cabi.org/isc/datasheet/45962 (accessed 14 August 2014).

Cabrera-Mederos, D., García-Hernández, D., González, J.E. and Portal, O. (2013) Manejo de epifitias del virus de la mancha anular de la papaya utilizando barreras de *Zea mays* L. en *Carica papaya* L. *Revista de Protección Vegetal* 28, 127–131.

Carvalho, F.A. and Renner, S.S. (2012) A dated phylogeny of the papaya family (Caricaceae) reveals the crop's closest relatives and the family's biogeographic history. *Molecular Phylogenetics and Evolution* 65, 46–53.

Chang, C.A. (1979) Isolation and comparison of two isolates of *Papaya ringspot virus* in Taiwan. *Journal of Agricultural Research in China* 28, 207–216.

Charoensilp, G., Attasart, P., Juricek, M., Panyim, S. and Kertbundit, S. (2003) Sequencing and characterization of Thai *Papaya ringspot virus* isolate type P (PRSVthP). *ScienceAsia* 29, 89–94.

Chen, K.C., Wang, C.H., Liu, F.L., Su, W.C. and Yeh, S.D. (2001) Construction of *in vitro* infectious clone of a type W strain of *Papaya ringspot virus* and analysis of host determinant for papaya infection. *Plant Pathology Bulletin* 10, 215–216.

Chen, K.-C., Chiang, C.-H., Raja, J.A.J., Liu, F.-L., Tai, C.-H., *et al.* (2008) A single amino acid of NIaPro of *Papaya ringspot virus* determines host specificity for infection of papaya. *Molecular Plant-Microbe Interactions* 21, 1046–1057.

Chiang, C.-H., Wang, J.-J., Jan, F.-J., Yeh, S.-D. and Gonsalves, D. (2001) Comparative reactions of recombinant papaya ringspot viruses with chimeric coat protein (CP) genes and wild-type viruses on CP-transgenic papaya. *Journal of General Virology* 82, 2827–2836.

Chiang, C.-H., Lee, C.-Y., Wang, C.-H., Jan, F.-J., Lin, S.-S., *et al.* (2007) Genetic analysis of an attenuated *Papaya ringspot virus* strain applied for cross-protection. *European Journal of Plant Pathology* 118, 333–348.

Chin, M., Ahmad, M.H. and Tennant, P. (2007a) *Momordica charantia* is a weed host reservoir for *Papaya ringspot virus* type P in Jamaica. *Plant Disease* 91, 1518.

Chin, M., Rojas, Y., Moret, J., Fermin, G., Tennant, P., *et al.* (2007b) Varying genetic diversity of *Papaya ringspot virus* isolates from two time-separated outbreaks in Jamaica and Venezuela. *Archives of Virology* 152, 2101–2106.

Chung, B.Y.-W., Miller, W.A., Atkins, J.F. and Firth, A.E. (2008) An overlapping essential gene in the Potyviridae. *Proceedings of the National Academy of Sciences USA* 105, 5897–5902.

Conover, R.A. (1964) *Papaya ringspot virus*. *Proceedings of the Florida State Horticultural Society* 77, 440.

Cruz, F.C., Tanada, J.M., Elvira, P.R., Dolores, L.M., Magdalita, P.M., *et al.* (2009) Detection of mixed virus infection with *Papaya ringspot virus* (PRSV) in papaya (*Carica papaya* L.) grown in Luzon, Philippines. *The Philippine Journal of Crop Science* 34, 62–74.

Desbiez, C., Justafre, I. and Lecoq, H. (2007) Molecular evidence that *Zucchini yellow fleck virus* is a distinct and variable Potyvirus related to *Papaya ringspot virus* and *Moroccan watermelon mosaic virus*. *Archives of Virology* 152, 449–455.

Drew, R.A., Siar, S.V., O'Brien, C.M., Magdalita, P.M. and Sajise, A.G.C. (2006) Breeding papaya ringspot virus resistance in *Carica papaya* via hybridization with *Vasconcellea quercifolia. Australian Journal of Experimental Agriculture* 46, 413–418.

Fermin, G., Inglessis, V., Garboza, C., Rangel, S., Dagert, M., *et al.* (2004) Engineered resistance against *Papaya ringspot virus* in Venezuelan transgenic papayas. *Plant Disease* 88, 516–522.

Fermin, G.A., Castro, L.T. and Tennant, P.F. (2010) CP-Transgenic and non-transgenic approaches for the control of papaya ringspot: current situation and challenges. *Transgenic Plant Journal* 4, 1–15.

Fermin, G., Keith, R.C., Suzuki, J.Y., Ferreira, S.A., Gaskill, D.A., *et al.* (2011) Allergenicity assessment of the *Papaya ringspot virus* coat protein expressed in transgenic rainbow papaya. *Journal of Agriculture and Food Chemistry* 59, 10006–10012.

Fernández-Rodríguez, T., Rubio, L., Carballo, O. and Marys, E. (2008) Genetic variation of *Papaya ringspot virus* in Venezuela. *Archives of Virology* 153, 343–349.

Fischer, H.U. and Lockhart, B.E.L. (1974) Serious losses in cucurbits caused by watermelon mosaic virus in Morocco. *Plant Disease Reports* 58, 143–146.

Gal-On, A., Antignus, Y., Rosner, A. and Raccah, B. (1992) A *Zucchini yellow mosaic virus* coat protein gene mutation restores aphid transmissibility but has no effect on multiplication. *Journal of General Virology* 73, 2183–2187.

Gao, L., Shen, W., Yan, P., Tuo, D., Li, X., *et al.* (2012a) NIa-Pro of *Papaya ringspot virus* interacts with papaya methionine sulfoxide reductase B1. *Virology* 434, 78–87.

Gao, L., Shen, W.T., Yan, P., Tuo, D.C., Li, X.Y., *et al.* (2012b) A set of host proteins interacting with *Papaya ringspot virus* NIa-Pro protein identified in a yeast two-hybrid system. *Acta Virologica* 56, 25–30.

Gibbs, A.J., Ohshima, K., Phillips, M.J. and Gibbs, M.J. (2008) The prehistory of potyviruses: their initial radiation was during the dawn of agriculture. *PLoS One* 3, e2523.

Gonsalves, D. (1994) *Papaya ringspot virus.* In: Ploetz, R.C., Zentmyer, G.A., Nishijima, W.T., Rohrbach, K.G. and Ohr, H.D. (eds) *Compendium of Tropical Fruit Diseases.* APS Press, St. Paul, Minnesota, pp. 67–68.

Gonsalves, D. (1998) Control of *Papaya ringspot virus* in papaya: a case study. *Annual Review of Phytopathology* 36, 415–437.

Gonsalves, D. and Ishii, M. (1980) Purification and serology of *Papaya ringspot virus. Phytopathology* 70, 1028.

Gonsalves, D., Suzuki, J.Y., Tripathi, S. and Ferreira, S.A. (2007) *Papaya ringspot virus (Potyviridae).* In: Mahy, B.W.J. and van Regenmortel, M.H.V. (eds) *Encyclopedia of Virology, 5 vols.* Elsevier Ltd, Oxford, UK.

Gonsalves, D., Tripathi, S., Carr, J.B. and Suzuki, J.Y. (2010) *Papaya ringspot virus. The Plant Health Instructor.* DOI: 10.1094/PHI-I-2010-1004-01. Available at: http://www.apsnet.org/edcenter/intropp/lessons/viruses/Pages/PapayaRingspotvirus.aspx (accessed 28 July 2014).

Gray, S.M. (1996) Plant virus proteins involved in natural vector transmission. *Trends in Microbiology* 4, 259–264.

Harrison, B.D. and Robinson, D.J. (1988) Molecular variation in vector-borne plant viruses: epidemiological significance. *Philosophical Transactions of the Royal Society of Biological Sciences* 321, 447–462.

Hernández-Castro, E., Riestra-Díaz, D., Villanueva-Jiménez, J.A. and Mosqueda-Vázquez, R. (2003) Análisis epidemiológico del virus de la mancha anular del papayo bajo diferentes densidades, aplicación de extractos acuosos de semillas de nim (*Azadirachta indica* A. Juss.) y eliminación de plantas enfermas del cv. Maradol Roja. *Revista Chapingo Serie Horticultura* 9, 55–68.

Hernández-Castro, E., Villanueva-Jiménez, J.A., Mosqueda-Vázquez, R. and Mora-Aguilera, J.A. (2004) Efecto de la erradicación de plantas enfermas por el PRSV-P en un sistema de manejo integrado del papayo (*Carica papaya* L.) en Veracruz, México. *Revista Mexicana de Fitopatología* 22, 382–388.

Hsieh, Y.-T. and Pan, T.-M. (2006) Influence of planting *Papaya ringspot virus* resistant transgenic papaya on soil microbial biodiversity. *Journal of Agriculture and Food Chemistry* 54, 130–137.

Husted, K. (1995) Reason for non-aphid transmissibility in a strain of Kalanchoë mosaic Potyvirus. *Virus Genes* 11, 59–61.

Inoue-Nagata, A.K., Mello-Franco, C., Martin, D.P., Rezende, J.A., Ferreira, G.B., *et al.* (2007) Genome analysis of a severe and a mild isolate of *Papaya ringspot virus*-type W found in Brazil. *Virus Genes* 35, 119–127.

Ivanov, K.I., Eskelin, K., Lõhmus, A. and Mäkinen, K. (2014) Molecular and cellular mechanisms underlying Potyvirus infection. *Journal of General Virology* 95, 1415–1429.

Jensen, D.D. (1949) Papaya virus diseases with special reference to papaya ringspot. *Phytopathology* 39, 191–211.

Kalam-Azad, M.A., Amin, L. and Sidik, N.M. (2014) Gene technology for *Papaya ringspot virus* disease management. *The Scientific World Journal* 2014, 768-038.

Kalleshwaraswamy, C.M. and Krishna Kumar, N.K. (2008) Transmission efficiency of *Papaya ringspot virus* by three aphid species. *Phytopathology* 98, 541–546.

Karl, E. and Schmelzer, K. (1971) Investigations on transmission of watermelon viruses by aphids. *Archives of PrallSchutz* 7, 3.

Kenganal, M.Y. (2009) Investigations on mild strains and transformation of PRSV-CP gene in papaya. PhD thesis, University of Agricultural Sciences, Dharwad, India.

Laney, A.G., Avanzato, M.V. and Tzanetakis, L.E. (2012) High incidence of seed transmission of *Papaya ringspot virus* and *Watermelon mosaic virus,* two viruses newly identified in *Robinia pseudoacacia. European Journal of Plant Pathology* 134, 227–230.

Lee, C.Y., Chiang, C.H. and Yeh, S.D. (2001) Analyses of the roles of P1 and HC-Pro genes of *Papaya ringspot virus* for the attenuated symptoms on papaya plants and local lesion formation on *Chenopodium quinoa. Plant Pathology Bulletin* 10, 211–212.

Li, Y., Peng, Y., Hallerman, E.M. and Wu, K. (2014) Biosafety management and commercial use of genetically modified crops in China. *Plant Cell Reports* 33, 565–573.

Lima, R.C.A., Lima, J.A.A., Souza, M.T. Jr, Pio-Ribeiro, G. and Andrade, G.P. (2001) Etiologia e estratégias de controle de viroses do mamoeiro no Brasil. *Fitopatologia Brasileira* 26, 689–702.

Lin, Y.-T., Jan, F.-J., Lin, C.-W., Chung, C.-H., Chen, J.-C., *et al.* (2013) Differential gene expression in response to *Papaya ringspot virus* infection in *Cucumis metuliferus* using cDNA-amplified fragment length polymorphism analysis. *PLoS One* 8, e68749.

Ling, K., Namba, S., Gonsalves, C., Slightom, J.L. and Gonsalves, D. (1991) Protection against detrimental effects of Potyvirus infection in transgenic tobacco plants expressing the *Papaya ringspot virus* coat protein gene. *Biotechnology* 9, 752–758.

Lu, Y.W., Shen, W.T., Zhou, P., Tang, Q.J., Niu, Y.M., *et al.* (2008) Complete genomic sequence of a *Papaya ringspot virus* isolate from Hainan Island, China. *Archives of Virology* 153, 991–993.

Mangrauthia, S.K., Parameswari, B., Jain, R.K. and Praveen, S. (2008) Role of genetic recombination in the molecular architecture of *Papaya ringspot virus. Biochemical Genetics* 46, 835–846.

Mangrauthia, S.K., Parameswari, B., Praveen, S. and Jain, R.K. (2009a) Comparative genomics of *Papaya ringspot virus* pathotypes P and W from India. *Archives of Virology* 154, 727–730.

Mangrauthia, S., Singh-Shakya, V., Jain, R. and Praveen, S. (2009b) Ambient temperature perception in papaya for *Papaya ringspot virus* interaction. *Virus Genes* 38, 429–434.

Melgarejo-Nárdiz, P., García-Jiménez, J., Jordá-Gutiérrez, M.C., López-González, M.M., Andrés-Yebes, M.F., *et al.* (2010) *Patógenos de Plantas Descritos en España, Coordinadores,* 2nd edn. Ministerio de Medio Ambiente y Medio Rural y Marino, Madrid.

Milne, K.S., Grogan, R.G. and Kimble, K.A. (1969) Identification of viruses infecting cucurbits in California. *Phytopathology* 59, 819–828.

Mora-Aguilera, G., Nieto-Angel, D., Lee-Campbell, C., Téliz, D. and García, E. (1996) Multivariate comparison of papaya ringspot epidemics. *Phytopathology* 86, 70–78.

Nakamura, K., Kondo, K., Kobayashi, T., Noguchi, A., Ohmori, K., *et al.* (2014) Identification and detection of genetically modified papaya resistant to *Papaya ringspot virus* strains in Thailand. *Biological and Pharmaceutical Bulletin* 37, 1–5.

Noa-Carrazana, J.C., González-de-León, D., Ruiz-Castro, B.S., Pinero, D. and Silva-Rosales, L. (2006) Distribution of *Papaya ringspot virus* and *Papaya mosaic virus* in papaya plants (*Carica papaya*) in Mexico. *Plant Disease* 90, 1004–1011.

Noa-Carrazana, J.C., González-de-León, D. and Silva-Rosales, L. (2007) Molecular characterization of a severe isolate of *Papaya ringspot virus* in Mexico and its relationship with other isolates. *Virus Genes* 35, 109–117.

Notomi, T., Okayama, H., Masubuchi, H., Yonekawa, T., Watanabe, K., *et al.* (2000) Loop–mediated isothermal amplification of DNA. *Nucleic Acids Research* 28, E63.

Ohmori, K., Tsuchiya, H., Watanabe, T., Akiyama, H., Maitani, T., *et al.* (2008) A DNA extraction method using a silica-base resin type. Kit for the detection of genetically modified papaya. *Journal of the Food Hygienic Society of Japan (Shokuhin Eiseigaku Zasshi)* 49, 63–69.

Olarte-Castillo, X.A., Fermin, G., Tabima, J., Rojas, Y., Tennant, P.F., *et al.* (2011) Phylogeography and molecular epidemiology of *Papaya ringspot virus. Virus Research* 159, 132–140.

Opina, O.S. (1991) Epidemiology of papaya ringspot. In: Bajet, N.B., Fabellar, N.G., Dizon, T.O., Talens, A.C.D. and Roperos, N.I. (eds.) *Proceedings of the First National Symposium/Workshop on Ringspot and Other Diseases of Papaya in the Philippines.* Bureau of Plant Industry, Manila, Philippines, pp. 46–51.

Parameswari, B., Mangrauthia, S.K., Praveen, S. and Jain, R.K. (2007) Complete genome sequence of an isolate of *Papaya ringspot virus* from India. *Archives of Virology* 152, 843–845.

Parida, M., Sannarangaiah, S., Dash, P.K. and Rao, P.V. (2008) Loop mediated isothermal amplification (LAMP): a new generation of innovative gene amplification technique; perspectives in clinical diagnosis of infectious diseases. *Reviews in Medical Virology* 18, 407–421.

Pirone, T.P. (1991) Viral genes and gene products that determine insect transmissibility. *Seminars in Virology* 2, 81–87.

Pirone, T.P. and Thornbury, D.W. (1983) Role of virion and helper component in regulating aphid transmission of *Tobacco etch virus*. *Phytopathology* 73, 872–875.

Provvidenti, R. and Gonsalves, D. (1982) Resistance to *Papaya ringspot virus* in *Cucumis metuliferus* and its relationship to resistance to *Watermelon mosaic virus* I. *Journal of Heredity* 73, 239–240.

Provvidenti, R. and Robinson, R. (1977) Inheritance of resistance to watermelon mosaic virus 1 in *Cucumis metuliferus*. *Journal of Heredity* 68, 56–57.

Pruss, G., Ge, X., Shi, X.M., Carrington, J.C. and Vance, V.B. (1997) Plant viral synergism: the potyviral genome encodes a broad range pathogenicity enhancer that transactivates replication of heterologous viruses. *Plant Cell* 9, 859–868.

Purcifull, D.E. and Hiebert, E. (1979) Serological distinction of watermelon mosaic virus isolates. *Phytopathology* 69, 112.

Purcifull, D.E., Edwardson, J.R., Hiebert, E. and Gonsalves, D. (1984) *Papaya ringspot virus*. In: Coronel, R.E. (ed.) *CMI/AAB Description of Plant Viruses*, no. 292 (Vol. 2), Wageningen University, Netherlands.

Rezende, J.A.M. and Costa, A.S. (1993) Papaya diseases caused by viruses and mycoplasmas. *Summa Phytopathologica* 19, 78–79.

Rivas-Valencia, P., Mora-Aguilera, G., Téliz-Ortiz, D. and Mora-Aguilera, A. (2003) Influencia de variedades y densidades de plantación de papayo (*Carica papaya* L.) sobre las epidemias de mancha anular. *Revista Mexicana de Fitopatología* 21, 109–116.

Rivas-Valencia, P., Mora-Aguilera, G., Téliz-Ortiz, D. and Mora-Aguilera, A. (2008) Evaluación de barreras vegetales en el manejo integrado de la mancha anular del papayo (PRSV-P) en Michoacán, México. *Summa Phytopathologica* 34, 307–312.

Romay, G., Lecoq, H. and Desbiez, C. (2014) Zucchini tigré mosaic virus is a distinct Potyvirus in the *Papaya ringspot virus* cluster: molecular and biological insights. *Archives of Virology* 159, 277–289.

Ruiz-Castro, S. and Silva-Rosales, L. (1997) Use of RT-PCR for *Papaya ringspot virus* detection in papaya (*Carica papaya*) plants from Veracruz, Tabasco and Chiapas. *Revista Mexicana de Fitopatologia* 15, 83–87.

Sahana, N., Kaur, H., Basavaraj, Tena, F., Jain, R.K., *et al.* (2012) Inhibition of the host proteasome facilitates *Papaya ringspot virus* accumulation and proteosomal catalytic activity is modulated by viral factor HcPro. *PLoS One* 7, e52546.

Sakuanrungsirikul, S., Sarindu, N., Prasartsee, V., Chaikiatiyos, S., Siriyan, R., *et al.* (2005) Update on the development of virus-resistant papaya: virus-resistant transgenic papaya for people in rural communities of Thailand. *Food and Nutrition Bulletin* 26, 422–426.

Shen, W., Yan, P., Gao, L., Pan, X., Wu, J., *et al.* (2010a) Helper component-proteinase (HC-Pro) protein of *Papaya ringspot virus* interacts with papaya calreticulin. *Molecular Plant Pathology* 1, 335–346.

Shen, W.T., Wang, M.Q., Yan, P., Gao, L. and Zhou, P. (2010b) Protein interaction matrix of *Papaya ringspot virus* type P based on a yeast two-hybrid system. *Acta Virologica* 54, 49–54.

Shen, W., Tuo, D., Yan, P., Yang, Y., Li, X., *et al.* (2014) Reverse transcription loop-mediated isothermal amplification assay for rapid detection of *Papaya ringspot virus*. *Journal of Virological Methods* 204, 93–100.

Shi, X.M., Miller, H., Verchot, J., Carrington, J.C. and Vance, V.B. (1997) Mutations in the region encoding the central domain of helper component-proteinase (HC-Pro) eliminate *Potato virus X*/potyviral synergism. *Virology* 231, 35–42.

Shulka, D., Ward, C. and Brunt, A. (1994) Coat protein structure and function. In: Shulka, D., Ward, C. and Brunt, A. (eds) *The Potyviridae*. CAB International, Wallingford, UK, pp. 113–148.

Sreenivasulu, M. and Sai-Gopal, D.V.R. (2010) Development of recombinant coat protein antibody based IC-RT-PCR and comparison of its sensitivity with other immunoassays for the detection of *Papaya ringspot virus* isolates from India. *Plant Pathology Journal* 26, 25–31.

Srinivasulu, M. and Sai-Gopal, D.V.R. (2011) Coat protein sequence comparison of south Indian isolates of *Papaya ringspot virus* with other Indian subcontinent isolates. *Phytopathologia Mediterranea* 50, 359–367.

Sta. Cruz, F.C., Tañada, J.M., Elvira, P.R.V., Dolores, L.M., Magdalita, P.M., *et al.* (2009). Detection of mixed virus infection with *Papaya ringspot virus* (PRSV) in papaya (*Carica papaya* L.) grown in Luzon, Philippines. *Philippine Journal of Crop Science* 34, 62–74.

Stokstad, E. (2008) GM papaya takes on ringspot virus and wins. *Science* 320, 472.

Suzuki, J.Y., Tripathi, S., Fermin, G.A., Jan, F.-J., Hou, S., *et al.* (2008) Characterization of insertion sites in Rainbow papaya, the first commercialized transgenic fruit crop. *Tropical Plant Biology* 1, 293–309.

Taylor, D.R. (2001) Virus diseases of *Carica papaya* in Africa – their distribution, importance, and control. In: Hughes, A.J. and Odu, B.O. (eds) *Plant Virology in Sub-Saharan Africa – Proceedings of a Conference Organized by IITA*. International Institute of Tropical Agriculture, Ibadan, Nigeria, pp. 25–32.

Tecson-Mendoza, E.M., Laurena, A.C. and Botella, J.R. (2008) Recent advances in the development of transgenic papaya technology. *Biotechnology Annual Review* 14, 423–462.

Tennant, P., Gonsalves, C., Ling, K., Fitch, M., Manshardt, R., *et al.* (1994) Differential protection against *Papaya ringspot virus* isolates in coat protein gene transgenic papaya and classically cross-protected papaya. *Phytopathology* 84, 1359–1366.

Tennant, P., Fermin, G., Fitch, M.M., Manshardt, R.M., Slightom, J.L., *et al.* (2001) *Papaya ringspot virus* resistance of transgenic Rainbow and SunUp is affected by gene dosage, plant development and coat protein homology. *European Journal of Plant Pathology* 107, 645–653.

Tennant, P.F., Fermin, G.A. and Roye, M.E. (2007) Viruses infecting papaya (*Carica papaya* L.): etiology, pathogenesis and molecular biology. *Plant Viruses* 1, 178–188.

Tripathi, S., Suzuki, J., Ferreira, S. and Gonsalves, D. (2008) *Papaya ringspot virus*-P: characteristics, pathogenicity, sequence variability and control. *Molecular Plant Pathology* 9, 269–280.

Urcuqui-Inchima, S., Haenni, A.L. and Bernardi, F. (2001) *Potyvirus* proteins: a wealth of functions. *Virus Research* 74, 157–175.

Usharani, T.R., Laxmi, V., Jalali, S. and Krishnareddy, M. (2013) Duplex PCR to detect both *Papaya ringspot virus* and *Papaya leaf curl virus* simultaneously from naturally infected papaya (*Carica papaya* L.). *Indian Journal of Biotechnology* 12, 269–272.

Vance, V.B., Berger, P.H., Carrington, J.C., Hunt, A.G. and Shi, X.M. (1995) 5′ proximal potyviral sequences mediate *Potato virus X*/potyviral synergistic disease in transgenic tobacco. *Virology* 206, 583–590.

Vegas, A., Trujillo, G., Pino, I., González, A., Mata, J., *et al.* (2000) Avances sobre el control integrado del virus de la mancha anillada del lechoso, mediante la implementación de la inoculación de cepas atenuadas y prácticas culturales. *Agronomía Tropical (UCV)* 50, 303–310.

Vijayapalani, P., Maeshima, M., Nagasaki-Takekuchi, N. and Miller, W.A. (2012) Interaction of the transframe Potyvirus protein P3N-PIPO with host protein PCaP1 facilitates Potyvirus movement. *PLoS Pathogens* 8, e1002639.

Wang, H.L., Wang, C.C., Chiu, R.J. and Sun, M.H. (1978) Preliminary study on *Papaya ringspot virus* in Taiwan. *Plant Protocol Bulletin* 20, 133–140.

Wang, H.L., Wang, Y., Cookmeyer, D.G., Lommel, S.A. and Lucas, W.J. (1998) Mutations in viral movement protein alter systemic infection and identify an intercellular barrier to entry into the phloem long distance transport system. *Virology* 245, 75–89.

Yakoubi, S., Desbiez, C., Fakhfakh, H., Wipf-Scheibel, C., Marrakchi, M., *et al.* (2008a) Biological characterization and complete nucleotide sequence of a Tunisian isolate of *Moroccan watermelon mosaic virus*. *Archives of Virology* 153, 117–125.

Yakoubi, S., Lecoq, H. and Desbiez, C. (2008b) *Algerian watermelon mosaic virus* (AWMV): a new Potyvirus species in the PRSV cluster. *Virus Genes* 37, 103–109.

Yeh, S., Jan, F., Chiang, C., Doong, T.J., Chen, M.C., *et al.* (1992) Complete nucleotide sequence and genetic organization of *Papaya ringspot virus* RNA. *Journal of General Virology* 73, 2531–2541.

Yeh, S., Bau, Y., Cheng, H., Yu, T., Yang, J., *et al.* (1998) Greenhouse and field evaluations of coat-protein transgenic papaya resistant to *Papaya ringspot virus*. *Acta Horticulturae* 461, 321–328.

Yeh, S.-D. and Gonsalves, D. (1985) Translation of *Papaya ringspot virus* RNA *in vitro*: detection of a possible polyprotein that is processed for capsid protein, cylindrical-inclusion protein, and amorphous-inclusion protein. *Virology* 143, 260–271.

Yeh, S.D., Gonsalves, D. and Provvidenti, R. (1984) Comparative studies on host range and serology of *Papaya ringspot virus* and *Watermelon mosaic virus* 1. *Phytopathology* 74, 1081–1085.

Yeh, S.D., Gonsalves, D., Wang, H.I., Namba, R. and Chiu, R.J. (1988) Control of *Papaya ringspot virus* by cross protection. *Plant Disease* 72, 375–380.

Zhang, Y., Yu, N., Huang, Q., Yin, G., Guo, A., *et al.* (2014) Complete genome of Hainan *Papaya ringspot virus* using small RNA deep sequencing. *Virus Genes* 48, 502–508.

Zettler, F.W., Edwardson, J.R. and Purcifull, D.E. (1968) Ultramicroscopic differences in inclusions of *Papaya mosaic virus* and *Papaya ringspot virus* correlated with differential aphid transmission. *Phytopathology* 58, 332–335.

Zitter, T.A. and Tsai, J.H. (1977) Transmission of three potyviruses by the leafminer *Liriomyza sativae* (Diptera: Agromyzidae). *Plant Disease Report* 61, 1025–1029.

12

Tomato Spotted Wilt

Tsung-Chi Chen[1,2] and Fuh-Jyh Jan[3,4]*

[1]*Department of Biotechnology, Asia University, Wufeng, Taichung, Taiwan;* [2]*Department of Medical Research, China Medical University Hospital, China Medical University, Taichung, Taiwan;* [3]*Department of Plant Pathology, National Chung Hsing University, Taichung, Taiwan;* [4]*Agricultural Extension Center, National Chung Hsing University, Taichung, Taiwan*

12.1 Introduction

Tomato spotted wilt disease was first reported in Australia in 1915 and later identified as a virus-infecting disease caused by *Tomato spotted wilt virus* (TSWV) (Brittlebank, 1919; Samuel *et al.*, 1930). Initially, geographically distinct TSWV isolates were classified in a particular 'group' on the basis of particle morphology, host range and transmission by thrips (Matthews, 1982). Until *Impatiens necrotic spot virus* (INSV) was discovered (Law and Moyer, 1990), this group was proposed as the genus *Tospovirus* and assigned to the family *Bunyaviridae* by the International Committee on Taxonomy of Viruses in 1991 based on virion morphology and genome organization (Francki *et al.*, 1991). Currently, tospoviruses have become a worldwide problem. Some tospoviruses, such as TSWV, *Iris yellow spot virus* (IYSV), *Groundnut bud necrosis virus* (GBNV) and *Watermelon silver mottle virus* (WSMoV), are of global importance. TSWV is the most important tospovirus with a worldwide distribution that includes South Africa, Asia, Australasia, Europe and North America, and can cause severe damage to various economically important crops such as tobacco, tomato, pepper, cucurbits, lettuce, cabbage and potato (Pappu *et al.*, 2009).

Tospoviruses are often referred to as generalist viruses given their wide host range,

including hosts from at least 1090 plant species in 15 monocotyledonous and 69 dicotyledonous families worldwide (Parrella *et al.*, 2003; Pappu *et al.*, 2009). These viruses usually cause lethal symptoms on host plants such as necrotic lesions, wilting and dieback, but the overall symptomatology varies depending on virus strains and host genotypes (Mumford *et al.*, 1996). Environmental factors and plant growth conditions can also affect the outcome of symptom expression. Additionally, host preference varies with the virus species. For example, analysis of host preference of WSMoV and Melon yellow spot virus (MYSV) reveals that these two serologically related tospoviruses share similar characteristics: they are vectored by *Thrips palmi* Karny (Chen *et al.*, 1990; Kato *et al.*, 1999), both viruses affect the same cucurbitaceous crops like watermelon, melon, wax gourd and cucumber (Yeh *et al.*, 1992; Okuda *et al.*, 2002; Chiemsombat *et al.*, 2008) and show the same geographic distribution pattern (Jan *et al.*, 2003; Chen *et al.*, 2008; Chiemsombat *et al.*, 2008). However, host preference between both tospoviruses differs; WSMoV prefers watermelon whereas MYSV is more widespread on melon, even though both watermelon and melon can be infected by WSMoV and MYSV naturally. Surprisingly, mixed infections with WSMoV and MYSV are rarely found in watermelon

*E-mail: fjjan@nchu.edu.tw

and melon plants (Peng *et al.*, 2011). Melon plants infected by MYSV at the early cultivated stage usually develop lethal symptoms, but when infected at later stages, infected plants can still produce fruits. MYSV has been detected within the exocarp, endocarp, fruit flesh and columella of melon fruits from infected plants, but no seed transmission of the virus has been demonstrated.

In this chapter, the currently updated taxonomic status of tospoviruses is addressed. Various detection methods developed for prompt identification of tospoviruses, the advanced next-generation sequencing (NGS) technology applied in tospovirus diagnosis, and molecular studies of virus–thrips interaction are also addressed.

12.2 Characterization of Tospoviruses

12.2.1 Morphology and genome organization

Virions of tospoviruses are quasi-spherical in shape, 80–110 nm in diameter, with a lipid envelope. The tospoviral genome consists of three segmented single-stranded RNA molecules, named large (L), medium (M) and small (S) RNAs based on their molecular sizes of 8700–9000, 4700–5000 and 2600–3600 nucleotides, respectively. All genomic RNAs have the same sequences of 5′-AGAGCAAU-3′ at the 5′ end and the complementary 5'-AU UGCUCU-3′ at the 3′ end. The L RNA is of negative sense and codes a large open reading frame (ORF) in the viral complementary strand for an RNA-dependent RNA polymerase (RdRp) of 310–330 kDa. Both M and S RNAs are ambisense; each RNA molecule consists of two ORFs oriented in opposite directions and an AU-rich non-coding intergenic region. The ORF in the viral (v) strand of the M RNA encodes a movement NSm protein (~35 kDa), and another ORF in the viral complementary strand encodes a glycoprotein precursor (*circa* 130 kDa) that is further proteolytically cleaved to yield the mature Gn and Gc proteins that will eventually form

spikes on the envelope surface. The ORF in the v strand of the S RNA encodes the NSs protein (*circa* 50 kDa) that functions as a suppressor of the host RNA silencing machinery (Takeda *et al.*, 2002; Bucher *et al.*, 2003). The ORF in the viral complementary strand of the S RNA encodes the structural nucleocapsid protein (NP) (28–31 kDa) that encapsidates the viral RNAs. Infectious ribonucleoproteins are formed by the association of the NP, RdRp and genomic RNA molecules (King *et al.*, 2012).

12.2.2 Taxonomy

Based on virion morphology and genome organization, all tospoviruses are placed in the genus *Tospovirus* of the family *Bunyaviridae*, which constitutes a diverse group of viruses that comprises, along with the aforementioned only plant virus genus, the four animal-infecting genera *Orthobunyavirus*, *Hantavirus*, *Nairovirus* and *Phlebovirus* (King *et al.*, 2012). *Tospovirus* was named after TSWV, the type species of the group. The disease caused by TSWV was first reported in Australia in 1915 (Brittlebank, 1919; Samuel *et al.*, 1930) and is currently a worldwide problem.

Previously, the geographically distinct TSWV isolates were classified in a particular 'group' on the basis of particle morphology, host range and transmission by thrips (Matthews, 1982). It was only after INSV was discovered (Law and Moyer, 1990) that the genus *Tospovirus* was proposed and assigned to the family *Bunyaviridae* by the International Committee on Taxonomy of Viruses in 1991 (Francki *et al.*, 1991). At present, the classification of tospoviruses is officially proposed by the International Committee on Taxonomy of Viruses according to the criteria of vector specificity (thrips species), host range, sequence identity and serological relationship of NP (King *et al.*, 2012). A threshold of 90% amino acid (aa) identity in NP is the most important criterion for the demarcation of a tospovirus at the species level (Goldbach and Kuo, 1996). As of 2014, at least 29 *Tospovirus* species have been identified (listed in

Table 12.1), including 11 officially accepted and 18 tentative species that can be divided into 5 major clades phylogenetically related to TSWV, WSMoV, *Groundnut yellow spot virus* (GYSV), Soybean vein necrosis-associated virus, and Lisianthus necrotic ringspot virus, which each represent the first characterized species in the respective clades (Fig. 12.1).

12.2.3   Serological grouping

Serological relationship is an important criterion for the classification of tospoviruses that, with the aid of NP-directed polyclonal antisera, provides a 'serogrouping' system of classification. Thus, serologically related tospoviruses are assigned to a serogroup. Initially, tospoviruses were classified into four serogroups (I to IV) based on the analysis

of a few species like TSWV in serogroup I, *Groundnut ringspot virus* (GRSV) and *Tomato chlorotic spot virus* (TCSV) in serogroup II, INSV in serogroup III, and WSMoV and GBNV in serogroup IV (Goldbach and Kuo, 1996). However, further investigations showed that since cross reactivity among TSWV, GRSV, TCSV and INSV isolates was frequent, the serological classification turned out to be inadequate (Bezerra *et al*., 1999; Chen *et al*., 2014). New serologically distinct tospoviruses reported at the same time also made the numeric system of classification quite confusing. Thus, a type member-based serological classification system was recommended (Jan *et al*., 2003). This way, most tospoviruses can be classified into four serogroups. First, GRSV, TCSV, INSV, *Zucchini lethal chlorosis virus*, Chrysanthemum stem necrosis virus (CSNV), Alstroemeria necrotic

Table 12.1. List of currently known tospovirus species

Species[a]	Acronym	References
Alstroemeria necrotic streak virus	ANSV	Hassani-Mehraban *et al*., 2010
Bean necrosis mosaic virus	BeNMV	de Oliveira *et al*., 2011
Capsicum chlorosis virus	CaCV	Lee *et al*., 2002
Calla lily chlorotic spot virus	CCSV	Chen *et al*., 2005a
Chrysanthemum stem necrosis virus	CSNV	Bezerra *et al*., 1999
Groundnut bud necrosis virus	GBNV	Reddy *et al*., 1992
Groundnut chlorotic fan-spot virus	GCFSV	Chen and Chiu, 1996
Groundnut ringspot virus	GRSV	Pang *et al*., 1993a
Groundnut yellow spot virus	GYSV	Satyanarayana *et al*., 1998
Hippeastrum chlorotic ringspot virus	HCRV	Dong *et al*., 2013
Impatiens necrotic spot virus	INSV	Law and Moyer, 1990
Iris yellow spot virus	IYSV	Cortes *et al*., 1998
Lisianthus necrotic ringspot virus	LNRV	Shimomoto *et al*., 2014
Melon severe mosaic virus	MeSMV	Ciuffo *et al*., 2009
Melon yellow spot virus	MYSV	Kato *et al*., 2000
Mulberry vein banding virus	MuVBV	Meng *et al*., 2013
Pepper chlorotic spot virus	PCSV	Cheng *et al*., 2014
Pepper necrotic spot virus	PNSV	Torres *et al*., 2012
Polygonum ringspot virus	PolRSV	Ciuffo *et al*., 2008
Soybean vein necrosis-associated virus	SVNaV	Zhou *et al*., 2011
Tomato chlorotic spot virus	TCSV	de Ávila *et al*., 1993
Tomato necrotic ringspot virus	TNRV	Seepiban *et al*., 2011
Tomato necrotic spot virus	TNSV	Yin *et al*., 2014
Tomato spotted wilt virus	TSWV	de Haan *et al*., 1990
Tomato yellow ring virus	TYRV	Hassani-Mehraban *et al*., 2005
Tomato zonate spot virus	TZSV	Dong *et al*., 2008
Watermelon bud necrosis virus	WBNV	Jain *et al*., 1998
Watermelon silver mottle virus	WSMoV	Yeh and Chang, 1995
Zucchini lethal chlorosis virus	ZLCV	Bezerra *et al*., 1999

[a]Official species names are in italics

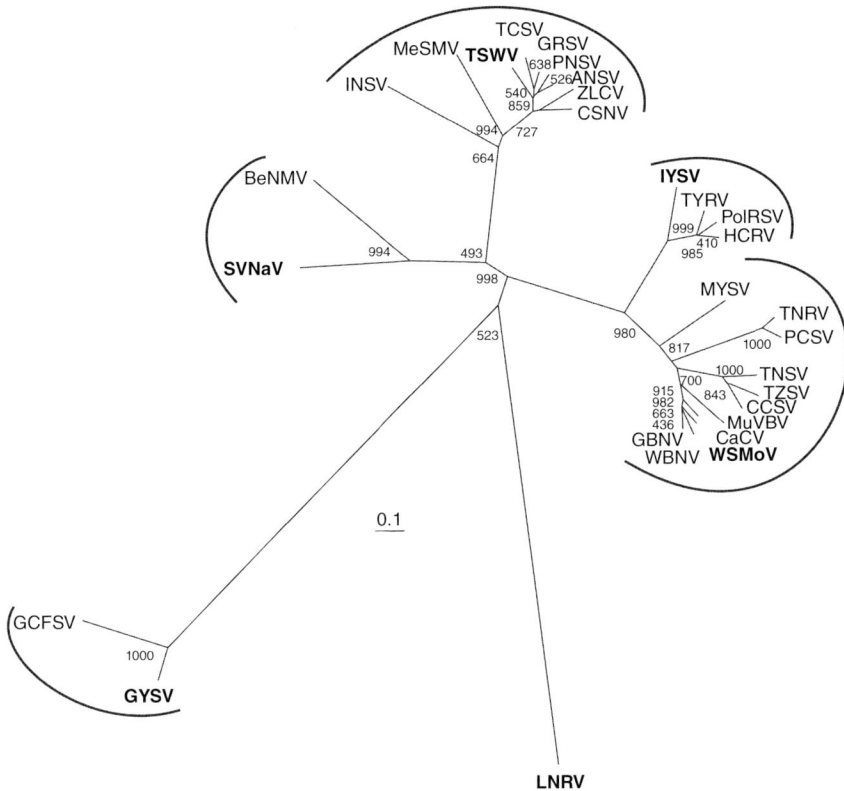

Fig. 12.1. Phylogenetic relationship of the currently known tospoviruses based on the comparison of the nucleocapsid protein gene sequence. The dendrogram was produced using the Neighbour-Joining algorithm with 1000 bootstrap replicates. Viruses with a close relationship are grouped together. The type species of each clade is indicated in bold. See Table 12.1 for details of virus acronyms.

streak virus, Pepper necrotic spot virus and Melon severe mosaic virus are serologically related to TSWV, and referred to as the TSWV serogroup (Ciuffo *et al.*, 2009; Hassani-Mehraban *et al.*, 2010; Chen *et al.*, 2014). Second, GBNV, Capsicum chlorosis virus (CaCV), Calla lily chlorotic spot virus (CCSV), MYSV, Pepper chlorotic spot virus, Tomato necrotic ringspot virus, Tomato necrotic spot virus, Tomato zonate spot virus and *Watermelon bud necrosis virus* (WBNV) are serologically related to WSMoV, and regarded as the WSMoV serogroup (Chen *et al.*, 2010; Cheng *et al.*, 2014; Yin *et al.*, 2014). Third, Tomato yellow ring virus (TYRV), Hippeastrum chlorotic ringspot virus and *Polygonum ringspot virus* are serologically related to IYSV, and cluster in the IYSV serogroup (Ciuffo *et al.*, 2008; Dong *et al.*, 2013). Finally,

the serological relatedness of Groundnut chlorotic fan-spot virus (GCFSV) and GYSV was recently verified, giving support to the fourth, or the GYSV serogroup (Kang *et al.*, 2014). The serological relationships of some tospoviruses, such as the newly identified Soybean vein necrosis-associated virus (Zhou *et al.*, 2011) or the Bean necrosis mosaic virus (de Oliveira *et al.*, 2012) and LNRV (Shimomoto *et al.*, 2014), are still in need of further clarification.

Noteworthy, the serological grouping of tospoviruses matches well with the corresponding phylogenetic clustering by which tospoviruses sharing more than 51.8% similarity at the NP sequence level are serologically related (Chen *et al.*, 2010; Seepiban *et al.*, 2011). NP serology is essential for the detection and characterization of tospoviruses.

12.3 Economic Importance of the Disease and Transmission of the Virus

12.3.1 Agricultural impact of tospoviruses

Based on the data collected between 1996 and 2006 in Georgia (USA) alone, the annual average losses were US$12.3 million in groundnut, US$11.3 million in tobacco, and US$9 million in tomato and pepper, which amounted to a total of US$32.6 million for the decade (Riley *et al.*, 2011). In 2004, the annual loss of groundnut due to an outbreak of TSWV in Georgia was estimated at more than US$100 million (Pearce, 2005). IYSV has been reported to mainly infect onion in South Africa, Asia, Australasia, Europe and North America. In recent years, outbreaks of IYSV in both onion bulb and seeds in the Treasure Valley of Oregon and Idaho and the Columbia Basin of Washington have resulted in total crop losses of up to US$480,000 per hectare (Pappu *et al.*, 2009). GBNV, on the other hand, is the most economically important virus affecting groundnut, potato, tomato and soybean in parts of China, India, Iran, Nepal, Sri Lanka and Thailand (Pappu *et al.*, 2009). Annual losses due to GBNV infection were estimated to reach over U$89 million in India (Reddy *et al.*, 1995). WSMoV mainly infects cucurbitaceous crops, including melon, watermelon and wax gourd, and it prevails in Asian countries such as Taiwan (Yeh and Chang, 1995), Japan (Iwaki *et al.*, 1984), Thailand (Chiemsombat *et al.*, 2008) and China (Rao *et al.*, 2013). In December 2006, WSMoV caused US$8 million losses in muskmelon production by infecting more than 500 ha of muskmelon in Tainan City, south Taiwan (Peng *et al.*, 2011).

12.3.2 Persistent transmission by thrips

Tospoviruses are transmitted persistently and propagatively by thrips. Review of the literature suggests that 14 thrips species belonging to the genera *Frankliniella*, *Thrips*, *Scirtothrips*, *Ceratothripoides* and *Dictyothrips* (Thysanoptera: Thripidae) have been reported as vectors of tospoviruses (listed in Table 12.2) (reviewed by Riley *et al.*, 2011). Thrips are important quarantine pests in international trade. Their tiny bodies, hidden behaviour and pesticide resistance make the control of thrips difficult. The relationship of a tospovirus with its thrips vector is well described for TSWV and *F. occidentalis*. Virus acquisition by thrips can only take place during the first and second larval stages. Transmission is successful at the late second stage of the larvae, but the most efficient transmission occurs with adults (Wijkamp and Peters, 1993; Wijkamp *et al.*, 1993; van de Wetering *et al.*, 1996).

When thrips are infected, tospoviruses tend to migrate from the midgut through the haemocoel to the salivary glands of adults (Whitfield *et al.*, 2005). The second pathway of translocation of a tospovirus was found via a thin ligament connecting the midgut and salivary glands of thrips (Nagata *et al.*, 2002). Many reports have demonstrated that the Gn and Gc proteins are essential determinants of thrips acquisition, but they may not be required for TSWV replication in plants (Sin *et al.*, 2005). The Gn protein may be involved in virus binding to the thrips cell receptors during virus entry, whereas the Gc protein may serve as a fusion protein essential for entry into the insect cells (Ullman *et al.*, 2005; Snippe *et al.*, 2007; Ribeiro *et al.*, 2009). A putative transcription factor of *F. occidentalis* has been found to interact with the TSWV RdRp (de Medeiros *et al.*, 2005). Recently, proteomic analysis of the response of *F. occidentalis* to TSWV infection revealed that a group of candidate proteins, including cyclophilin, heat shock proteins, and many others, are differentially expressed upon TSWV infection (Badillo-Vargas *et al.*, 2012). Nonetheless, the actual mechanism of tospovirus-thrips interaction remains unclear.

12.4 Diagnosis and Detection of Tospoviruses

12.4.1 Serological assays

Symptomatology is not sufficient for virus identification due to the fact that similar

Table 12.2. List of thrips species reported to vector tospoviruses

Thrip species	Tospovirus[a]	Reference
Ceratothripoides claratris	CaCV	Premachandra *et al.*, 2005
Dictyothrips betae	PolRSV	Ciuffo *et al.*, 2010
Frankliniella occidentalis	CSNV	Nagata and de Ávila, 2000
	GRSV	Wijkamp *et al.*, 1995
	INSV	Deangelis *et al.*, 1993
	TCSV	Wijkamp *et al.*, 1995
	TSWV	Allen and Broadbent, 1986
F. cephalica	TSWV	Ohnishi *et al.*, 2006
F. gemina	GRSV	de Borbón *et al.*, 1999
	TSWV	de Borbón *et al.*, 1999
F. schultzei	CSNV	Nagata and de Ávila, 2000
		Wijkamp *et al.*, 1995
	GRSV	Meena *et al.*, 2005
	GBNV	Wijkamp *et al.*, 1995
	TCSV	Sakimura, 1969
	TSWV	
F. fusca	TSWV	Sakimura, 1963
F. intonsa	GRSV	Wijkamp *et al.*, 1995
	INSV	Sakurai *et al.*, 2004
	TCSV	Wijkamp *et al.*, 1995
	TSWV	Wijkamp *et al.*, 1995
F. bispinosa	TSWV	Moyer, 2000
F. zucchini	ZLCV	Nakahara and Monteiro, 1999
Thrips tabaci	IYSV	Cortes *et al.*, 1998
	TSWV	Pittman, 1927
	TYRV	Golnaraghi *et al.*, 2007
T. setosus	TSWV	Kobatake *et al.*, 1984
T. palmi	CCSV	Chen *et al.*, 2005a
	GBNV	Vijayalakshmi, 1994
	MYSV	Kato *et al.*, 2000
	TSWV	Moyer, 2000
	WSMoV	Iwaki *et al.*, 1984
Scirtothrips dorsalis	GBNV	German *et al.*, 1992
	GCFSV	Chen and Chiu, 1996
	GYSV	Reddy *et al.*, 1990

[a]See Table 12.1 for details of virus acronyms.

symptoms may be induced by different to-spoviruses. Conventionally, serological (antibody-based) or molecular (nucleic acid-based) techniques are used to specifically detect and identify diverse tospoviruses. The type of antibody, however, is very important for successful serological detection assays. In general, antiserum consisting of polyclonal antibodies against the tospoviral NP is used in ELISA and western blotting for virus diagnosis. As mentioned earlier, the NP-derived antiserum of a tospovirus may cross-react with other tospovirus species of the same serogroup. For instance, the antiserum derived from the NP of WSMoV can react not only with WSMoV, but also with CaCV, CCSV, GBNV, WBNV and MYSV (Chen *et al.*, 2005b; Lin *et al.*, 2005; Chen *et al.*, 2010). An antiserum raised against the NP of TSWV, that cross-reacted with the heterologous CSNV and INSV proteins, was recently reported by Chen *et al.* (2014). With regard to the serological relationship of tospoviruses, the contiguous conserved amino acid residues can be found within the antibody-recognized epitopes (Kang *et al.*, 2014). Thus, the NP antiserum is insufficient to distinguish tospoviruses clustered in the same serogroup.

Empirically, monoclonal antibody targeting a specific epitope is highly specific and can be used to differentiate tospovirus species in the same serogroup. However, tospoviruses such as CaCV, GBNV, WBNV and WSMoV, which show similarity values greater than 80% at the amino acid sequence level of NP, are still indistinguishable when specific monoclonal antibodies are used (Chen *et al.*, 2005b).

A novel detection idea developed for detecting all tospoviruses has been proposed. The 24-aa conserved peptide 'YFLSK-TLEVLPKNLQTMSYLDSIQC', which locates at the C-terminal region of the NSs proteins of serogroup I to IV of tospoviruses, was artificially synthesized to produce rabbit antisera. The antisera successfully reacted with TSWV, GRSV, TCSV and INSV (serogroup I to III), but failed to react with WSMoV (serogroup IV) (Heinze *et al.*, 2000). Thereafter, the NSs protein of WSMoV was used to prepare polyclonal antiserum and monoclonal antibodies (Chen *et al.*, 2006). Surprisingly, the NSs-derived monoclonal antibody, denoted MAb-WNSs, can broadly react with WSMoV, GBNV, WBNV, CaCV, CCSV and MYSV, all clustered in the WSMoV serogroup, and the IYSV and TYRV belonging to the IYSV serogroup (Chen *et al.*, 2011). The epitope 'VRK-PGVKNTGCKFTMHNQIFNPN' of MAb-WNSs was mapped to the N-terminal position (amino acid 98–120) of the NSs protein of WSMoV; the epitope is conserved within the NSs proteins of the aforementioned tospovirus species and is designated as WNSscon (Chen *et al.*, 2006). The NSs protein has become an ideal candidate for producing serogroup-universal antibodies.

12.4.2 Nucleic acid amplification

The primer pair designed from the N gene is used in conventional reverse transcription (RT)-PCR assays to identify tospovirus species when polyclonal and monoclonal antibodies are unavailable or unable to distinguish among various species. However, the category of viruses occurring in the field is usually complex; thus, the application of a single species-specific primer pair for the detection of field samples is insufficient. Multiplex

RT-PCR using a mixture of various primer sets is recommended to simultaneously detect different viruses in a certain geographic area or crop (Uga and Tusda, 2005). An improved method termed multi-PCR system, using a universal degenerate primer coupled with five tospovirus species-specific primers has been further developed to amplify size-differentiated DNA fragments corresponding to CaCV, CSNV, INSV, IYSV and TSWV infecting plants of the families Solanaceae and Asteraceae (Kuwabara *et al.*, 2010). Although multiplex RT-PCR can be applied to simultaneously detect various tospoviruses, the exploration for most characterized and uncharacterized tospovirus species is still limited. In previous studies, degenerate primers that recognize the consensus sequences of L and M RNAs of tospoviruses were designed based on the determination and comparison of full-length virus genome sequences (Chu *et al.*, 2001; Chen *et al.*, 2012). These primer sets can be successfully used to detect at least 12 tospovirus species, including CaCV, CCSV, MYS, GBNV, WBNV, WSMoV, GRSV, TCSV, TSWV, IYSV, TYRV and INSV (Chen *et al.*, 2012). It also provides a useful approach for the diagnosis of new tospoviruses. However, these described methods can only detect the virus qualitatively, not quantitatively.

Real-time RT-PCR, also called quantitative RT-PCR (qRT-PCR), is a highly specific and sensitive method that can be used to quantify target nucleic acid molecules. There are two systems developed for qRT-PCR: (i) the TaqMan system using fluorescent-labelled specific probes that hybridize with the target DNA molecule; and (ii) the non-specific fluorescent dyes, such as SYBR Green I, that intercalate with any newly synthesized double-stranded DNA molecules. The TaqMan-based qRT-PCR has been developed to identify and quantify TSWV and INSV in host plants and their thrip vectors (Roberts *et al.*, 2000; Boonham *et al.*, 2002; Debreczeni *et al.*, 2011; Chen *et al.*, 2013). A SYBR Green I-based qRT-PCR system has been developed to detect IYSV (Pappu *et al.*, 2008). Accompanying melting curve analysis is recommended in order to corroborate PCR results derived from the SYBR Green I qRT-PCR system via the determination of the T_m values of amplicons (Dai *et al.*, 2012). The SYBR Green I qRT-PCR system, which is as sensitive as the probe-based TaqMan system, but without the need of a specialized probe and with relatively lower cost, is more suitable for diagnosis of tospoviruses in the field.

12.4.3 Deep sequencing analysis

A new technique, NGS, has been developed recently to facilitate genome sequencing of any organism. Unlike the conventional Sanger's method, template nucleic acid molecules are randomly fragmented into small sizes and then clonally amplified by emulsion PCR (Dressman *et al.*, 2003) and solid-phase amplification (Fedurco *et al.*, 2006). Commercial NGS systems such as Illumina, SOLiD and Roche 454 are commonly used to generate large masses of nucleotide sequences, so-called reads. Millions of reads are *de novo* assembled to generate contigs, which are aligned to known reference sequences. Bioinformatics methods are necessary for the analyses of NGS-produced high-throughput data. NGS has become a powerful tool in the study of the complete genome sequence of any organism (Wheeler *et al.*, 2008), transcriptome analysis (Wilhelm *et al.*, 2008), small RNA profiling (Morin *et al.*, 2008), single-nucleotide polymorphisms (Brockman *et al.*, 2008), and many others.

NGS is also used in deep sequencing analyses of plant viruses. Small-interfering RNAs (siRNAs) and polyadenylated RNAs are usually used as the template for NGS depending on the genome category of viruses (reviewed by Barba *et al.*, 2014). A recent study reported that the tospoviruses TSWV and CaCV can be diagnosed from host plants at the early infection stage before the onset of symptoms using siRNAs for sequencing (Hagen *et al.*, 2011). It provides an efficient approach for detecting viruses at extreme low concentration that normally cannot be detected by conventional methods. In addition, the transcriptome data of viruliferous insects, which are obtained from sequencing of mRNA pools, can be analyzed for the study of virus–vector interaction (Xu *et al.*, 2012). Comparison of gene expression between TSWV-infected and non-infected *F. occidentalis* populations revealed that TSWV modulates cellular processes and the immune response of *F. occidentalis*, which might lead to lower virus titres in thrip cells and result in no detrimental effects to the vector (Zhang *et al.*, 2013). An understanding of the molecular interaction between TSWV and *F. occidentalis* will provide insight into the strategic development of disease control measures.

12.5 Management of Tospoviral Diseases

Control of thrip vectors is typically applied to the management of tospoviral diseases. However, there are no effective measures to sufficiently control the tiny thrips with chemical insecticides. Two well-known natural resistance genes, *Sw-5* gene identified in tomato (Stevens *et al.*, 1992; Rosello *et al.*, 1998) and *Tsw* gene in pepper (Jahn *et al.*, 2000), are widely used in traditional breeding programs against TSWV. The *Sw-5* gene also provides resistance against TCSV and GRSV, which are phylogenetically related to TSWV (Boiteux and Giordano, 1993). Unfortunately, the resistance conferred by either *Sw-5* or *Tsw* gene can be broken down under field conditions by natural resistance-breaking TSWV strains (Margaria *et al.*, 2007; López *et al.*, 2011). The high divergence of tospoviruses and the scarcity of natural resistance resources in crops make traditional breeding difficult.

Since the concept of pathogen-derived resistance was proposed in 1985 (Sanford and Johnston, 1985), transgenic resistance has become a crucial approach for controlling viral diseases. Previously, the protein-mediated resistance strategy that results in the accumulation of high-levels of TSWV NP in transgenic *Nicotiana benthamiana* plants was shown to confer broad-spectrum resistance not only against the homologous virus, but also against distantly related INSV isolates (Pang *et al.*, 1994). However, the NP-mediated resistance can be overcome with increasing inoculum strength (Schwach *et al.*, 2004). The RNA-mediated resistance approach, on the other hand, is more effective than protein-mediated resistance (Pang *et al.*, 1993b; Prins and Goldbach, 1996). The N gene has been extensively used against TSWV and various investigations report on efficient resistance against TSWV infections in transgenic plants carrying segments of the gene from TSWV (Prins and Goldbach, 1996). An early report showed that a large segment (longer than 387 base pairs) of the N gene of TSWV confers virus resistance through sense post-transcriptional gene silencing, whereas a smaller segment (smaller than 235 base pairs) does not (Pang *et al.*, 1997). However, N gene sequences as short as 110 base pairs are sufficient in conferring virus resistance, when the small segment is fused with a silencer DNA such as the green fluorescent protein (Jan *et al.*, 2000a). An artificial microRNA approach, developed to trigger RNA silencing in *Arabidopsis thaliana* plants, also confers RNA-mediated resistance against tospoviruses (Niu *et al.*, 2006). Using this approach, artificial microRNAs targeting sequence elements within conserved RdRp motifs of the WSMoV L gene successfully provide high degrees of transgenic resistance against the homologous virus (Kung *et al.*, 2012).

RNA-mediated resistance is highly sequence homology-dependent and only specific to the target virus. Thus several attempts have been made to develop transgenic plants with multiple virus resistance using a single transgene construct. Among these is a chimera consisting of the N genes isolated from TSWV, TCSV and GRSV, which on transfer to tobacco, introduced broad resistance against the three tospoviruses (Prins *et al.*, 1995). Similarly, a composite transgene containing small, partial fragments of N genes from WSMoV, TSWV, GRSV and TCSV in a hairpin construct triggered RNA silencing against the corresponding viruses (Bucher *et al.*, 2006). Transgenic plants expressing an NP-interacting peptide derived from the N ORF of TSWV also exhibited high degrees of broad-spectrum resistance against TSWV, GRSV, TCSV and CSNV (Rudolph *et al.*, 2003). Nonetheless, the resistance derived from

chimeric transgenes is seemingly limited to the progenitor viruses. Moreover, regardless of their origin (from N transgenes or from other sources (e.g. peptides) targeting N gene sequences), the resistance is generally not expected to be broad-spectrum and/or durable, because of the high degree of variation among the N gene sequences of tospoviral species and strains.

Recent evidence suggests that a new approach that employs a single fragment corresponding to the conserved region of the L RNA of WSMoV in building translatable, sense non-translatable, antisense and hairpin transgene constructs may be more effective in protecting against tospoviruses. So far, the constructs have successfully provided broad-spectrum transgenic resistance against all challenged tospoviruses, including WSMoV, IYSV, GCFSV, TSWV, GRSV and INSV in *N. benthamiana* plants. This approach has also proved effective in tomato plants. Resistance was exhibited against WSMoV, TSWV and GRSV infections (Peng *et al.*, 2014). To develop transgenic plants resistant to tospoviruses and viruses belonging to other genera, Jan *et al.* (2000a) used a single chimeric gene composed of linked viral segments to confer resistance against two RNA viruses, *Turnip mosaic virus* and TSWV (Jan *et al.*, 2000b). Recently, the strategy was applied to tobacco for concurrent transgenic resistance against a DNA geminivirus and a RNA tospovirus (Lin *et al.*, 2011; Yang *et al.*, 2014).

12.6 Concluding Remarks

Since the first tospovirus TSWV was reported in 1915, 29 tospoviruses have been characterized as of 2014. The numbers of newly discovered tospoviruses continue to increase, suggesting that more unidentified tospoviruses remain at large in nature. The development of efficient detection methods is crucial not only for the diagnosis of threatening tospoviruses, but also for the identification of newly unexplored species. Additionally, the new NGS technology will promote increased and widened genetic studies of tospoviruses, which invariably will contribute to further exploration of the molecular interactions between tospoviruses and their plant hosts and thrips vectors. The recently developed approaches that generate multivirus transgenic resistance should provide useful and versatile tools to combat tospovirus disease infections in various crops.

References

Allen, W.R. and Broadbent, A.B. (1986) Transmission of *Tomato spotted wilt virus* in Ontario greenhouses by the western flower thrips, *Frankliniella occidentalis* (Pergande). *Canadian Journal of Plant Pathology* 8, 33–38.

Badillo-Vargas, I.E., Rotenberg, D., Schneweis, D.J., Hiromasa, Y., Tomich, J.M., *et al.* (2012) Proteomic analysis of *Frankliniella occidentals* and differentially expressed proteins in response to *Tomato spotted wilt virus* infection. *Journal of Virology* 86, 8793–8809.

Barba, M., Czosnek, H. and Hadidi, A. (2014) Historical perspective, development and applications of next-generation sequencing in plant virology. *Viruses* 6, 106–136.

Bezerra, I.C., Resende, R.O., Pozzer, L., Nagata, T., Kormelink, R., *et al.* (1999) Increase of tospoviral diversity in Brazil with the identification of two new tospovirus species, one from chrysanthemum and one from zucchini. *Phytopathology* 89, 823–830.

Boiteux, L.S. and Giordano, L.D.B. (1993) Genetic basis of resistance against two Tospovirus species in tomato (*Lycopersicon esculentum*). *Euphytica* 71, 151–154.

Boonham, N., Smith, P., Walsh, K., Tame, J., Morris, J., *et al.* (2002) The detection of *Tomato spotted wilt virus* (TSWV) in individual thrips using real time fluorescent RT-PCR (TaqMan). *Journal of Virological Methods* 101, 37–48.

Brittlebank, C.C. (1919) Tomato diseases. *Journal of Agriculture Victoria* 17, 213–235.

Brockman, W., Alvarez, P., Young, S., Garber, M., Giannoukos, G., *et al.* (2008) Quality scores and SNP detection in sequencing-by-synthesis systems. *Genome Research* 18, 763–770.

Bucher, E., Sijen, T., de Haan, P., Goldbach, R. and Prins, M. (2003) Negative-strand tospoviruses and tenuiviruses carry a gene for a suppressor of gene silencing at analogous genomic positions. *Journal of Virology* 77, 1329–1336.

Bucher, E., Lohuis, D., Pieter, M., van Poppel, J.A., Geerts-Dimitriadou, C., *et al.* (2006) Multiple virus resistance at a high frequency using a single transgene construct. *Journal of General Virology* 87, 3697–3701.

Chen, C.C. and Chiu, R.J. (1996) A tospovirus infecting peanut in Taiwan. *Acta Horticulturae* 431, 57–67.

Chen, C.C., Shy, J.F. and Yeh, S.D. (1990) Thrips transmission of *Tomato spotted wilt virus* from watermelon. *Plant Protection Bulletin* 32, 331–332.

Chen, C.C, Chen, T.C., Lin, Y.H., Yeh, S.D. and Hsu, H.T. (2005a) A chlorotic spot disease on calla lilies (*Zantedeschia* spp.) is caused by a tospovirus serologically but distantly related to *Watermelon silver mottle virus*. *Plant Disease* 89, 440–445.

Chen, S.M., Wang, Y.C., Wu, P.R. and Chen, T.C. (2014) Production of antiserum against the nucleocapsid protein of *Tomato spotted wilt virus* and investigation of its serological relationship with other tospoviruses. *Plant Protection Bulletin* 56, 55–74.

Chen, T.C., Hsu, H.T., Jain, R.K., Huang, C.W., Lin, C.H., *et al.* (2005b) Purification and serological analyses of tospoviral nucleocapsid proteins expressed by *Zucchini yellow mosaic virus* vector in squash. *Journal of Virological Methods* 129, 113–124.

Chen, T.C., Huang, C.W., Kuo, Y.W., Liu, F.L., Hsuan-Yuan, C.H., *et al.* (2006) Identification of common epitopes on a conserved region of NSs proteins among tospoviruses of *Watermelon silver mottle virus* serogroup. *Phytopathology* 96, 1296–1304.

Chen, T.C., Lu, Y.Y., Cheng, Y.H., Chang, C.A. and Yeh, S.D. (2008) Melon yellow spot virus in watermelon: a first record from Taiwan. *Plant Pathology* 54, 765.

Chen, T.C., Lu, Y.Y., Cheng, Y.H., Li, J.T., Yeh, Y.C., *et al.* (2010) Serological relationship between Melon yellow spot virus and *Watermelon silver mottle virus* and differential detection of the two viruses in cucurbits. *Archives of Virology* 155, 1085–1095.

Chen, T.C., Lu, Y.Y., Kang, Y.C., Li, J.T., Yeh, Y.C., *et al.* (2011) Detection of eight different tospovirus species by a monoclonal antibody against the common epitope of NSs protein. *Acta Horticulturae* 901, 61–66.

Chen, T.C., Li, J.T., Lin, Y.P., Yeh, Y.C., Kang, Y.C., *et al.* (2012) Genomic characterization of Calla lily chlorotic spot virus and design of broad-spectrum primers for detection of tospoviruses. *Plant Pathology* 61, 183–194.

Chen, X., Xu, X., Li, Y. and Liu, Y. (2013) Development of a real-time fluorescent quantitative PCR assay for detection of *Impatiens necrotic spot virus*. *Journal of Virological Methods* 189, 299–304.

Cheng, Y.H., Zheng, Y.X., Tai, C.H., Yen, J.H., Chen, Y.K., *et al.* (2014) Identification, characterization and detection of a new tospovirus on sweet pepper. *Annals of Applied Biology* 164, 107–115.

Chiemsombat, P., Gajanandana, O., Warin, N., Hongprayoon, R., Bhunchoth, A., *et al.* (2008) Biological and molecular characterization of tospoviruses in Thailand. *Archives of Virology* 153, 571–577.

Chu, F.H., Chao, C.H., Chung, M.H., Chen, C.C. and Yeh, S.D. (2001) Completion of the genome sequence of *Watermelon silver mottle virus* and utilization of degenerate primers for detecting tospoviruses in five serogroups. *Phytopathology* 91, 361–368.

Ciuffo, M., Tavella, L., Pacifico, D., Masenga, V. and Turina, M. (2008) A member of a new *Tospovirus* species isolated in Italy from wild buckwheat (*Polygonum convolvulus*). *Archives of Virology* 153, 2059–2068.

Ciuffo, M., Kurowski, C., Vivoda, E., Copes, B., Masenga, V., *et al.* (2009) A new *Tospovirus* sp. in cucurbit crops in Mexico. *Plant Disease* 93, 467–474.

Ciuffo, M., Mautino, G.C., Bosco, L., Turina, M. and Tavella, L. (2010) Identification of *Dictyothrips betae* as the vector of *Polygonum ring spot virus*. *Annals of Applied Biology* 157, 299–307.

Cortes, I., Livierations, I.C., Derks, A., Peters, D. and Kormelink, R. (1998) Molecular and serological characterization of Iris yellow spot virus, a new and distinct tospovirus species. *Phytopathology* 88, 1276–1282.

Dai, H., Yang, C., Zhou, Q. and Yu, C. (2012) Development and comparative study of several PCR-based approaches for detection of Iris yellow spot virus (IYSV). *Acta Phytopathologica Sinica* 42, 126–130.

de Ávila, A.C., de Haan, P., Kormelink, R., Resende, R.O., Goldback, R.W., *et al.* (1993) Classification of tospoviruses based on phylogeny of nucleoprotein gene sequences. *Journal of General Virology* 74, 153–159.

de Borbón, C.M., Gracia, O. and De Santis, L. (1999) Survey of Thysanoptera occurring on vegetable crops as potential *Tospovirus* vectors in Mendoza, Argentina. *Revista de la Sociedad Entomológica Argentina* 58, 59–66.

de Haan, P., Wagemakers, L., Peters, D. and Goldbach, R.W. (1990) The S RNA segment of *Tomato spotted wilt virus* has an ambisense character. *Journal of General Virology* 71, 1001–1007.

de Medeiros, R.B., Figueiredo, J., Resende, R.O. and de Ávila, A.C. (2005) Expression of a viral polymerase-bound host factor turns human cell lines permissive to a plant- and insect-infecting virus. *Proceedings of the National Academy of Sciences USA* 102, 1175–1180.

de Oliveira, A.S., Bertran, A.G.M., Inoue-Nagata, A.K., Nagata, T., Kitajima, E.W., *et al.* (2011) An RNA-dependent RNA polymerase gene of a distinct Brazilian tospovirus. *Virus Genes* 43, 385–389.

de Oliveira, A.S., Melo, F.L., Inoue-Nagata, A.K., Nagata, T., Kitajima, E.W., *et al.* (2012) Characterization of Bean necrotic mosaic virus: a member of a novel evolutionary lineage within the genus Tospovirus. *PLoS One* 7, e38634.

Deangelis, J.D., Sether, D.M. and Rossignol, P.A. (1993) Survival, development, and reproduction in western flower thrips (Thysanoptera: Thripidae) exposed to *Impatiens necrotic spot virus*. *Environmental Entomology* 22, 1308–1312.

Debreczeni, D.E., Ruiz-Ruiz, S., Aramburu, J., Lopez, C., Belliure, B., *et al.* (2011) Detection, discrimination and absolute quantitation of *Tomato spotted wilt virus* isolates using real time RT-PCR with TaqMan MGB probes. *Journal of Virological Methods* 176, 32–37.

Dong, J.H., Cheng, X.F., Yin, Y.Y., Fang, Q., Ding, M., *et al.* (2008) Characterization of Tomato zonate spot virus, a new tospovirus in China. *Archives of Virology* 153, 855–864.

Dong, J.H., Yin, Y.Y., Fang, Q., McBeath, J.H., and Zhang, Z.K. (2013) A new tospovirus causing chlorotic ringspot on *Hippeastrum* sp. in China. *Virus Genes* 46, 567–570.

Dressman, D., Yan, H., Traverso, G., Kinzler, K.W. and Vogelstein, B. (2003) Transforming single DNA molecules into fluorescent magnetic particles for detection and enumeration of genetic variations. *Proceedings of the National Academy of Sciences USA* 100, 8817–8822.

Fedurco, M., Romieu, A., Williams, S., Lawrence, I. and Turcatti, G. (2006) BTA, a novel reagent for DNA attachment on glass and efficient generation of solid-phase amplified DNA colonies. *Nucleic Acids Research* 34, e22.

Francki, R.I.B., Fauquent, C.M., Knudson, D.L. and Brown, F. (1991) Classification and nomenclature of viruses. Fifth Report of the International Committee on Taxonomy of Viruses. *Archives of Virology Supplement* 2, 450.

German, T.L., Ullman, D.E. and Moyer, J.W. (1992) Tospoviruses: diagnosis, molecular biology, phylogeny, and vector relationships. *Annual Review in Phytopathology* 30, 315–348.

Goldbach, R. and Kuo, G. (1996) Introduction of tospoviruses and thrips of floral and vegetable crops. *Acta Horticulturae* 431, 21–26.

Golnaraghi, A.R., Pourrahim, R., Farzadfar, S., Ohshima, K., Shahraeen, N., *et al.* (2007) Incidence and distribution of Tomato yellow fruit ring virus on soybean in Iran. *Journal of Plant Pathology* 6, 14–21.

Hagen, C., Frizzi, A., Kao, J., Jia, L., Huang, M., *et al.* (2011) Using small RNA sequences to diagnose, sequence, and investigate the infectivity characteristics of vegetable-infecting viruses. *Archives of Virology* 156, 1209–1216.

Hassani-Mehraban, A., Saaijer, J., Peters, D., Goldbach, R.W. and Kormelink, R. (2005) A new tomato-infecting tospovirus from Iran. *Phytopathology* 95, 852–858.

Hassani-Mehraban, A., Botermans, M., Verhoeven, J.T.J., Meekes, E., Saaijer, J., *et al.* (2010) A distinct tospovirus causing necrotic streak on *Alstroemeria* sp. in Colombia. *Archives of Virology* 155, 423–428.

Heinze, C., Roggero, P., Sohn, M., Vaira, A.-M., Masenga, V., *et al.* (2000) Peptide-derived broad-reacting antisera against tospovirus NSs-protein. *Journal of Virological Methods* 89, 137–146.

Iwaki, M., Honda, Y., Hanada, K., Tochahara, H., Yonaha, T., *et al.* (1984) Silver mottle disease of watermelon caused by *Tomato spotted wilt virus*. *Plant Disease* 68, 1006–1008.

Jain, R.K., Pappu, R.H., Pappu, S.S., Krishna Reddy, M. and Vani, A. (1998) Watermelon bud necrosis tospovirus is a distinct virus species belonging to serogroup IV. *Archives of Virology* 143, 1637–1644.

Jahn, M., Paran, I., Hoffmann, K., Radwanski, E.R., Livingstone, K.D., *et al.* (2000) Genetic mapping of the *Tsw* locus for resistance to the *Tospovirus Tomato spotted wilt virus* in *Capsicum* spp. and its relationship to the *Sw-5* gene for resistance to the same pathogen in tomato. *Molecular Plant-Microbe Interactions* 13, 673–682.

Jan, F.J., Fagoaga, C., Pang, S.Z. and Gonsalves, D. (2000a) A minimum length of N gene sequence in transgenic plants is required for RNA-mediated tospovirus resistance. *Journal of General Virology* 81, 235–242.

Jan, F.J., Fagoaga, C., Pang, S.Z. and Gonsalves, D. (2000b) A single chimeric transgene derived from two distinct viruses confers multi-virus resistance in transgenic plants through homology-dependent gene silencing. *Journal of General Virology* 81, 2103–2109.

Jan, F.J., Chen, T.C. and Yeh, S.D. (2003) Occurrence, importance, taxonomy, and control of thrips-borne tospoviruses. In: Haung, H. and Acharya, S.N. (eds) *Advances in Plant Disease Management*. Research Signpost, Kerala, India, pp. 391–411.

Kang, Y.C., Yeh, S.D., Liao, C.H., Chou, W.C., Liu, F.L., *et al.* (2014) Verification of serological relationship between two phylogenetically related peanut-infecting *Tospovirus* species. *European Journal of Plant Pathology* 140, 815–828.

Kato, K., Hanada, K. and Kameya-Iwaki, M. (1999) Transmission mode, host range and electron microscopy of a pathogen causing a new disease of melon (*Cucumis melo*) in Japan. *Annals of the Phytopathological Society of Japan* 65, 624–627.

Kato, K., Hanada, K. and Kameya-Iwaki, M. (2000) Melon yellow spot virus: a distinct species of the genus *Tospovirus* isolated from melon. *Phytopathology* 90, 422–426.

King, A.M.Q., Lefkowitz, E., Adams, M.J. and Carstens, E.B. (2012) *Virus Taxonomy – 9th Report of the International Committee on Taxonomy of Viruses*. Academic Press, Elsevier, New York, pp. 737–739.

Kobatake, H., Osaki, T. and Inouye, T. (1984) The vector and reservoirs of *Tomato spotted wilt virus* in Nara prefecture. *Annals of the Phytopathological Society of Japan* 50, 541–544.

Kung, Y.J., Lin, S.S., Huang, Y.L., Chen, T.C., Harish, S.S., *et al.* (2012) Multiple artificial microRNAs targeting conserved motifs of the replicase gene confer robust transgenic resistance to negative-sense single-stranded RNA plant virus. *Molecular Plant Pathology* 13, 303–317.

Kuwabara, K., Yokoi, N., Ohki, T. and Tsuda, S. (2010) Improved multiplex reverse transcription-polymerase chain reaction to detect and identify five tospovirus species simultaneously. *Journal of General Plant Pathology* 76, 273–277.

Law, M.D., and Moyer, J.W. (1990) A tomato spotted wilt-like virus with a serologically distinct N protein. *Journal of General Virology* 71, 933–938.

Lee, A.M., Persley, D.M. and Thomas, J.E. (2002) A new tospovirus serogroup IV species infecting capsicum and tomato in Queensland, Australia. *Australasian Plant Pathology* 31, 231–239.

Lin, Y.H., Chen, T.C., Hsu, H.T., Liu, F.L., Chu, F.H., *et al.* (2005) Serological comparison and molecular characterization for verification of Calla lily chlorotic spot virus as a new tospovirus species belonging to *Watermelon silver mottle virus* serogroup. *Phytopathology* 95, 1482–1488.

Lin, C.Y., Ku, H.M., Tsai, W.S., Green, S.K. and Jan, F.J. (2011) Resistance to a DNA and a RNA virus in transgenic plants by using a single chimeric transgene construct. *Transgenic Research* 20, 261–270.

López, C., Aramburu, J., Galipienso, L., Soler, S., Nuez, F., *et al.* (2011) Evolutionary analysis of tomato *Sw-5* resistance-breaking isolates of *Tomato spotted wilt virus. Journal of General Virology* 92, 210–215.

Margaria, P., Ciuffo, M., Pacifico, D. and Turina, M. (2007) Evidence that the nonstructural protein of *Tomato spotted wilt virus* is the avirulence determinant in the interaction with resistant pepper carrying the *Tsw* gene. *Molecular Plant-Microbe Interactions* 20, 547–558.

Matthews, R.E.F. (1982) Classification and nomenclature of viruses. *Intervirology* 17, 1–199.

Meena, R.L., Venkatesan, T.R.S. and Mohankumar, S. (2005) Molecular characterization of *Tospovirus* transmitting thrips populations from India. *American Journal of Biochemistry and Biotechnology* 1, 167–172.

Meng, J.R., Liu, P.P., Zou, C.W., Wang, Z.Q., Liao, Y.M., *et al.* (2013) First report of a Tospovirus in mulberry. *Plant Disease* 97, 1001.

Morin, R.D., O'Connor, M.D., Griffith, M., Kuchenbauer, F., Delaney, A., *et al.* (2008) Application of massively parallel sequencing to microRNA profiling and discovery in human embryonic stem cells. *Genome Research* 18, 610–621.

Moyer, J.W. (2000) Tospoviruses. *Encyclopedia of Microbiology* 4, 592–597.

Mumford, R.A., Barker, I. and Wood, K.R. (1996) The biology of the tospoviruses. *Annals of Applied Biology* 128, 159–183.

Nagata, T. and de Ávila, A.C. (2000) Transmission of Chrysanthemum stem necrosis virus, a recently discovered *Tospovirus*, by two thrips species. *Journal of Phytopathology* 148, 123–125.

Nagata, T., Inoue-Nagata, A.K., van Lent, J., Goldbach, R. and Peter, D. (2002) Factors determining vector competence and specificity for transmission of *Tomato spotted wilt virus. Journal of General Virology* 83, 663–671.

Nakahara, S. and Monteiro, R.C. (1999) *Frankliniella zucchini* (Thysanoptera: Thripidae), a new species and vector of tospovirus in Brazil. *Proceedings of the Entomological Society of Washington* 101, 290–294.

Niu, Q.W., Lin, S.S., Reyes, J.L., Chen, K.C., Wu, H.W., *et al.* (2006) Expression of artificial microRNAs in transgenic *Arabidopsis thaliana* confers virus resistance. *Nature Biotechnology* 24, 1420–1428.

Ohnishi, J., Katsuzaki, H., Tsuda, S., Sakurai, T., Akutsu, K., *et al.* (2006) *Frankliniella cephalica*, a new vector for *Tomato spotted wilt virus. Plant Disease* 90, 685.

Okuda, M., Takeuchi, S., Taba, S., Kato, K. and Hanada, K. (2002) Melon yellow spot virus and *Watermelon silver mottle virus*: outbreak of cucurbit infecting tospovirus in Japan. *Acta Horticulturae* 588, 143–148.

Pang, S.Z., Slightom, J.L. and Gonsalves, D. (1993a) The biological properties of a distinct tospovirus and sequence analysis of its S RNA. *Phytopathology* 83, 728–733.

Pang, S.Z., Slightom, J.L. and Gonsalves, D. (1993b) Different mechanisms protect transgenic tobacco against tomato spotted wilt and impatiens necrotic spot tospoviruses. *Nature Biotechnology* 11, 819–824.

Pang, S.Z., Bock, J.H., Gonsalves, C., Slightom, J.L. and Gonsalves, D. (1994) Resistance of transgenic *Nicotiana benthamiana* plants to tomato spotted wilt and impatiens necrotic spot tospoviruses: evidence of involvement of the N protein and N gene RNA in resistance. *Phytopathology* 84, 243–249.

Pang, S.Z., Jan, F.J. and Gonsalves, D. (1997) Nontarget DNA sequences reduce the transgene length necessary for RNA-mediated tospovirus resistance in transgenic plants. *Proceedings of the National Academy of Sciences USA* 94, 8261–8266.

Pappu, H.R., Rosales, I.M. and Druffel, K.L. (2008) Serological and molecular assays for rapid and sensitive detection of Iris yellow spot virus infection of bulb and seed onion crops. *Plant Disease* 92, 588–594.

Pappu, H.R., Jones, R.A. and Jain, R.K. (2009) Global status of tospovirus epidemics in diverse cropping systems: successes gained and challenges that lie ahead. *Virus Research* 141, 219–236.

Parrella, G., Gognalons, P., Gebre-Selassiè, K., Vovlas, C. and Marchoux, G. (2003) An update of the host range of *Tomato spotted wilt virus. Plant Pathology* 85, 227–264.

Pearce, M. (2005) *2004 Georgia Plant Disease Loss Estimates*. University of Georgia Cooperative Extension Service, 24 pp.

Peng, J.C., Yeh, S.D., Huang, L.H., Li, J.T., Cheng Y.F., *et al.* (2011) Emerging threat of thrips-borne Melon yellow spot virus on melon and watermelon in Taiwan. *European Journal of Plant Pathology* 130, 205–214.

Peng, J.C., Chen, T.C., Raja, J.A.J., Yang, C.F., Chien, W.C., *et al.* (2014) Broad-spectrum transgenic resistance against distinct tospovirus species at the genus level. *PLoS One* 9, e96073. doi: 10.1371/journal. pone.0096073.

Pittman, H.A. (1927) Spotted wilt of tomatoes: preliminary note concerning the transmission of the 'spotted wilt' of tomatoes by an insect vector (*Thrips tabaci* Lind.). *Journal of the Council of Science and Industrial Research* 1, 74–77.

Premachandra, W.T., Borgemeister, C., Maiss, E., Knierim, D. and Poehling, H.M. (2005) *Ceratothripoides claratris*, a new vector of a Capsicum chlorosis virus isolate infecting tomato in Thailand. *Phytopathology* 95, 659–663.

Prins, M. and Goldbach, R. (1996) RNA-mediated virus resistance in transgenic plants. *Archives of Virology* 141, 2259–2276.

Prins, M., de Haan, P., Luyten, R., van Veller, M., van Grinsven, M.Q., *et al.* (1995) Broad resistance to tospoviruses in transgenic tobacco plants expressing three tospoviral nucleoprotein gene sequences. *Molecular Plant-Microbe Interactions* 8, 85–91.

Rao, X., Wu, Z. and Li, Y. (2013) Complete genome sequence of a *Watermelon silver mottle virus* isolate from China. *Virus Genes* 46, 576–580.

Reddy, D.V.R., Sudarshana, M.R., Ratna, A.S., Reddy, A.S., Amin, P.W., *et al.* (1990) The occurrence of yellow spot virus, a member of tomato spotted wilt virus group, on peanut (*Arachis hypogaea* L.) in India. *U.S. Department of Agriculture-Agricultural Research Service Bulletin* 87, 77–88.

Reddy, D.V.R., Ranta, A.S., Suudarshana, M.R., Poul, F. and Kumar, I.K. (1992) Serological relationships and purification of bud necrosis virus, a tospovirus occurring in peanut (*Arachis hypogaea* L.) in India. *Annals of Applied Biology* 120, 279–286.

Reddy, D.V.R., Buiel, A.A.M., Satyanarayana, T., Dwivedi, S.L., Reddy, A.S., *et al.* (1995) Peanut bud necrosis virus disease: an overview. In: Buiel, A.A.M., Parlevliet, J.E. and Lenne, J.M. (eds) *Recent Studies on Peanut Bud Necrosis Disease: Proceedings of a Meeting*. March 20, 1995, ICRISAT Asia Centre, Andhra Pradesh, India, pp. 3–7.

Ribeiro, D., Borst, J.W., Goldbach, R. and Kormelink, R. (2009) *Tomato spotted wilt virus* nucleocapsid protein interacts with both viral glycoproteins Gn and Gc *in planta*. *Virology* 383, 121–130.

Riley, D.G., Joseph, S.V., Srinivasan, R. and Diffie, S. (2011) Thrips vectors of tospoviruses. *Journal of Integrated Pest Management* 2, 1–10.

Roberts, C.A., Dietzgen, R.G., Heelan, L.A. and Maclean, D.J. (2000) Real-time RT-PCR fluorescent detection of *Tomato spotted wilt virus*. *Journal of Virological Methods* 88, 1–8.

Rosello, S., Diez Maria, J. and Nuez, F. (1998) Genetics of *Tomato spotted wilt virus* resistance coming from *Lycopersicon peruvianum*. *European Journal of Plant Pathology* 104, 499–509.

Rudolph, C., Schreier, P.H. and Uhrig, J.F. (2003) Peptide-mediated broad-spectrum plant resistance to tospoviruses. *Proceedings of the National Academy of Sciences USA* 100, 4429–4434.

Sakimura, K. (1963) *Frankliniella fusca*, an additional vector for the *Tomato spotted wilt virus*, with notes on *Thrips tabaci*, a thrips vector. *Phytopathology* 53, 412–415.

Sakimura, K. (1969) A comment on the color forms of *Frankliniella schultzei* (Thysanoptera: Thripidae) in relation to transmission of the *Tomato spotted wilt virus*. *Pacific Insects* 11, 761–762.

Sakurai, T., Inoue, T. and Tsuda, S. (2004) Distinct efficiencies of *Impatiens necrotic spot virus* transmission by five thrips vector species (Thysanoptera: Thripidae) of tospoviruses in Japan. *Applied Entomology and Zoology* 39, 71–78.

Samuel, G., Blad, J.G. and Pittman, H.A. (1930) Investigation on "spotted wilt" of tomatoes. *Australasian Council of Science and Industrial Research Bulletin* 44, 1–46.

Sanford, J.C. and Johnston, S.A. (1985) The concept of parasite-derived resistance-deriving resistance genes from the parasite's own genome. *Journal of Theoretical Biology* 113, 395–405.

Satyanarayana, T., Gowda, S., Reddy, K.L., Mitchell, S.E., Dawson, W.O., *et al.* (1998) Peanut yellow spot virus is a member of a new serogroup of *Tospovirus* genus based on small (S) RNA sequence and organization. *Archives of Virology* 143, 353–364.

Schwach, F., Adam, G. and Heinze, C. (2004) Expression of a modified nucleocapsid-protein of *Tomato spotted wilt virus* (TSWV) confers resistance against TSWV and *Groundnut ringspot virus* (GRSV) by blocking systemic spread. *Molecular Plant Pathology* 5, 309–316.

Seepiban, C., Gajanandana, O., Attathom, T. and Attathom, S. (2011) Tomato necrotic ringspot virus, a new tospovirus isolated in Thailand. *Archives of Virology* 156, 263–274.

Shimomoto, Y., Kobayashi, K. and Okuda, M. (2014) Identification and characterization of Lisianthus necrotic ringspot virus, a novel distinct tospovirus species causing necrotic disease of lisianthus (*Eustoma grandiflorum*). *Journal of General Plant Pathology* 80, 169–175.

Sin, S.H., McNulty, B.C., Kennedy G.G. and Moyer, J.W. (2005) Viral genetic determinants for thrips transmission of *Tomato spotted wilt virus*. *Proceedings of the National Academy of Sciences USA* 102, 5168–5173.

Snippe, M., Borst, J.W., Goldbach, R. and Kormelink, R. (2007) *Tomato spotted wilt virus* Gc and N proteins interact in vivo. *Virology* 357, 115–123.

Stevens, M.R., Scott, S.J. and Gergerich, R.C. (1992) Inheritance of a gene for resistance to *Tomato spotted wilt virus* (TSWV) from *Lycopersicon esculentum* Mill. *Euphytica* 59, 9–17.

Takeda, A., Sugiyama, K., Nagano, H., Mori, M., Kaido, M., *et al.* (2002) Identification of a novel RNA silencing suppressor, NSs protein of *Tomato spotted wilt virus*. *FEBS Letters* 532, 75–79.

Torres, R., Larenas, J., Fribourg, C. and Romero, J. (2012) Pepper necrotic spot virus, a new tospovirus infecting solanaceous crops in Peru. *Archives of Virology* 157, 609–615.

Uga, H. and Tusda, S. (2005) A one-step reverse transcription-polymerase chain reaction system for the simultaneous detection and identification of multiple tospovirus infections. *Phytopathology* 95, 166–171.

Ullman, D.E., Whitfield, A.E. and German, T.L. (2005) Thrips and tospoviruses come of age: mapping determinants of insect transmission. *Proceedings of the National Academy of Sciences USA* 102, 4931–4932.

van de Wetering, F., Goldbach, R. and Peters, D. (1996) Tomato spotted wilt tospovirus ingestion by first instar larvae of *Frankliniella occidentalis* is a prerequisite for transmission. *Phytopathology* 86, 900–905.

Vijayalakshmi, K. (1994) Transmission and ecology of *Thrips palmi* Karny, the vector of peanut bud necrosis virus. PhD thesis, Andhra Pradesh Agricultural University, Hyderabad, India.

Wheeler, D.A., Srinivasan, M., Egholm, M., Shen, Y., Chen, L., *et al.* (2008) The complete genome of an individual by massively parallel DNA sequencing. *Nature* 452, 872–876.

Whitfield, A.E., Ullman, D.E. and German, T.L. (2005) Tospovirus-thrips interactions. *Annual Review of Phytopathology* 43, 459–489.

Wijkamp, I. and Peters, D. (1993) Determination of the median latent period of two tospoviruses in *Frankliniella occidentalis*, using a novel leaf disk assay. *Phytopathology* 83, 986–991.

Wijkamp, I., van Lent, J., Kormelink, R., Goldbach, R. and Peters, D. (1993) Multiplication of *Tomato spotted wilt virus* in its vector, *Frankliniella occidentalis*. *Journal of General Virology* 74, 341–349.

Wijkamp, I., Almarza, N., Goldbach, R. and Peters, D. (1995) Distinct levels of specificity in thrips transmission of tospoviruses. *Phytopathology* 85, 1069–1074.

Wilhelm, B.T., Marguerat, S., Watt, S., Schubert, F., Wood, V., *et al.* (2008) Dynamic repertoire of a eukaryotic transcriptome surveyed at single-nucleotide resolution. *Nature* 453, 1239–1243.

Xu, Y., Zhou, W., Zhou, Y., Wu, J. and Zhou, X. (2012) Transcriptome and comparative gene expression analysis of *Sogatella furcifera* (Horváth) in response to Southern rice black-streaked dwarf virus. *PLoS One* 7, e36238.

Yang, C.F., Chen, K.C., Cheng, Y.H., Raja, J.A.J., Huang, Y.L., *et al.* (2014) Generation of marker-free transgenic plants concurrently resistant to a DNA geminivirus and a RNA tospovirus. *Scientific Reports* 4, 5717. doi: 10.1038/srep05717.

Yeh, S.D. and Chang, T.F. (1995) Nucleotide sequence of the N gene of *Watermelon silver mottle virus*, a proposed new member of the genus *Tospovirus*. *Phytopathology* 85, 58–64.

Yeh, S.D., Lin, Y.C., Cheng, Y.H., Jih, C.L., Chen, M.J. and Chen, C.C. (1992) Identification of tomato spotted wilt-like virus infecting watermelon in Taiwan. *Plant Disease* 76, 835–840.

Yin, Y., Zheng, K., Dong, J., Fang, Q., Wu, S., *et al.* (2014) Identification of a new tospovirus causing necrotic ringspot on tomato in China. *Virology Journal* 11, 213.

Zhang, Z., Zhang, P., Li, W., Zhang J., Huang, F., *et al.* (2013) *De novo* transcriptome sequencing in *Frankliniella occidentalis* to identify genes involved in plant virus transmission and insecticide resistance. *Genomics* 101, 296–305.

Zhou, J., Kantartzi, S.K., Wen, R.H., Newman, M., Hajimorad, M.R., *et al.* (2011) Molecular characterization of a new tospovirus infecting soybean. *Virus Genes* 43, 289–295.

13 Tomato Yellow Leaf Curl

Cindy-Leigh Hamilton,[1] Sudeshna Mazumdar-Leighton,[2] Icolyn Amarakoon[3] and Marcia Roye[1]*

[1]*Biotechnology Centre, The University of the West Indies, Mona Campus, Jamaica; [2]Department of Botany, Delhi University, Delhi, India; [3]Department of Basic Medical Sciences, The University of the West Indies, Mona Campus, Jamaica*

13.1 Introduction

Tomato yellow leaf curl virus (TYLCV), a geminivirus of the genus *Begomovirus* and the family *Geminiviridae*, has impacted tomato (*Solanum lycopersicum*) cultivation worldwide in tropical and subtropical regions for many years (Picó *et al.*, 1996). The virus was first reported in the Jordan Valley, Israel, in the 1940s. Years later, it was isolated (Czosnek *et al.*, 1988) and sequenced (Navot *et al.*, 1991), and was among the first begomoviruses shown to consist of a single genomic DNA molecule. TYLCV also infects several other economically important crop plants including pepper (*Capsicum* spp.), bean (*Phaseolus vulgaris*) and tobacco (*Nicotiana* spp.), as well as numerous weed species (Roye *et al.*, 1999; Martínez-Zubiaur *et al.*, 2002; Polston *et al.*, 2006). This chapter summarizes the characteristics of TYLCV as well as the integrated approaches used to manage the disease it induces in various crops.

13.2 Characteristic Symptoms

Symptoms caused by TYLCV include severely stunted growth and reduced leaf size, which typically results in a bushy appearance of infected plants. The leaves of the infected plants curl upwards and are usually mottled and chlorotic at the margins. Flower abscission is common in tomato plants infected during the early stages of seedling development, leading to the production of small unmarketable fruits or limited fruit production and reduced yields (Polston *et al.*, 1999; Gilbertson *et al.*, 2007). Infected plants generally occur in random distribution patterns or clusters in the field. Although TYLCV can induce severe disease symptoms in tomato, the virus is capable of establishing symptomless infections in other hosts, such as peppers and cucurbits. These plants serve as reservoirs for the virus and despite the lack of symptom development, the whitefly vector (*Bemisia tabaci*) is capable of acquiring and transmitting TYLCV to other plants, including tomato (Polston *et al.*, 2006; Czosnek and Ghanim, 2012).

13.3 Economic Impact

TYLCV is one of the factors that severely limits tomato production (Hanssen *et al.*, 2010). Tomato is considered one of the most economically valuable vegetable crops, constituting 72% of the value of all vegetable crops. According to the Food and Agriculture Organization (FAO), global tomato production in 2012 reached 161 million metric tonnes valued at US$ 62.5 billion. The Americas in total produced 24.58 million tonnes of tomatoes while the Caribbean

*E-mail: marcia.roye@uwimona.edu.jm

produced 1.04 million tonnes with Jamaica producing just above 29 thousand tonnes (FAOSTAT Database, 2015). TYLCV incidence can reach up to 100% in both protected and open fields and can cause economic losses of between 50% and 90%, especially when the virus infects the crop during the early growing stages. Reduction in production by TYLCV can have a devastating economic impact, resulting in a reduction in income for crop producers and distributors and higher prices for consumers (Lapidot and Friedmann, 2002; Polston and Lapidot, 2007).

13.4 Geographic Distribution

The geographical range of TYLCV is among the widest of plant viruses known to cause epidemics (Czosnek and Laterrot, 1997). The primary centre of TYLCV diversity is the tomato-growing regions in the Middle-East/Mediterranean region/Africa, which includes Iran as the putative center of origin (Lefeuvre *et al.*, 2010; Hosseinzadeh *et al.*, 2014). Diverse isolates of the TYLCV from South-East Asia/Far East and Australia include reports from Japan, South Korea, China and Taiwan. The third region is the continental Americas and the Caribbean (Polston and Anderson, 1997; Morales and Anderson, 2001; Zambrano *et al.*, 2007). Multiple introductions of TYLCV isolates have contributed to the spread of tomato yellow leaf curl disease (TYLCD) in this region (Duffy and Holmes, 2007). At least 43 isolates are reported for TYLCV (Fauquet *et al.*, 2008). Most of these reports are from surveys of symptomatic hosts and epidemiological studies following disease outbreaks (Czosnek and Laterrot, 1997; Moriones and Navas-Castillo, 2000; Cathrin and Ghanim, 2014). Expansion of the geographic distribution and host range of TYLCV has been attributed to introductions of infected and/or whitefly-infested planting materials. Upsurges in whitefly populations, changes in cropping patterns and climatic conditions have presumably contributed to the prevalence and global spread of the virus (Navas-Castillo *et al.*, 2011).

13.5 Genome Organization

The family *Geminiviridae* contains viruses possessing circular, single-stranded DNA (ssDNA) genomes and is divided into seven genera based on their insect vector, host range and genome organization (Fauquet *et al.*, 2008; Varsani *et al.*, 2014). TYLCV is a member of the genus *Begomovirus*. The genome of most begomoviruses is bipartite, with their DNA separated into two components, DNA-A and DNA-B. Six open reading frames are shared between both DNA components. TYLCV is unique as it is monopartite and all six open reading frames are on a single component of 2700–2800 base pairs (Lazarowitz, 1992; Fig. 13.1).

The TYLCV genes run in opposite directions from an approximate 200 nucleotide intergenic region (IR). The virion sense genes are V1 and V2, and the complementary strand genes are C1, C2, C3, and C4. Flanking the ends of the IR are conserved repeats with the motif TAATATTAC, which function as the origin of replication (*ori*). The IR also contains transcriptional promoters for V1, V2, C1 and C4. The promoters for C2 and C3 are found within C1 (Lazarowitz, 1992; Fig. 13.1).

V1 encodes the capsid protein (CP, 30 kDa) that is responsible for the encapsidation of the ssDNA genome. The TYLCV CP is believed to be a functional homolog of the nuclear shuttle protein and movement protein (MP) encoded by the DNA–B of begomoviruses (Rojas *et al.*, 2001). TYLCV CP tends to localize near the nucleus and is necessary for the establishment of systemic infection (Wartig *et al.*, 1997). CP also determines insect transmission specificity by interacting with GroEL proteins produced by endosymbiotic bacteria found with *B. tabaci* (Morin *et al.*, 1999).

The V2 or movement protein (13 kDa) works along with the CP and C4 gene products to ensure movement of virions (Wartig *et al.*, 1997). TYLCV infection appears to trigger host RNA silencing as part of the host defence mechanism; V2 counter silences this by producing RNA-silencing suppressor proteins (Bar-Ziv *et al.*, 2012). C1 codes for the replication initiation protein (Rep,

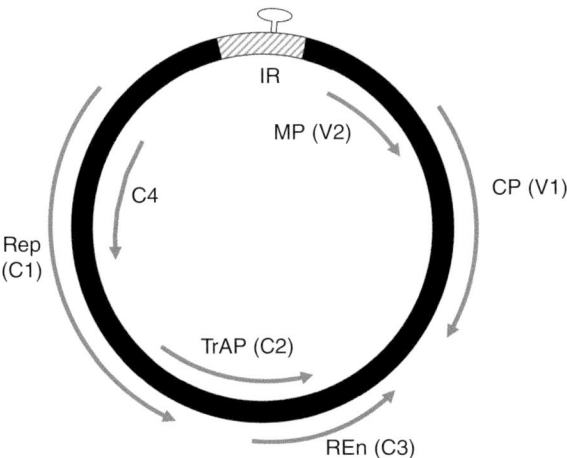

Fig. 13.1. Genome organization of *Tomato leaf curl virus* (TYLCV). V1, capsid protein (CP); V2, movement protein (MP); C1, replication initiation protein (Rep); C2, transcriptional activator protein (TrAP); C3, replication enhancer (Ren); C4 is responsible for symptom expression and viral movement.

40–41 kDa) and is the only viral protein involved in the replication of its DNA. During replication, the circular ssDNA is first converted to circular double-stranded DNA, and amplification occurs via a rolling circle mechanism. Rep triggers replication by binding to the *ori* site and cleaves at specific DNA sequences. The 5′ end of the nick remains linked to Rep while the 3′ end acts as a primer for the DNA polymerase to synthesize a new strand (Desbiez *et al.*, 1995).

C2 encodes the transcription activator protein (TrAP, 15 kDa) and, as in bipartite geminiviruses, TrAP has been shown to enhance transcription from DNA-A, V1 and V2 promoters and DNA-B C1 and V1 promoters (Sunter and Bisaro, 1992; Noris *et al.*, 1996). In addition to its transactivation activity, TrAP is involved in suppressing RNA-silencing (Chellappan *et al.*, 2004).

C3 codes for a replication enhancer (16 kDa); its presence increases the viral ssDNA and dsDNA concentration during infection, which is linked to the degree of symptom expression in plants (Gutiérrez, 1999). Tomato plants infected with TYLCV having a C4 mutation do not exhibit symptoms of infection; this is because C4 mutants are unable to systemically infect plants.

These findings suggest that C4 gene products play an important role in viral movement (Jupin *et al.*, 1994).

13.6   Main Ecotypes

TYLCD is transmitted by whiteflies belonging to the *B. tabaci* Gennadius species complex (Czosnek, 2008). TYLCV includes viral isolates from Israel sequenced in the 1990s that caused severe (TYLCV-IL) and mild (TYLCV-Mld) symptoms (Navot *et al.*, 1991; Antignus and Cohen, 1994). The TYLCV-IL and TYLCV-Mld strains are particularly widespread in their distribution (Czosnek, 2008; Cathrin and Ghanim, 2014). There are seven known strains of TYLCV that include TYLCV-IL , TYLCV-Mld, TYLCV-Gezira (Sudan), TYLCV-Oman, and three strains from Iran (TYLCV-Iran, TYLCV-Ker and TYLCV-Bou) (Lefeuvre *et al.*, 2010; Hosseinzadeh *et al.*, 2014). Sequences of TYLCV strains from afflicted countries can provide valuable spatial and temporal information on trajectories of virus emergence, introduction, persistence and evolution. Partial and/or complete genome sequences of TYLCV isolates are available from Iran (Bananej *et al.*, 2004), Egypt (Nakhla *et al.*, 1993), Sudan (Idris and Brown, 2005), Portugal

(Navas-Castillo *et al.*, 2000), Spain (Noris *et al.*, 1994), Japan (Kato *et al.*, 1998), China (Wu and Zhou, 2006), South Korea (Kim *et al.*, 2011); Mexico (Ascencio-Ibáñez *et al.*, 1999), Morocco (Boukhatem *et al.*, 2008), Puerto Rico (Bird *et al.*, 2001), Tunisia (Chouchane *et al.*, 2007), Turkey (Köklü *et al.*, 2006), Oman (Khan *et al.*, 2008), Iraq (Al-Kuwaiti *et al.*, 2013), Jordan (Anfoka *et al.*, 2009); Dominican Republic (Nakhla *et al.*, 1994), Reunion islands (Delatte *et al.*, 2005), Cuba (Gómez and Gonzáles, 1993), Jamaica and the Caribbean region (McGlashan *et al.*, 1994; Roye *et al.*, 1999), the USA (Polston *et al.*, 1999) and Australia (Stonor *et al.*, 2003). Isolates of TYLCV-IL and TYLCV-Mld were recently detected in tomato fruits in Northern Europe (Just *et al.*, 2014).

In addition, at least ten species of begomoviruses cause TYLCD (Díaz-Pendón *et al.*, 2010). Isolates of this complex of TYLCV-like begomoviruses have been reported from Europe (Spain, Italy), Africa (Mali) and Asia (China, Indonesia, Thailand and Vietnam). These begomoviruses cause devastating disease outbreaks when they recombine with each other or with strains of TYLCV (Monci *et al.*, 2002; Davino and Accotto, 2009; Díaz-Pendón *et al.*, 2010; Kim *et al.*, 2011).

In general, DNA-B genomes and satellite molecules have not been reported for TYLCV strains, which mostly have a monopartite genome (Czosnek, 2008). TYLCV-like viruses are also monopartite, except for begomoviruses from Thailand causing TYLCD (Rochester *et al.*, 1994; Díaz-Pendón *et al.*, 2010). Recently, a TYLCV-IR strain from Oman has been found associated with a satellite DNA molecule (Khan *et al.*, 2008).

13.7 Transmission

Whiteflies belonging to the species complex (biotype) of *B. tabaci* are the sole vectors for begomoviruses causing TYLCD (Cohen and Nitzany, 1966; Ghanim, 2014). TYLCD outbreaks often accompany upsurges in vector populations (Czosnek and Ghanim, 2011). *B. tabaci* has a wide host range including cotton, cassava, sweet potato, tobacco and tomato that are vulnerable to begomoviruses.

About 300 plant species belonging to 60 families can be infested by *B. tabaci* (Mound and Halsey, 1978).

Analysis of mitochondrial *cytochrome oxidase I* gene sequences suggest the existence of at least 24 putative species associated with spread of begomoviruses (Dinsdale *et al.*, 2010; Navas-Castillo *et al.*, 2011). Further, examination of 16S ribosomal RNA gene sequences indicates the existence of five geographical groups of *B. tabaci* from the (i) New World; (ii) South-East Asia; (iii) Mediterranean basin; (iv) Indian subcontinent; and (v) Equatorial Africa (Cathrin and Ghanim, 2014). Insect biotypes from different geographical regions have been differentiated as well by using isozymes (Brown *et al.*, 1995a), and various molecular markers (Gill and Brown, 2010).

Biotypes of *B. tabaci* can differ in phenotypic traits regarding utilization of different host plants, transmissibility of geminiviruses, developmental rate, resistance to pesticides, response to heat stress and parasitoids, and nature/composition of endosymbionts (Bedford *et al.*, 1994; Brown and Bird, 1995; Wang and Tsai, 1996; Díaz-Pendón *et al.*, 2010). In areas of TYLCD prevalence, both local and invasive biotypes frequently co-exist. But invasive biotypes frequently outperform local biotypes in their ability to transmit TYLCD. TYLCV is vectored solely by the biotype B in the Middle-Eastern–Asia-Minor region (Cathrin and Ghanim, 2014). This biotype or *Bemisia argentifolii* has been implicated with begomovirus invasions, outcompeting local insect biotypes and causing outbreaks of TYLCD (Polston *et al.*, 2014). The biotype B (silverleaf whitefly) emerged in the 1980s (Brown *et al.*, 1995b). Another invasive *B. tabaci* is biotype Q from the Mediterranean (Guirao *et al.*, 1997). A closely related biotype Q in the USA is currently restricted to greenhouses (McKenzie *et al.*, 2009). TYLCV has been vectored by biotype B and Q in Spain, Italy, Israel and China (Noris *et al.*, 1994; Jiang *et al.*, 2004; Horowitz *et al.*, 2005; Polston *et al.*, 2014). A local biotype Asia II-1 in China is only half as efficient as the invasive biotypes B and Q in transmission of TYLCV (Li *et al.*, 2010). The biotypes B and Q do not seem to mate, and viruliferous females of one biotype do not pass TYLCV to males of the other biotype and vice versa (Pascual and Callejas, 2004; Ghanim et al., 2007). Biotype NW-1 from the New World cannot mate with biotype B (De Barro et al., 2011). These results further strengthen the species complex concept for *B. tabaci*. Horizontal transmission of TYLCV and *Tomato yellow leaf curl China virus* was reported among the invasive biotypes B and Q at very low frequency (Wang et al., 2009).

After acquisition from phloem cells of an infected host plant, TYLCV circulates in the adult insect moving from stylet and food canal, via the oesophagus, to the gut. The virions traverse the midgut and enter the haemolymph through membranes of the filter chamber (Ghanim, 2014). In the haemolymph, the virions escape the insects' immune response and proteolytic cascades by binding to a 63-kDa GroEL chaperone encoded by symbiotic bacteria (Morin *et al.*, 1999; Gottlieb *et al.*, 2010). The virus can be retained in the vector for a few hours or a lifetime (Ghanim, 2014). From the haemolymph, the virions reach the primary salivary glands (Brown and Czosnek, 2002; Czosnek and Ghanim, 2002) and are subsequently transmitted to host plant phloem tissues via steps involving insect alighting, probing and stylet insertion, salivation and inoculation of virus (Hogenhout *et al.*, 2008; Fereres and Moreno, 2009).

Biotype B can begin transmitting TYLCV-IL 8 hours after an access acquisition phase (AAP) of 10 minutes (Ghanim et al., 2001). The inoculation acquisition phase can be as low as 1.8 minutes upon reaching phloem tissue (Jiang et al., 2000). The inoculation acquisition phase varies according to the virus, vector, host plant type and experimental conditions (Ghanim, 2014). Accumulation of viral transcripts and an increase in viral DNA within the insect after AAP have been demonstrated (Czosnek et al., 2001; Sinisterra et al., 2005). Nearly 100% efficiency of transmission (and a 24 hour AAP) can be achieved with 5–15 viruliferous insects (Cohen and Nitzany, 1966; Mehta et al., 1994). The AAP can vary from 15–60 minutes for TYLCV-IL and other

Mediterranean isolates. Transmission efficiency can vary between different biotypes (Polston et al., 2014).

In general, the virus is not transmitted sexually. However, transmission of TYLCV-IL between males and females of biotype B (if no other source of inoculum is available) and the ability of the recipient fly to infect a host plant have been documented (Ghanim and Czosnek, 2000; Ghanim et al., 2001, 2007). TYLCV is transmitted vertically and the virus can be detected in reproductive tissues of *B. tabaci* (Ghanim et al., 1998). TYLCV was shown as a negative load with fitness costs for the vector in terms of reduced fertility (Rubinstein and Czosnek, 1997; Jiu et al., 2007; Matsuura and Hoshino, 2009; De Barro et al., 2011), suggesting that the virus may also be an insect pathogen (Czosnek and Ghanim, 2012). Reports document the effects of TYLCV on insect behaviour regarding sluggishness, increased salivation and prolonged sap ingestion (Fang et al., 2013). Behavioural responses in viruliferous insects have been associated with improved feeding and enhanced nutrition in *B. tabaci* (Moreno-Delafuente et al., 2013). Viruliferous biotype B isolates show a preference for healthy host plants, whereas those of the biotype Q prefer feeding on infected hosts. This altered host plant preference of the vectors has been implicated in the successful spread and persistence of biotype Q in China (Fang et al., 2013).

13.8 Control Strategies

Since TYLCV is not transmitted mechanically or by seeds, effective control and management of TYLCV transmission relies almost entirely on the successful control of the vector. Before planting, it is important to select a suitable growing site. The area chosen should be free of weeds and separated from fields containing other TYLCV and whitefly host crops. Plant residues from the previous growing season should be completely destroyed. It may even be necessary to strategically rotate periods in which host plants and non-host plants are grown. These measures minimize the chances of host crops and weeds acting as reservoirs for TYLCV and whitefly proliferation. To be effective, all farmers must adhere to similar growing practices. This initiative was put forward in the Dominican Republic between 1995 and 1996. The Ministry of Agriculture established a mandatory country-wide 3 month host-free period. Planters were restricted from growing common bean, cucurbits, eggplant, melon, okra, pepper and tomato. This implementation resulted in a drastic reduction in TYLCV (Polston and Anderson, 1997).

The soil mineral content must also be considered before planting. The soil must provide the optimal mineral concentrations in order for the plant to thrive and withstand infection. Poor nitrogen levels may stress the plant causing them to succumb to disease; increased nitrogen and phosphorus levels are associated with increased viral replication (Pennazio and Roggero, 1997).

Once an ideal site has been selected and prepared, it is time to transfer transplants that should be free of virus infection and whitefly infestation. To achieve this, transplants are usually grown in areas away from production sites within protected greenhouses. For protection during the first week in the field, transplants may be treated with a neonicotinoid insecticide (such as imidacloprid or thiamethoxam) a week prior to field transplant.

The use of TYLCV-resistant cultivars is an effective way of ensuring virus-free transplants. TYLCV-resistant cultivars have been developed by identifying tomato genes that naturally confer resistance. Identifying such genes within the domesticated tomato plant *S. lycopersicum* was unsuccessful. Instead, several wild tomato plants have been noted for possessing resistant genes including *S. pimpinellifolium*, *S. chmielewskii*, *Lycianthes glandulosa* (previously *S. glandulosum*), *S. lycopersicoides*, *S. habrochaites*, *S. chilense* and *S. peruvianum*. Five major resistant genes have been identified, designated Ty-1 to Ty-5. Several commercial varieties of TYLCV-resistant cultivars are available. These varieties were developed as a result of the introgression breeding of several accessions displaying resistance within the same species (Czosnek, 2007).

Limitations, however, exist with the use of TYLCV-resistant cultivars. If these plants are still susceptible to other geminiviral infections, they are likely to succumb to infection if exposed to highly viruliferous whiteflies shortly after being transplanted, and they may possess reduced fruit quality and increased susceptibility to biotic and abiotic stresses (Tal and Shannon, 1983; Polston and Anderson, 1997; Lapidot and Friedmann, 2002). There are also concerns of how well resistant cultivars will maintain tolerance in different regions. For example, the resistant variety 'TY52', derived from *S. chilense* accession LA1969, displays resistance in Israel, but is susceptible in Guatemala, and this is possibly due to the prevalence of bipartite begomoviruses and the high viral pressure that exists in Guatemala. Commercially tolerant varieties that have been proven effective in Jamaica include 'Gemstar', 'Gempair', 'Gempride' and 'Adonis', all from the Seminis Seed Company.

Efforts have also focused on the assessment of *L. pennellii,* which although susceptible to TYLCV, demonstrates a high level of resistance to *B. tabaci*. This resistance to the virus vector is due to a sticky material secreted by granular trichomes present on the leaves and stems of the plant (Berlinger and Dahan, 1987). However, to date there are no commercially available *B. tabaci* resistant cultivars.

Reflective polyethylene mulches have been shown to delay whitefly infestation. Some mulches may be partially aluminized and reflect both white and UV light. This disorients whiteflies and thus affects their ability to land on plants. Yellow mulches attract whiteflies; the high temperature of these mulches dehydrates the whiteflies that have landed on them. Yellow mulches are more effective in dry, arid conditions where temperatures remain high, such as in Israel. In Florida, where atmospheric humidity tends to be high, aluminium mulches are preferred as the high humidity counters the dehydrating nature of yellow mulches and may only serve to attract whiteflies. The effectiveness of these mulches decreases once canopy size increases; it is advisable to use them early in the growing season (Cohen and Melamed-Madjar, 1978).

Once plants are growing in the fields, it becomes necessary to regularly inspect plants for the presence of the whitefly and TYLCV symptoms. Growers must be familiar with the physical features exhibited during each stage of whitefly life cycle. The underside of leaves may be examined with a hand lens to identify either the adult or their eggs. Alternatively, sticky yellow traps may be hung vertically on tomato plants and examined. When plant leaves are disturbed, the whitefly will be attracted to the yellow traps, stick to them and eventually die.

If infestation or infection has been detected early in the growing season, affected plants must be removed, bagged and discarded. Controlling infestation may be achieved by the use of chemical insecticides. Various classes of insecticides are available, each targeting different stages of the whitefly life cycle and using different mechanisms of action. These differences further influence the method of application used. Common classes of insecticides include pyrethroids, organophosphates, neonicotinoids, pyridine-azomethines, carbamates, chlorinated hydrocarbons, insect growth regulators, oils and soaps (Perring et al., 1999). Of these, the most effective and widely used are the neonicotinoids. However, frequent neonicotinoid use has been associated with the destruction of important pollinators such as the honey bee (van der Sluijs et al., 2013). Once insecticides are being used, there is the potential for development of resistance (Polston and Anderson, 1997). To reduce this risk, it is necessary to ensure complete coverage when applying insecticides, especially on the underside of leaves, and routinely rotating the classes of insecticides used. Soaps and oils do not result in resistance and should therefore be considered as alternatives.

The concerns of insecticide resistance have led to explorations of alternative methods of managing the disease; namely RNA interference and biological agents. The verdict is still out on whether RNA interference will be able to deliver effective resistance against geminiviruses, which is durable in the field. Most efforts so far have been less than successful and produced tolerant transgenic varieties, presumably because of the counter defence mechanisms of this group of viruses. Similar results have been reported with biological control agents. Biological control focuses on the use of natural predators or parasitoids against the TYLCV vector. Several natural enemies of B. tabaci have been identified; these include beetles (Coccinellidae), true bugs (Miridae, Anthocoridae), lace wings (Chrysopidae, Coniopterygidae), mites (Phytoseiidae) and spiders (Araneae), a few of which are commercially available. Parasites belonging to the genera Encarsia and Eretmocerus have also been successfully used (Gerling et al., 2001). Fungal agents such as, Cordyceps confragosa (previously Verticillium lecanii), Paecilomyces fumosoroseus and Beauveria bassiana seem to have potential for success; however, their commercialization has been hindered by several limitations (Faria and Wraight, 2001). The greenhouse whitefly, in particular Trialeurodes vaporariorum, has responded positively to biological control methods. The control of B. tabaci is more complex, partly due to their extensive host range coupled with their high reproductive capacity often resulting in high intercrop infestation and rapid population increase. Other problematic factors include the varying climatic conditions which require the use of a combination of various predators, the lethal effects of insecticides on identified predators and the inconsistent production of certain annual crops which makes it difficult to establish a stable environment for predators (Gerling et al., 2001).

13.9 Concluding Remarks

Despite significant progress in disease diagnosis, molecular characterization and development of tolerant or resistant tomato cultivars, TYLCV continues to be among the viruses that are causing great economic losses in field- and greenhouse-grown tomatoes.

Controlling the whitefly vector and limiting transmission of the virus have been met with limited success. Challenges to controlling outbreaks caused by TYLCD include limited number of cultivars with tolerance or resistance to the different virus complexes, the adaptability of these cultivars to the varied ecosystems in which tomatoes are grown, and the susceptibility of these cultivars to other begomoviruses. Also of notable concern is the continued evolution of the virus through recombination and mutation which may increase adaptability to the host or to virus transmission.

References

Al-Kuwaiti, N., Otto, B., Collins, C., Seal, S. and Maruthi, M. (2013) Molecular characterization and first complete genome sequence of *Tomato yellow leaf curl virus* (TYLCV) infecting tomato in Iraq. *New Disease Reports* 27, 17.

Anfoka, G., Haj-Ahmad, F., Abhary, M. and Hussein, A. (2009) Detection and molecular characterization of viruses associated with tomato yellow leaf curl disease in cucurbit crops in Jordan. *Plant Pathology* 58, 754–762.

Antignus, Y. and Cohen, S. (1994) Complete nucleotide sequence of an infectious clone of a mild isolate of *Tomato yellow leaf curl virus* (TYLCV). *Phytopathology* 84, 707–712.

Ascencio-Ibáñez, J.T., Díaz-Plaza, R., Méndez-Lozano, J., Monsalve-Fonnegra, Z.I., Argüello-Astorga, G.R., *et al.* (1999) First report of Tomato yellow leaf curl geminivirus in Yucatán, México. *Plant Disease* 83, 1178.

Bananej, K., Kheyr-Pour, A., Hosseini Salekdeh, G. and Ahoonmanesh, A. (2004) Complete nucleotide sequence of Iranian tomato yellow leaf curl virus isolate: further evidence for natural recombination amongst begomoviruses. *Archives of Virology* 149, 1435–1443.

Bar-Ziv, A., Levy, Y., Hak, H., Mett, A., Belausov, E., *et al.* (2012) The *Tomato yellow leaf curl virus* (TYLCV) V2 protein interacts with the host papain-like cysteine protease CYP1. *Plant Signaling and Behavior* 8, 983–989.

Bedford, I.D., Briddon, R.W., Brown, J.K., Rosell, R.C. and Markham, P.G. (1994) Geminivirus transmission and biological characterisation of *Bemisia tabaci* (Gennadius) biotypes from different geographic regions. *Annals of Applied Biology* 125, 311–325.

Berlinger, M.J. and Dahan, R. (1987) Breeding for resistance to virus transmission by whiteflies in tomatoes. *Insect Science and its Application* 8, 783–784.

Bird, J., Idris, A.M., Rogan, D. and Brown, J.K. (2001) Introduction of the exotic *Tomato yellow leaf curl virus*-Israel in tomato to Puerto Rico. *Plant Disease* 85, 1028.

Boukhatem, N., Jdaini, S., Muhovsky, Y., Jacquemin, J.M., Del Rincóne, C.L., *et al.* (2008) Molecular characterization of *Tomato yellow leaf curl virus* Alm in Morocco: complete sequence and genome organization. *Journal of Plant Pathology* 9, 109–112.

Brown, J.K. and Bird, J. (1995) Variability within the *Bemisia tabaci* species complex and its relation to new epidemics caused by geminivirus. *CEIBA* 36, 73–80.

Brown, J.K. and Czosnek, H. (2002) Whitefly transmission of plant viruses. In: Plumb, R.T. (ed.) *Advances in Botanical Research: Plant Virus Vector Interactions*. Academic Press, New York, pp. 65–100.

Brown, J.K., Coats, S.A., Bedford, I.D., Markham, P.G., Bird, J., *et al.* (1995a) Characterization and distribution of esterase electromorphs in the whitefly, *Bemisia tabaci* (Genn.) (Homoptera: Aleyrodidae). *Biochemical Genetics* 33, 205–214.

Brown, J.K., Frohlich, D.R. and Rosell, R.C. (1995b) The sweetpotato or silverleaf whiteflies: biotypes of *Bemisia tabaci* or a species complex? *Annual Review of Entomology* 40, 511–534.

Cathrin, P.B. and Ghanim, M. (2014) Recent advances on the interactions between the whitefly *Bemisia tabaci* and begomoviruses, with emphasis on *Tomato yellow leaf curl virus*. In: Gaur, P.K., Hohn, T. and Sharma, P. (eds) *Plant Virus-Host Interactions: Molecular Approach and Viral Evolution*. Elsevier Ltd., Oxford, UK, pp. 79–86.

Chellappan, P., Vanitharani, R. and Fauquet, C.M. (2004) Short interfering RNA accumulation correlates with host recovery in DNA virus-infected hosts, and gene silencing targets specific viral sequences. *Journal of Virology* 78, 7465–7477.

Chouchane, S.G., Gorsane, F., Nakhla, M.K., Maxwell, D.P., Marrakchi, M., *et al.* (2007) First report of *Tomato yellow leaf curl virus*-Israel species infecting tomato, pepper and bean in Tunisia. *Journal of Phytopathology* 155, 236–240.

Cohen, S. and Melamed-Madjar, V. (1978) Prevention by soil mulching of the spread of *Tomato yellow leaf curl virus* transmitted by *Bemisia tabaci* (Gennadius) (Hemiptera: Aleyrodidae) in Israel. *Bulletin of Entomological Research* 68, 465–470.

Cohen, S. and Nitzany, F.E. (1966) Transmission and host range of *Tomato yellow leaf curl virus*. *Phytopathology* 56, 1127–1131.

Czosnek, H. (2007) *Tomato Yellow Leaf Curl Virus. Management, Molecular Biology, Breeding for Resistance*. Springer, Dordrecht, The Netherlands.

Czosnek, H. (2008) *Tomato Yellow Leaf Curl Virus.* In: Mahy, B.W.J. and van Regenmortel, M.H.V. (eds) *Encyclopedia of Virology*, 5 vols. Elsevier Ltd., Oxford, UK.

Czosnek, H. and Ghanim, M. (2002) Circulative pathway of begomoviruses in the whitefly vector *Bemisia tabaci* – insights from studies with *Tomato yellow leaf curl virus*. *Annals of Applied Biology* 140, 215–231.

Czosnek, H. and Ghanim, M. (2011) *Bemisia tabaci* – *Tomato yellow leaf curl virus* interaction causing world-wide epidemics. In: Thompson, W.M.O. (ed.) *The Whitefly, Bemisia tabaci (Homoptera: Aleyrodidae) Interaction with Geminivirus-Infected Host Plants.* Springer, Dordrecht, The Netherlands, pp. 51–67.

Czosnek, H. and Ghanim, M. (2012) Back to basics: are begomoviruses whitefly pathogens? *Journal of Integrative Agriculture* 11, 225–234.

Czosnek, H. and Laterrot, H. (1997) A worldwide survey of Tomato yellow leaf curl viruses. *Archives of Virology* 142, 1391–1406.

Czosnek, H., Beer, R., Antignus, Y., Cohen, S., Navot, N., *et al.* (1988) Isolation of *Tomato yellow leaf curl virus*, a geminivirus. *Phytopathology* 78, 508–512.

Czosnek, H., Ghanim, M., Rubinstein, G., Morin, S., Fridman, V., *et al.* (2001) Whiteflies: vectors – or victims ? – of geminiviruses. *Advances in Virus Research* 57, 291–322.

Davino, M. and Accotto, G.P. (2009) Two new natural begomovirus recombinants associated with tomato yellow leaf curl disease co-exist with parental viruses in tomato epidemics in Italy. *Virus Research* 143, 15–23.

De Barro, P.J., Liu, S.-S., Boykin, L.-M. and Dinsdale, A.-B. (2011) *Bemisia tabaci*: a statement of species status. *Annual Review of Entomology* 56, 1–19.

Delatte, H., Holota, H., Naze, F., Peterschmitt, M., Reynaud, B., *et al.* (2005) South West Indian Ocean islands tomato begomovirus populations represent a new major monopartite begomovirus group. *Journal of General Virology* 86, 1533–1542.

Desbiez, C., David, C., Mettouchi, A., Laufs, J. and Gronenborn, B. (1995) Rep protein of *Tomato yellow leaf curl geminivirus* has an ATPase activity required for viral DNA replication. *Proceedings in National Academy of Science USA* 92, 5640–5644.

Díaz-Pendón, J.A., Cañizares, M.C., Moriones, E., Bejarano, E.R., Czosnek, H., *et al.* (2010) Tomato yellow leaf curl viruses: *ménage à trois* between the virus complex, the plant and the whitefly vector. *Molecular Plant Pathology* 11, 441–450.

Dinsdale, A., Cook, L., Riginos, C., Buckley, Y.M. and De Barro, P. (2010) Refined global analysis of *Bemisia tabaci* (Hemiptera: Sternorrhyncha: Aleyrodoidea: Aleyrodidae) Mitochondrial *cytochrome oxidase I* to identify species level genetic boundaries. *Annals of the Entomological Society of America* 103, 196–208.

Duffy, S. and Holmes, E.C. (2007) Multiple introductions of the Old World begomovirus *Tomato yellow leaf curl virus* into the New World. *Applied Environmental Microbiology* 73, 7114–7117.

Fang, Y., Jiao, X., Xie, W., Wang, S., Wu, Q., *et al.* (2013) *Tomato yellow leaf curl virus* alters the host preferences of its vector *Bemisia tabaci*. *Scientific Reports* 3, 2876.

FAOSTAT Database (2015) http://faostat3.fao.org/download/Q/QC/E (accessed 15 July 2015).

Faria, M. and Wraight, S.P. (2001) Biological control of *Bemisia tabaci* with fungi. *Crop Protection* 20, 767–778.

Fauquet, C., Briddon, R., Brown, J., Moriones, E., Stanley, J., *et al.* (2008) Geminivirus strain demarcation and nomenclature. *Archives of Virology* 153, 783–821.

Fereres, A. and Moreno, A. (2009) Behavioural aspects influencing plant virus transmission by homopteran insects. *Virus Research* 141, 158–168.

Gerling, D., Alomar, O. and Arno, J. (2001) Biological control of *Bemisia tabaci* using predators and parasitoids. *Crop Protection* 20, 779–799.

Ghanim, M. (2014) A review of the mechanisms and components that determine the transmission efficiency of *Tomato yellow leaf curl virus* (*Geminiviridae*; *Begomovirus*) by its whitefly vector. *Virus Research* 186, 47–54.

Ghanim, M. and Czosnek, H. (2000) Tomato yellow leaf curl geminivirus (TYLCV-Is) is transmitted among whiteflies (*Bemisia tabaci*) in a sex-related manner. *Journal of Virology* 74, 4738–4745.

Ghanim, M., Morin, S., Zeidan, M. and Czosnek, H. (1998) Evidence for transovarial transmission of *Tomato yellow leaf curl virus* by its vector the whitefly *Bemisia tabaci. Virology* 240, 295–303.

Ghanim, M., Morin, S. and Czosnek, H. (2001) Rate of *Tomato yellow leaf curl virus* pathway at its vector, the whitefly *Bemisia tabaci. Phytopathology* 91, 188–196.

Ghanim, M., Sobol, I., Ghanim, M. and Czosnek, H. (2007) Horizontal transmission of begomoviruses between *Bemisia tabaci* biotypes. *Arthropod-Plant Interactions* 1, 195–204.

Gilbertson, R., Rojas, M.R., Kon, T. and Jaquez, J. (2007) Introduction of *Tomato yellow leaf curl virus* into the Dominican Republic: the development of a successful integrated pest management strategy. In: Czosnek, H. (ed.) *Tomato Yellow Leaf Curl Virus Disease*. Springer, Dordrecht, The Netherlands, pp. 279–303.

Gill, R.J. and Brown, J.K. (2010) Systematics of *Bemisia* and *Bemisia* relatives: can molecular techniques solve the *Bemisia tabaci* complex conundrum – a taxonomist's viewpoint. In: Stansly, P.A. and Naranjo, S.E. (eds) *Bemisia: Bionomics and Management of a Global Pest*. Springer, Dordrecht, The Netherlands, pp. 5–29.

Gómez, G. and Gonzáles, G. (1993) Genetic approach for resistance to the Cuban geminiviruses transmitted by whiteflies. *Tomato Leaf Curl News* 3, 3.

Gottlieb, N., Zchori-Fein, E., Mozes-Daube, N., Kontsedalov, S., Skaljac, M., *et al.* (2010) The transmission efficiency of *Tomato yellow leaf curl virus* by the whitefly *Bemisia tabaci* is correlated with presence of a specific symbiotic bacterial species. *Journal of Virology* 84, 9310–9317.

Guirao, P., Beitia, F. and Cenis, J.C. (1997) Biotype determination of Spanish populations of *Bemisia tabaci* (Hemiptera: Aleyrodidae). *Bulletin of Entomological Research* 87, 587–593.

Gutiérrez, C. (1999) Geminivirus DNA replication. *Cellular and Molecular Life Science* 56, 313–329.

Hanssen, I.M., Lapidot, M. and Thomma, B.P. (2010) Emerging viral diseases of tomato crops. *Molecular Plant-Microbe Interactions* 23, 539–548.

Hogenhout, S.A., Ammar, E.D., Whitfield, A.E. and Redinbaugh, M.G. (2008) Insect vector interactions with persistently transmitted viruses. *Annual Review of Phytopathology* 46, 327–359.

Horowitz, A.R., Kontsedalov, S., Khasdan, V. and Ishaaya, I. (2005) Biotypes B and Q of *Bemisia tabaci* and their relevance to neonicotinoid and pyriproxyfen resistance. *Archives of Insect Biochemistry and Physiology* 58, 216–225.

Hosseinzadeh, M., Shams-Bakhsh, M., Osaloo, S.K. and Brown, J.K. (2014) Phylogenetic relationships, recombination analysis, and genetic variability among diverse variants of *Tomato yellow leaf curl virus* in Iran and the Arabian Peninsula: further support for a TYLCV center of diversity. *Archives of Virology* 159, 485–497.

Idris, A.M. and Brown, J.K. (2005) Evidence for inter-specific recombination for three monopartite begomoviral genomes associated with the tomato leaf curl disease from central Sudan. *Archives of Virology* 150, 1003–1012.

Jiang, Y.X., de Blas, C., Barrios, L., Fereres, A. (2000) Correlation between whitefly (Homoptera: Aleyrodidae) feeding behavior and transmission of *Tomato yellow leaf curl virus. Annals of the Entomology Society of America* 93, 573–579.

Jiang, Y.X., de Blas, C., Bedford, I.D., Nombela, G. and Muñiz, M. (2004) Effect of *Bemisia tabaci* biotype in the transmission of *Tomato yellow leaf curl Sardinia virus* (TYLCSV-ES) between tomato and common weeds. *Spanish Journal of Agricultural Research* 2, 115–119.

Jiu, M., Zhou, X.P., Tong, L., Xu, J., Yang, X., *et al.* (2007) Vector-virus mutualism accelerates population increase of an invasive whitefly. *PLoS One* 2, e182.

Jupin, I., De Kouchkovsky, F., Jouanneau, F. and Gronenborn, B. (1994) Movement of Tomato yellow leaf curl geminivirus (TYLCV): involvement of the protein encoded by ORF C4. *Virology* 204, 82–90.

Just, K., Leke, W.N., Sattar, M.N., Luik, A. and Kvarnheden, A. (2014) Detection of *Tomato yellow leaf curl virus* in imported tomato fruit in northern Europe. *Plant Pathology* 63, 1454–1460.

Kato, K., Onuki, M., Fuji, S. and Hanada, K. (1998) The first occurrence of *Tomato yellow leaf curl virus* in tomato (*Lycopersicon esculentum* Mill.) in Japan. *Annals of the Phytopathological Society of Japan* 64, 552–559.

Khan, A.J., Idris, A.M., Al-Saady, N.A., Al-Mahruki, M.S., Al-Subhi, A.M., *et al.* (2008) A divergent isolate of *Tomato yellow leaf curl virus* from Oman with an associated DNA beta satellite: an evolutionary link between Asian and the Middle Eastern virus-satellite complexes. *Virus Genes* 36, 169–176.

Kim, S.H., Oh, S., Oh, T.-K., Park, J.S., Kim, S.C., *et al.* (2011) Genetic diversity of tomato-infecting *Tomato yellow leaf curl virus* (TYLCV) isolates in Korea. *Virus Genes* 42, 117–127.

Köklü, G., Rojas, A. and Kvamheden, A. (2006) Molecular identification and the complete nucleotide sequence of a *Tomato yellow leaf curl virus* isolate from Turkey. *Journal of Plant Pathology* 88, 61–66.

Lapidot, M. and Friedmann, M. (2002) Breeding for resistance to whitefly-transmitted geminiviruses. *Annals of Applied Biology* 140, 109–127.

Lazarowitz, S.G. (1992) Geminiviruses: genome structure and gene function. *Critical Reviews in Plant Science* 11, 327–349.

Lefeuvre, P., Martin, D.P., Harkins, G., Lemey, P., Gray, A.J.A., *et al*. (2010) The spread of *Tomato yellow leaf curl virus* from the Middle East to the World. *PLoS Pathogens* 6, e1001164.

Li, M., Hu, J., Xu, F.-C. and Liu, S.S. (2010) Transmission of *Tomato yellow leaf curl virus* by two invasive biotypes and a Chinese indigenous biotype of the whitefly *Bemisia tabaci*. *International Journal of Pest Management* 56, 275–280.

Martínez-Zubiaur, J., Quiñones, M., Fonseca, D., Potter, J.L. and Maxwell, D.P. (2002) First report of *Tomato yellow leaf curl virus* associated with beans, *Phaseolus vulgaris*, in Cuba. *Plant Disease* 86, 814.

Matsuura, S. and Hoshino, S. (2009) Effect of tomato yellow leaf curl disease on reproduction of *Bemisia tabaci* Q biotype (Hemiptera: Aleyrodidae) on tomato plants. *Applied Entomology and Zoology* 44, 143–148.

McGlashan, D., Polston, J.E. and Bois, D. (1994) Tomato yellow leaf curl geminivirus in Jamaica. *Plant Disease* 78, 1219.

McKenzie, C.L., Hodges, G.S., Osborne, L., Byrne, F.J, and Shatters, R.G. (2009) Distribution of *Bemisia tabaci* (Hemiptera: Aleyrodidae) biotypes in Florida – investigating the Q invasion. *Journal of Economic Entomology* 102, 670–676.

Mehta, P., Wyman, J.A., Nakhla, M.K. and Maxwell, D.P. (1994) Transmission of tomato yellow leaf curl geminivirus by *Bemisia tabaci* (Homoptera: Aleyrodidae). *Journal of Economic Entomology* 87, 1291–1297.

Monci, F., Sánchez-Campos, S., Navas-Castillo, J. and Moriones, E. (2002) A natural recombinant between the geminiviruses *Tomato yellow leaf curl Sardinia virus* and *Tomato yellow leaf curl virus* exhibits a novel pathogenic phenotype and is becoming prevalent in Spanish populations. *Virology* 303, 317–326.

Morales, F.J. and Anderson, P.K. (2001) The emergence and dissemination of whitefly-transmitted geminiviruses in Latin America. *Archives of Virology* 146, 415–441.

Moreno-Delafuente, A., Garzo, E., Moreno, A. and Fereres, A. (2013) A plant virus manipulates the behavior of its whitefly vector to enhance its transmission efficiency and spread. *PLoS One* 8, e61543.

Morin, S., Ghanim, M., Zeidan, M., Czosnek, H., Verbeek, M., *et al*. (1999) A GroEL homologue from endosymbiotic bacteria of the whitefly *Bemisia tabaci* is implicated in the circulative transmission of *Tomato yellow leaf curl virus*. *Virology* 30, 75–84.

Moriones, E. and Navas-Castillo, J. (2000) *Tomato yellow leaf curl virus*, an emerging virus complex causing epidemics worldwide. *Virus Research* 71, 123–134.

Mound, L.A. and Halsey, S.H. (1978) *Whiteflies of the World, a Systematic Catalogue of the Aleyrodidae (Homoptera) with Host Plant and Natural Enemy Data*. British Museum (Natural History), London.

Nakhla, M.K., Mazyad, H.M. and Maxwell, D.P. (1993) Molecular characterization of four *Tomato yellow leaf curl virus* isolates from Egypt and development of diagnostic methods. *Phytopathologia Mediterranea* 32, 163–173.

Nakhla, M.K., Maxwell, D.P., Martinez, R.T., Carvalho, M.G. and Gilbertson, R.L. (1994) Widespread occurrence of the Eastern Mediterranean strain of tomato yellow leaf curl geminivirus in tomatoes in the Dominican Republic. *Plant Disease* 78, 926.

Navas-Castillo, J., Sánchez-Campos, S., Noris, E., Louro, D., Accotto, G.P., *et al*. (2000) Natural recombination between *Tomato yellow leaf curl virus*-Is and *Tomato leaf curl virus*. *Journal of General Virology* 81, 2797–2801.

Navas-Castillo, J., Fiallo-Olivé, E. and Sánchez-Campos, S. (2011) Emerging virus diseases transmitted by whiteflies. *Annual Review of Phytopathology* 49, 219–248.

Navot, N., Pichersky, E., Zeidan, M., Zamir, D. and Czosnek, H. (1991) *Tomato yellow leaf curl virus*: a whitefly-transmitted geminivirus with a single genomic component. *Virology* 185, 151–161.

Noris, E., Hidalgo, E., Accotto, G.P. and Moriones, E. (1994) High similarity among the *Tomato yellow leaf curl virus* isolates from the West Mediterranean Basin: the nucleotide sequence of an infectious clone from Spain. *Archives of Virology* 135, 165–170.

Noris, E., Jupin, I., Accotto, G.P. and Gronenborn, B. (1996) DNA-binding activity of the C2 protein of *Tomato yellow leaf curl* geminivirus. *Virology* 217, 607–612

Pascual, S. and Callejas, C. (2004) Intra- and interspecific competition between biotypes B and Q of *Bemisia tabaci* (Hemiptera: Aleyrodidae) from Spain. *Bulletin of Entomological Research* 94, 369–375.

Pennazio, S. and Roggero, P. (1997) Mineral nutrition and systemic virus infections in plants. *Phytopathologia Mediterranea* 36, 54–66.

Perring, T., Gruehagen, N.M. and Farrar, C. (1999) Management of plant viral disease through chemical control of insect vectors. *Annual Review of Entomology* 44, 457–481.

Picó, B., Díez, M.J. and Nuez, F. (1996) Viral diseases causing the greatest economic losses to the tomato crop. II. The *Tomato yellow leaf curl virus* – a review. *Scientia Horticulturae* 67, 151–196.

Polston, J.E. and Anderson, P.K. (1997) The emergence of whitefly-transmitted geminiviruses in tomato in the Western Hemisphere. *Plant Disease* 81, 1358–1369.

Polston, J. and Lapidot, M. (2007) Management of *Tomato yellow leaf curl virus*: US and Israel perspectives. In: Czosnek, H. (ed.) *Tomato Yellow Leaf Curl Virus Disease*. Springer, Dordrecht, The Netherlands, pp. 251–262.

Polston, J.E., McGovern, R.J. and Brown, L.G. (1999) Introductions of *Tomato yellow leaf curl virus* in Florida and implications for the spread of this and other geminiviruses of tomato. *Plant Disease* 83, 984–988.

Polston, J.E., Cohen, L., Sherwood, T.A., Ben-Joseph, R. and Lapidot, M. (2006) *Capsicum* species: symptomless hosts and reservoirs of *Tomato yellow leaf curl virus*. *Phytopathology* 96, 447–452.

Polston, J.E., De Barro, P. and Boykin, L.M. (2014) Transmission efficiencies of plant viruses transmitted by the newly identified species of the *Bemisia tabaci* species complex. *Pest Management Science* 70, 1547–1552.

Rochester, D.E., Depaulo, J.J., Fauquet, C.M. and Beachy, R.N. (1994) Complete nucleotide sequence of the geminivirus *Tomato yellow leaf curl virus,* Thailand isolate. *Journal of General Virology* 75, 477–485.

Rojas, M.R., Jiang, H., Salati, R., Xoconostle-Cázares, B., Sudarshana, M.R., *et al.* (2001) Functional analysis of proteins involved in movement of the monopartite begomovirus *Tomato yellow leaf curl virus*. *Virology* 291, 110–125.

Roye, M.E., Wernecke, M.E., McLaughlin, W.A., Nakhla, M.K. and Maxwell, D.P. (1999) Tomato dwarf leaf curl virus, a new bipartite geminivirus associated with tomatoes and peppers in Jamaica and mixed infection with *Tomato yellow leaf curl virus*. *Plant Pathology* 48, 370–378.

Rubinstein, G. and Czosnek, H. (1997) Long-term association of *Tomato yellow leaf curl virus* (TYLCV) with its whitefly vector *Bemisia tabaci*: effect on the insect transmission capacity, longevity and fecundity. *Journal of General Virology* 78, 2683–2689.

Sinisterra, X.H., McKenzie, C.L., Hunter, W.B., Powell, C.A. and Shatters, R.G. Jr. (2005) Differential transcriptional activity of plant-pathogenic begomoviruses in their whitefly vector (*Bemisia tabaci,* Gennadius: Hemiptera Aleyrodidae). *Journal of General Virology* 86, 1525–1532.

Stonor, J., Hart, P., Gunther, M., DeBarro, P. and Rezaian, M.A. (2003) Tomato leaf curl geminivirus in Australia: occurrence, detection, sequence diversity and host range. *Plant Pathology* 52, 379–388.

Sunter, G. and Bisaro, D.M. (1992) Transactivation of geminivirus AR1 and BR1 gene expression by the viral AL2 gene product occurs at the level of transcription. *Plant Cell* 4, 1321–1331.

Tal, M. and Shannon, M.C. (1983) Salt tolerance in the wild relatives of the cultivated tomato: response of *Lycopersicon esculentum, L. cheesmanii, L. peruvianum, Solanum pennellii* and Fl hybrids to high salinity. *Australian Journal of Plant Physiology* 10, 109–117.

van der Sluijs, J.P., Simon-Delso, N., Goulson, D., Maxim, L., Bonmatin, J., *et al.* (2013) Neonicotinoids, bee disorders and the sustainability of pollinator services. *Current Opinion in Environmental Sustainability* 5, 293–305.

Varsani, A., Navas-Castillo, J., Moriones, E., Hernández-Zepeda, C., Idris, A., *et al.* (2014) Establishment of three new genera in the family *Geminiviridae*: *Becurtovirus, Eragrovirus* and *Turncurtovirus*. *Archives of Virology* 159, 2193–2203.

Wang, K. and Tsai, J.H. (1996) Development and reproduction of *Bemisia argentifolii* (Homoptera: Aleyrodidae) on five host plants. *Annals of Entomological Society of America* 89, 375–384.

Wang, J., Zhao, H., Jian, L., Jiu, M., Qian, Y.-J., *et al.* (2009) Low frequency of horizontal and vertical transmission of two begomoviruses through whitefly *Bemisia tabaci* biotype B and Q. *Annals of Applied Biology* 157, 125–133.

Wartig, L., Kheyr-Pour, A., Noris, E., De Kouchkovsky, F., Jouanneau, F., *et al.* (1997) Genetic analysis of the monopartite *Tomato yellow leaf curl* geminivirus: roles of V1, V2, and C2 ORFs in viral pathogenesis. *Virology* 228, 132–140.

Wu, J.B. and Zhou, X.P. (2006) First report of *Tomato yellow leaf curl virus* in China. *Plant Disease* 90, 1359.

Zambrano, S., Carballo, O., Geraud, F., Chirinos, D., Fernandez, C., *et al.* (2007) First report of *Tomato yellow leaf curl virus* in Venezuela. *Plant Disease* 91, 768.

14 Tristeza

Latanya C. Fisher,[1] Paula Tennant[2] and Vicente J. Febres[1]*

[1]Horticultural Sciences Department, *University of Florida, Florida, USA;* [2]*Department of Life Sciences, The University of the West Indies, Mona Campus, Jamaica*

14.1 Introduction

Citrus tristeza virus (CTV) is the most economically devastating viral plant pathogen and is responsible for the death and loss of productivity of approximately 100 million citrus trees worldwide (Timmer *et al.*, 2000; Moreno *et al.*, 2008; Saponari *et al.*, 2013). Reports of the CTV epidemics have been recorded in several countries, including Argentina, Brazil, Cuba, Cyprus, the Dominican Republic, Israel, Italy, Mexico, South Africa, Spain, the USA (California and Florida) and Venezuela (Bar-Joseph *et al.*, 1989; Kyriakou *et al.*, 1996; Gottwald *et al.*, 1998; Durán-Vila and Moreno, 2000; Garnsey *et al.*, 2000; Timmer *et al.*, 2000; Gottwald *et al.*, 2002; Davino *et al.*, 2003). Presently, CTV is widely distributed (Fig. 14.1) as it is found on all continents where citrus is cultivated (CABI/EPPO, 2010).

The origin of CTV and its co-evolution with citrus is speculated to be in South-East Asia and the Malayan archipelago (Moreno *et al.*, 2008). Presumably the dawn of commercial citriculture in the 19th century also involved the initial movement of CTV-infected citrus germplasm from Asia to different parts of the world. CTV subsequently became a problem as the new environmental and climatic conditions allowed for interactions with new host varieties. Additionally, the citrus industry was fighting another pathogen, an oomycete of the *Phytophthora* species, which led to the widespread use of the resistant sour orange rootstock in the Mediterranean and the Americas. This particular rootstock had excellent agronomic qualities such as cold hardiness, enhancement of fruit quality and high adaptability to adverse soil conditions (Moreno *et al.*, 2008). However, CTV outbreaks grew problematic to the citrus industry due to: (i) the loss of trees and production that occurred from CTV-susceptible genotypes grafted on the widely used sour orange rootstock (death of trees, low yield and poor quality of fruits); (ii) the indirect cost of replacing sour orange as the main rootstock; and (iii) the introduction of CTV-tolerant rootstocks which have a number of associated agronomic problems.

CTV is a *Closterovirus* of the family *Closteroviridae*, the members of which possess the largest and most complex of the plant virus RNA genomes (Dolja *et al.*, 2006). The virus is phloem restricted and consists of flexuous, rod-shaped particles with dimensions of 2000 nm × 12 nm (Febres *et al.*, 1996). Two capsid proteins make up the protein coat. The monopartite, positive sense, single-stranded RNA genome ranges in size from 19,226 to 19,302 nucleotides, depending on the isolate (Karasev *et al.*, 1995; Karasev, 2000; Ruiz-Ruiz *et al.*, 2006). Closteroviruses represent a large diverse group of plant viruses that affect vegetable and fruit crops. They are economically important as they keep emerging and contribute to production losses in important agricultural crops, due to invasions and spread by new vectors and

*E-mail: vjf@ufl.edu

Fig. 14.1. Present distribution of *Citrus tristeza virus* (CTV). Dark grey areas highlighted indicate the presence of CTV in a particular country. Map generated based on information from CABI/EPPO (2010).

changes in agricultural practices (Karasev, 2000). CTV still remains a concern and ongoing research is relevant for developing effective control strategies for management and regulatory purposes. This chapter presents an overview of CTV, the disease it causes and its current status and biotechnological applications as a vector for controlling viral or other diseases in plant pathosystems.

14.2 Disease

14.2.1 Host range

All economically important, fruit-producing species of *Citrus* are susceptible to CTV, although individual genotypes may be tolerant or resistant to particular strains (Yoshida, 1996; Timmer *et al.*, 2000; Moreno *et al.*, 2008). Other species such as *C. glauca* (syn. *Eremocitrus glauca*) and *C. australis* (syn. *Microcitrus australis*) are also susceptible (Yoshida, 1996). Certain genotypes of *C. trifoliata* (syn. *Poncirus trifoliata*) such as 'Pomeroy', 'Rubidoux' and 'Flying Dragon' are considered resistant; however, CTV strains capable of breaking this resistance have been found in New Zealand and are known as NZRB for

New Zealand resistance breaking (Harper *et al.*, 2010). Outside of the Rutaceae, CTV has been experimentally transmitted to a few species of *Passiflora* and *Nicotiana benthamiana* (Timmer *et al.*, 2000; Ambrós *et al.*, 2011).

14.2.2 Symptomatology

The virus causes a wide range of symptoms, depending on the CTV strain–citrus genotype combination. Generally, symptoms fall within four major categories:

1. Mild strains cause visible symptoms of vein clearing or yellowing only in 'Mexican' lime, *C. aurantiifolia* (Roistacher, 1991; Timmer *et al.*, 2000).
2. Seedling yellows (SY) or severe chlorosis and stunting on seedlings. 'Duncan' grapefruit (*C. x paradisi*) and sour orange (*C. aurantium*) are normally used to identify Seedling yellows strains (Roistacher, 1991; Timmer *et al.*, 2000).
3. Decline is induced by more severe strains on sweet orange (*C. sinensis*), grapefruit and mandarins grafted on sour orange rootstock (Timmer *et al.*, 2000; Moreno *et al.*, 2008). When decline occurs within a few weeks, the disease is referred to as 'quick decline',

and is one of the most devastating and eco- nomically important symptoms produced by CTV. Trees may appear yellow, wilted or exhibit thinning canopies (Fig. 14.2a). In the most severe cases, the leaves completely fall off and the tree collapses with the fruit still hanging from the branches. Apparently, it is to this symptom that CTV owes its name, tristeza or 'sadness'.

4. Stem pitting (SP) is also produced by severe strains and is best visualized by strip- ping the bark off of affected branches expos- ing grooves or pits along the twigs (Fig. 14.2b). SP can affect many genotypes of citrus,

including rootstocks. 'Duncan' grapefruit is considered very susceptible to SP strains and is used in their diagnosis. More severe SP strains can be identified using 'Madam Vinous' sweet orange (Roistacher, 1991). SP strains are also economically important since they affect all major cultivated citrus types and lead to reduced production over time.

It is of note that citrus trees can be infected by CTV populations of multiple strains due to repeated inoculations, the longevity of citrus trees and the inability of certain strains to exclude superinfection by other strains,

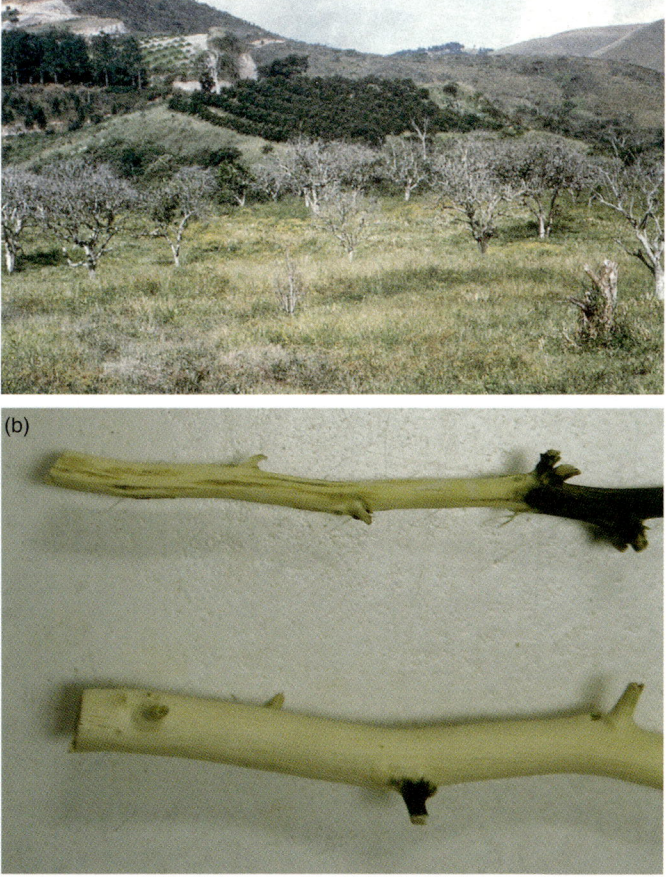

Fig. 14.2. (a) Decline symptoms induced by *Citrus tristeza virus* (CTV) on infected sweet orange trees (foreground) in comparison to healthy trees (upper center). Picture taken in Carabobo state, Venezuela. (b) Stem pitting symptoms observed in CTV-infected citrus stem, notice absence of pits in the lower healthy citrus stem.

all this contributing to the wide range of phenotypes observed within and between citrus types (Moreno *et al.*, 2008; Folimonova *et al.*, 2010).

14.2.3 *Transmission*

In nature, CTV is transmitted by several species of aphids: *Aphis citricidus* (syn. *Toxoptera citricida*), *A. gossypii*, *A. spiraecola* and *A. aurantii* (syn. *T. aurantii*) in a semi-persistent manner. By far the most efficient vector of the virus is *A. citricidus*. Rates of transmission that are 6 to 25 times more efficient than *A. gossypii* have been recorded and depend on the CTV strain. Nonetheless, *A. citricidus* greatly facilitates the spread of the virus and the onset of epidemics (Yokomi and Damsteegt, 1991; Yokomi *et al.*, 1994; Gottwald *et al.*, 1996). CTV is also transmitted very efficiently through grafting with infected plant tissue such as buds, bark segments and leaf midribs as long as there is phloem to phloem contact. This is the preferred experimental method of transmission given its simplicity and high rate of success. The technique is also common in commercial citrus propagation and has presumably contributed to the worldwide distribution of the virus before effective diagnostic methods and certification programs existed. Additionally, CTV can be mechanically transmitted by slash-inoculation (Muller and Garnsey, 1984).

14.3 Molecular Biology

14.3.1 Replication and gene function

Like most closteroviruses, the genome of CTV is uniquely organized and contains 12 open reading frames (ORFs) that potentially code for at least 19 protein products (Dolja *et al.*, 2006). Replication and expression of its genes occurs via three mechanisms: a +1 translational frameshift, polyprotein processing and subgenomic mRNAs (sgRNAs) expression (Fig. 14.3a–d). ORF 1a–1b, located in the 5′ terminal region, contains the replication

machinery: the papain-like proteases L1 and L2, an RNA methyl transferase (MET), a large interdomain region (IDR), an RNA helicase (HEL) and an RNA-dependent RNA polymerase (POL). L1 and L2 have accessory and processing functions and are also responsible for the activation and enhancement of viral RNA amplification, viral cell-to-cell movement and viral invasion; they are possibly aphid transmission factors (Dolja *et al.*, 1994, 2006; Peng *et al.*, 2001). During replication, translation of the CTV genomic RNA yields two polyproteins (Fig. 14.3b,c), MET-HEL-POL (400 kDa) and MET-HEL (349 kDa). The former is translated from ORF 1b presumably by a +1 translational frameshift that occurs as a result of the two overlapping ORFs 1a and 1b, while the second one is translated from ORF 1a (Dolja *et al.*, 1994, 2006; Karasev, 2000). These polyproteins include L1 and L2, along with the large IDR between the MET and HEL domains, an unusual characteristic of closteroviral replicases. Later, these two polyproteins are apparently processed by the two tandem papain-like proteases, L1 and L2, by an unknown mechanism to generate the functional MET, HEL and POL products (Karasev *et al.*, 1995; Dolja *et al.*, 2006).

The sgRNA expression of ORFs 2–11 produces several proteins, including p6, p65 (Hsp70h homologue), p61, major coat protein (CP) and minor coat protein (CPm). The function of this gene set is associated with virion assembly, virus cell-to-cell movement and other aspects of viral replication (Febres *et al.*, 1996; Alzhanova *et al.*, 2000, 2001; Dolja *et al.*, 2006). Proteins p18, p13 and p33 have been recently found to extend viral host range (Tatineni *et al.*, 2011). CP, p20 and p23 have been identified as RNA silencing suppressors (Lu *et al.*, 2004). The p23 protein also regulates the asymmetrical accumulation of viral RNA strands (Satyanarayana *et al.*, 2002b). There are also untranslatable regions (UTR) at the 5′ and 3′ terminal ends of the CTV genome; the latter region is highly conserved, and both are required for viral replication and assembly (Pappu *et al.*, 1997; Satyanarayana *et al.*, 2002a). The process of sgRNA expression involves the formation of ten 3′ co-terminal (+) sgRNAs acting as mRNA

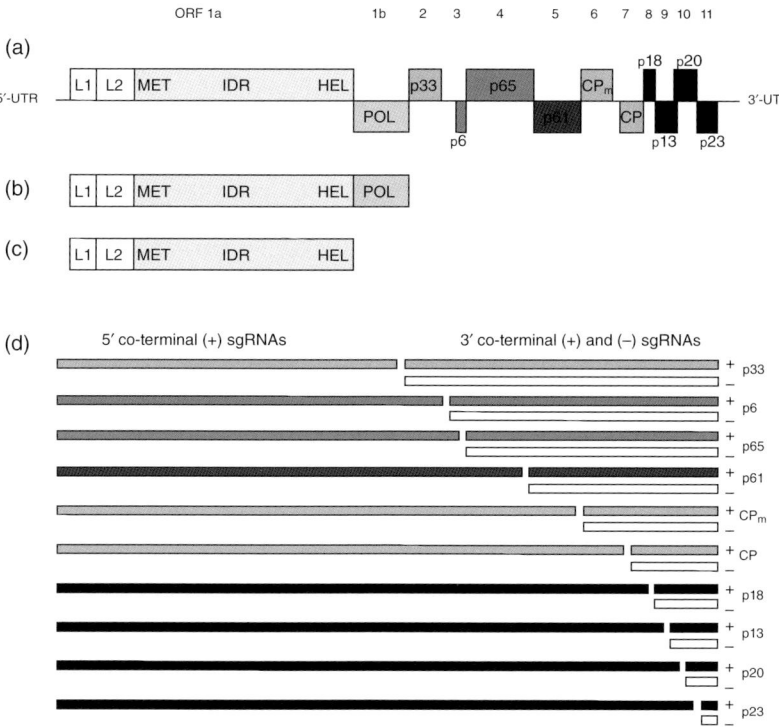

Fig. 14.3. (a) Schematic representation of *Citrus tristeza virus* (CTV) genome organization (adapted from Dolja *et al*., 2006). Replication and expression of CTV via (b) +1 translational frameshift, (c) polyprotein processing, and (d) sgRNA expression (ORFs are depicted as boxes; RNA species (+/−) produced in sgRNA expression are also indicated).

templates to translate the particular 5′ proximal ORF, another ten 3′ co-terminal (−) sgR-NAs produced at lower amounts and ten 5′ co-terminal (+) sgRNAs (Fig. 14.3d) (Hilf *et al*., 1995). The 5′ co-terminal sgRNAs terminate at variable positions upstream of the initiation site of the corresponding sgRNAs and are less abundant than the 3′ sgRNAs. Controller elements contain hairpin structures that function as terminators during the transcription of the 5′ and 3′ sgRNAs; they regulate gene expression and are found upstream of each of the internal CTV ORFs (Gowda *et al*., 2001). sgRNA synthesis is regulated (at the transcriptional level by the virus) in both amounts and timing (Hilf *et al*., 1995; Ayllón *et al*., 2003). CTV contains 10 to 11 controller elements and each controller element produces 3 sgRNAs (5′ co-terminal positive strand RNA, 3′ co-terminal negative and positive strand

RNA); hence, the virus contains 30 to 33 RNA species at any time in infected cells. These RNA species possibly promote recombination events between sgRNAs and their counterparts from the 5′-terminal region producing defective RNAs (D-RNAs) (Rubio *et al*., 2000; Che *et al*., 2003). Large amounts of positive and negative stranded D-RNAs can be observed in CTV-infected tissues. D-RNAs are 2.0–12.0 kb in size and have been described as possessing a common 3′ component containing the 3′-most gene, 3′ and 5′ or 5′ and 3′ termini of various lengths and others that retain ORFs 1a and 1b, 3′ and 5′ termini of the genomic RNA lacking variable portions of the central region (Che *et al*., 2002, 2003). The biological role of D-RNAs is unknown, but has been linked to viral evolution and symptom regulation.

14.3.2   Plant–pathogen interactions

The interaction between CTV and their hosts to establish infection is not well understood (Gandía *et al.*, 2007). CTV movement throughout the plant is thought to be systemic, which involves a certain degree of cell-to-cell and long-distance movement. Long-distance movement involves viral entry into the sieve elements, whereas in cell-to-cell movement the virus moves from adjacent cells to fill clusters of multiple cells. Infection is limited to phloem-associated cells and movement is citrus host dependent since movement mechanisms appear to vary based on the citrus species. Studies have demonstrated CTV infection in citrus phloem cells with green fluorescent protein-labelled CTV in less susceptible citrus hosts ('Duncan' grapefruit and sour orange), more susceptible hosts (Mexican lime and Alemow) and an intermediate host, like sweet orange (Folimonova *et al.*, 2008). The less susceptible citrus hosts show little to no cell-to-cell movement, whereas the more susceptible ones show small clusters of infected phloem cells. For long-distance movement of CTV in the more susceptible citrus species, 10–20% infected phloem-associated cells was observed; the intermediate citrus species showed less, while the less susceptible species showed single cells being infected or fewer infection sites. Clearly CTV mostly uses the long-distance movement mechanism for systemic infection.

14.3.3   Phylogeny of strains

Classification of CTV isolates into various strains is more accurately done genetically than phenotypically (Harper, 2013). Based on complete genome sequences, CTV can be categorized into several phylogenetically distinct strains, than if partial gene or gene regions such as ORF1a/b, CP or other 3′ genes are used. A strain here means that in a single phylogenetic lineage there is shared evolutionary ancestry (homology) and, thus, a high degree of sequence similarity. Existing CTV strains based on phylogenetic analysis have been denoted as T3, T30, T36 and T68 (Florida; Karasev *et al.*, 1995; Albiach-Marti *et al.*, 2000; Harper, 2013), VT (Israel; Mawassi *et al.*, 1996), RB (New Zealand; Harper *et al.*, 2010), and recently HA16-5 (Hawaii; Melzer *et al.*, 2010), that may represent a new CTV strain type. Of these existing strains, RB, T68 and HA16-5 are considered recombinant strains, whereas VT is the most diverse strain. A link between CTV genotypes and phenotypes has not been determined as these strains show a range in their phenotypic characteristics. Diversity is observed more in the 5′ half than the 3′ half of the genome. Sequence analysis also revealed increased degree of similarity within, but not between strains (Harper, 2013). The diversity in CTV can be attributed to mutation, recombination, selection, genetic drift and gene flow between regions, cultural and natural dispersal of the virus, and continuous re-inoculation with different CTV strains by aphid vectors (Moreno *et al.*, 2008).

14.4   Diagnostics

Traditionally, diagnosis of CTV employs a combination of two approaches: biological tests and tests based on serological or molecular properties of the virus. For the most part, detection of CTV infections has relied on biological indexing, which involves graft inoculation onto sensitive citrus indicator plants, mainly Mexican lime, and noting the time and type of symptom development (Roistacher, 1991).Typically, Mexican lime seedlings exhibit vein clearing, leaf cupping, reduced internodes and stem pitting, depending on the severity of the virus isolate. Differentiation between mild, decline and stem pitting strains is generally achieved by inoculation on other species including grapefruit, sour orange and sweet orange, respectively. Diagnostic symptoms on these plants include pitting in the stems of grapefruit and sweet orange, and stunting and small yellow leaves in grapefruit and sour orange. Some 11 reaction types are associated with reference CTV isolates worldwide (Garnsey *et al.*, 2005). Serological detection using ELISA and other techniques (e.g. direct tissue immunoassay, dot immunoblotting and Western blot assay) with polyclonal antibodies is also routinely used in the diagnosis of CTV (Garnsey and Cambra, 1991). Serological methods have several practical advantages over traditional biological tests; namely reduced costs, reduced analysis time and large scale application. Moreover, the production of monoclonal antibodies against CTV has greatly improved the specificity and sensitivity of these tests. The monoclonal antibody MCA-13 to the T-36 isolate, for example, reacts selectively with T-36 and isolates associated with decline, seedling yellows and stem pitting diseases (Permar et al., 1990). A mixture of the monoclonal antibodies along with their recombinant versions generally achieves broad spectrum detection (Cambra et al., 2000). New insights into the antigenic structure of the CTV CP are expected to contribute to further improvements (Wu et al., 2014). Successful virus purification and sequencing of the virus genome from different regions have resulted in the development of sensitive detection procedures based on molecular hybridization with complementary DNA probes to specific CTV groups (Narváez et al., 2000; Niblett et al., 2000) or reverse transcription followed by PCR amplification (RT-PCR) (Olmos et al., 1999; Hilf et al., 2005; Roy and Brlansky, 2010). More recent quantitative real-time RT-PCR assays provide a reliable and reproducible estimation of virus accumulation in various tissues and, depending on the primers, broad spectrum as well as genotype specific strain differentiation (Ruiz-Ruiz et al., 2007; Ananthakrishnan et al., 2010; Saponari et al., 2013).

14.5 Management Strategies

Effective management of CTV diseases requires an integrated approach aimed at preventing or delaying infection that is tailored to the prevailing ecological and epidemiological conditions. Education of citrus growers and the general public on the potential impact of the virus and its vector are also important, and together with the prompt response by scientists and government officials, contribute to delaying the occurrence of epidemics (Rocha-Peña et al., 1995).

14.5.1 Control strategies

In regions where CTV incidence is low or where only mild strains are present, management efforts typically focus on quarantine of incoming materials, to limit and/or avoid introduction of decline-inducing or stem pitting isolates. Other precautionary measures include the introduction of official virus-free certification programmes for monitoring local distribution of citrus materials, prohibition of importation of citrus varieties from countries with CTV, and treatment of exported fruits for pests in order to prevent the introduction of the vector. Surveys for CTV using ELISA are encouraged at regular intervals to detect new outbreaks of the disease (Navarro et al., 1984; Navarro et al., 1988). Removal of CTV-infected trees in areas having low virus incidence and where T. citricida is not present, has proven useful and has extended productivity when there is replacement with trees budded on a recommended tolerant rootstock for the region (Costa and Muller, 1980). Eradication and suppression, on the other hand, are not useful in regions where CTV is established, the disease pressure is high, and T. citricida is the primary vector (Rocha-Peña et al., 1995). Disease caused by decline strains may be managed by replanting trees on a tolerant rootstock. The combined strategies of mild strain cross-protection and replanting with nursery stock certified as CTV-free are reported to contribute to several years of productivity in some regions where stem pitting strains predominate (Costa and Muller, 1980; van Vuuren et al., 1993). However, the protection can be variable and dependent on the citrus scion varieties, prevailing CTV strains and environmental conditions (Cox et al., 1976; da Graca et al., 1984; Leki and Yamaguchi, 1988; Folimonova et al., 2010).

14.5.2 Resistance

Although genetic resistance is the most effective way to control the disease, the complex reproductive biology of citrus has impeded the genetic improvement of the crop through conventional breeding methods (Peña *et al.*, 2007). Additionally, useful genetic resistance to CTV strains within the genus is limited. Resistance to CTV has been observed in close relatives of *Citrus,* including *P. trifoliata, Severinia buxifolia* and *Swinglea glutinosa* (Yoshida, 1985; Gmitter *et al.*, 1996). In *P. trifoliata*, there are two dominant CTV resistance genes, *Ctm* and *Ctv*, for which it is heterozygous at both loci (Rai, 2006). *Ctv* has been mapped and is restricted to a 121 kb region consisting of ten genes, as a single dominant locus relying on a gene-to-gene recognition of CTV (Gmitter *et al.*, 1996; Rai, 2006). The resistance is constitutive and results in the suppression of viral replication, movement and accumulation. Resistance based on activating the endogenous RNA silencing machinery has been applied to citrus and appears to depend on the successful attenuation of the complex suppression mechanism of the virus. Genetic transformations with p23, POL, CP (translatable, untranslatable or truncated versions) or sense or antisense constructs of the 3′ terminus of the virus genome have failed to provide durable resistance (Domínguez *et al.*, 2002a,b; Febres *et al.*, 2003; Fagoaga *et al.*, 2006; López *et al.*, 2010). However, transformations with constructs consisting of full untranslatable versions of genes coding the three suppressors, p23, p20 and CP, along with the 3′-UTR in sense and antisense orientations separated by an intron (i.e. sense–intron–antisense), show promise for disease resistance (Soler *et al.*, 2012). Whether or not transgenic plants could be an alternative strategy for managing CTV in the field remains to be tested.

14.6 From Foe to Friend: *Citrus tristeza virus* Vectors

A rewarding paradox has emerged from years of detailed study of CTV's replication and understanding its life cycle and pathogenicity determinants: the use of this virus as a vector for gene expression and directed RNA interference (RNAi) for the purpose of developing systems for disease and pest control, for furthering our knowledge of virus–host interactions, and in general as an instrument for plant biotechnology. Two recent major developments have improved the use of such vectors and made this system a reality. Firstly, the development of a flexible and stable CTV vector (Folimonov *et al.*, 2007), and secondly, a much improved infection method based on agroinfiltration and an alternative host (Ambrós *et al.*, 2011). These two developments made it possible to work with a simpler system (*N. benthamiana*), while at the same time improved and facilitated the infection of citrus hosts. When the first CTV cDNA-derived clone was developed (Satyanarayana *et al.*, 1999), only infection of *N. benthamiana* protoplasts was achieved. Protoplast transfection relied on *in vitro*-generated RNA transcripts synthesized with SP6 RNA polymerase and linearized plasmid-cloned CTV cDNA as template. Subsequent improvements, including the use of virions as inoculum (either purified virions or crude protoplast 'sap') instead of RNA and three to seven consecutive passages in protoplasts to increase CTV titre, made it possible to infect citrus plants by slash-inoculation with 90% efficiency (Satyanarayana *et al.*, 2001). Despite its success, this system was laborious and time consuming; however, it enabled several studies on the biology and gene function of CTV, its interaction with various hosts and the testing and improvement of different viral vectors.

Based on these accomplishments, a stable CTV vector was developed in which an extra gene (the gene of interest) could be added between the CPm and the CP under the transcriptional control of either native (CTV CP) or heterologous (*Beet yellows virus* CP) controller elements (Folimonov *et al.*, 2007). Trees inoculated with such a vector and GFP as the extra gene maintained fluorescence for several years (Folimonov *et al.*, 2007). The extra gene could also be successfully expressed when added between p13 and p20, downstream of p23 or by replacing p13, but not when added between p20 and p23 (El-Mohtar and Dawson, 2014). The p13 substitution worked best when expressing longer genes (El-Mohtar and Dawson, 2014). Another important advance was the development of an infection method based on *Agrobacterium* binary vectors (Ambrós *et al.*, 2011). This system uses either pCAMBIA or BAC plasmid vectors and drives CTV transcription via a CaMV 35*S* promoter and a NOS terminator, eliminating the need for the production of *in vitro* transcribed RNA as the source of inoculum. *A. tumefaciens* containing either plasmid vector could be directly infiltrated into *N. benthamiana* leaves and the plants became CTV positive days after. Co-inoculation with silencing suppressors also improved the long-distance movement and infectivity, although it was not necessary for successful infection (Ambrós *et al.*, 2011). CTV reached high titres in *N. benthamiana*, and purified virions or crude sap could be used to slash-inoculate citrus plants, with the former giving the best results. These improvements make for a simpler, shorter and more effective CTV vector system; in particular, they eliminate the need for protoplast cultures that were quite involved previously.

Unfortunately, direct agroinfection of citrus hosts has not been achieved (Ambrós *et al.*, 2013). This latest, improved version of the CTV vector system has recently been used to induce RNAi in the insect vector of another important disease of citrus, Huanglongbing, in an attempt to control the spread of this disease. An Asian citrus psyllid (*Diaphorina citri*, a vector of Huanglongbing) gene that controls wing development (truncated abnormal wing disc, t*Awd*) was targeted for silencing using a CTV vector engineered into a pCAMBIA plasmid and containing the t*Awd* downstream of p23 (Hajeri *et al.*, 2014). *D. citri* nymphs fed on plants infected with the transgenic CTV-t*Awd* gave rise to 30% of adults with malformed wings (Hajeri *et al.*, 2014). Under field conditions such adults would theoretically be limited in their ability to fly, and thus to spread the bacteria.

CTV-based vector systems have come a long way since the first infectious clone was developed and they have proven to be important research tools. They have demonstrated application to exotic gene expression and silencing. At the very least they are an alternative for testing genes of interest without going through the more elaborate citrus stable transformation procedure. Due to concerns with transgenic organisms and potential for recombination and escape they may not be as suitable for commercial or field release. Additionally, their long-term effect and durability in a perennial crop such as citrus has not been evaluated. Despite the present limitations, they still have a place in biotechnology and molecular biology – and represent a great example of turning an enemy into a friend.

14.7 Concluding Remarks

CTV has had disastrous effects on the citrus industry, and today its incidence and distribution remain a concern in many countries. This unique closterovirus, having a large and complex RNA genome, will continually be of interest due to questions that have remain unanswered. These include the full understanding of its plant–pathogen interaction, viral evolution, symptom regulation and the link between phylogenetic and genotypic relationships among strains. Further research is imperative to give insight into such queries that could improve on diagnostic techniques and provide information necessary to assist with more effective control methods. This will also help the advancement of education and certification programs aimed at limiting spread and/or management of the disease. Although CTV has been mostly known for its negative impact on citrus agriculture, recent advances in plant biotechnology have led to the use of the virus as a vector and its potential use in engineering resistance mechanisms against itself, other pathogens and pests.

References

Albiach-Marti, M.R., Guerri, J., Cambra, M., Garnsey, S.M. and Moreno, P. (2000) Differentiation of citrus tristeza virus isolates by serological analysis of p25 coat protein peptide maps. *Journal of Virological Methods* 88, 25–34.

Alzhanova, D.V., Hagiwara, Y., Peremyslov, V.V. and Dolja, V.V. (2000) Genetic analysis of the cell-to-cell movement of beet yellows closterovirus. *Virology* 268, 192–200.

Alzhanova, D.V., Napuli, A.J., Creamer, R. and Dolja, V.V. (2001) Cell-to-cell movement and assembly of a plant closterovirus: roles for the capsid proteins and Hsp70 homolog. *The EMBO Journal* 20, 6997–7007.

Ambrós, S., El-Mohtar, C., Ruiz-Ruiz, S., Peña, L., Guerri, J., *et al.* (2011) Agroinoculation of *Citrus tristeza virus* causes systemic infection and symptoms in the presumed nonhost *Nicotiana benthamiana*. *Molecular Plant-Microbe Interactions* 24, 1119–1131.

Ambrós, S., Ruiz-Ruiz, S., Peña, L. and Moreno, P. (2013) A genetic system for *Citrus tristeza virus* using the non-natural host *Nicotiana benthamiana*: an update. *Frontiers in Microbiology* 4, 165.

Ananthakrishnan, G., Venkataprasanna, T., Roy, A. and Brlansky, R.H. (2010) Characterization of the mixture of genotypes of a *Citrus tristeza virus* isolate by reverse transcription-quantitative real-time PCR. *Journal of Virological Methods* 164, 75–82.

Ayllón, M.A., Gowda, S., Satyanarayana, T., Karasev, A.V., Adkins, S., *et al.* (2003) Effects of modification of the transcription initiation site context on *Citrus tristeza virus* subgenomic RNA synthesis. *Journal of Virology* 77, 9232–9243.

Bar-Joseph, M., Marcus, R. and Lee, R.F. (1989) The continuous challenge of *Citrus tristeza virus* control. *Annual Review of Phytopathology* 27, 291–316.

CABI/EPPO (2010) *Citrus tristeza virus. Distribution Maps of Plant Diseases*. CAB International, Wallingford, UK.

Cambra, M., Gorris, M.T., Román, M.P., Terrada, E., Garnsey, S.M., *et al.* (2000) Routine detection of *Citrus tristeza virus* by direct immunoprinting-ELISA method using specific monoclonal and recombinant antibodies. In: da Graça, J.V., Lee, R.F. and Yokomi, R.K. (eds) *Proceedings of the 14th Conference of the International Organization of Citrus Virologists, Riverside, California,* pp. 34–41.

Che, X., Mawassi, M. and Bar-Joseph, M. (2002) A novel class of large and infectious defective RNAs of *Citrus tristeza virus*. *Virology* 298, 133–145.

Che, X., Dawson, W.O. and Bar-Joseph, M. (2003) Defective RNAs of *Citrus tristeza virus* analogous to *Crinivirus* genomic RNAs. *Virology* 310, 298–309.

Costa, A.S. and Muller, G.W. (1980) Tristeza control by cross protection. *Plant Disease* 64, 538–541.

Cox, J., Fraser, L.R. and Broadbent, P. (1976) Stem pitting of grapefruit: field protection by the use of mild strains, an evaluation of trials in two climatic districts. In: Calavan, E.C. (ed) *Proceedings of the 7th Conference of the International Organization of Citrus Virologists, Riverside, California,* pp. 68–70.

da Graca, J.V., Marais, L.J. and von Broemsen, L.A. (1984) Severe tristeza stem pitting decline of young grapefruit in South Africa. In: Garnsey, S.M., Timmer, L.W. and Dodds, J.A. (eds) *Proceedings of the 9th Conference of the International Organization of Citrus Virologists, Riverside, California,* pp. 62–65.

Davino, S., Davino, M., Sambade, A., Guardo, M. and Caruso, A. (2003) The first *Citrus tristeza virus* outbreak found in a relevant citrus producing area of Sicily, Italy. *Plant Disease* 87, 314–314.

Dolja, V.V., Karasev, A.V. and Koonin, E.V. (1994) Molecular biology and evolution of closteroviruses: sophisticated build-up of large RNA genomes. *Annual Review of Phytopathology* 32, 261–285.

Dolja, V.V., Kreuze, J.F. and Valkonen, J.P.T. (2006) Comparative and functional genomics of closteroviruses. *Virus Research* 117, 38–51.

Domínguez, A., de Mendoza, A., Guerri, J., Cambra, M., Navarro, L., *et al.* (2002a) Pathogen-derived resistance to *Citrus tristeza virus* (CTV) in transgenic Mexican lime (*Citrus aurantiifolia* (Christm.) Swing.) plants expressing its p25 coat protein gene. *Molecular Breeding* 10, 1–10.

Domínguez, A., Fagoaga, C., Navarro, L., Moreno, P. and Peña, L. (2002b) Constitutive expression of untranslatable versions of the p25 coat protein gene in Mexican lime (*Citrus aurantiifolia* (Christm.) Swing.) transgenic plants does not confer resistance to *Citrus tristeza virus* (CTV). In: Duran-Vila, N., Milne, R.G. and da Graca, J.V. (eds) *Proceedings of the 15th Conference of the International Organization of Citrus Virologists, Riverside, California,* pp. 341–344.

Durán-Vila, N. and Moreno, P. (2000) *Enfermedades de los Cítricos*. Monografía de la Sociedad Española de Fitopatología No. 2. Ediciones Mundi–Prensa, Madrid.

El-Mohtar, C. and Dawson, W.O. (2014) Exploring the limits of vector construction based on *Citrus tristeza virus*. *Virology* 448, 274–283.

Fagoaga, C., López, C., de Mendoza, A.H., Moreno, P., Navarro, L., *et al.* (2006) Post-transcriptional gene silencing of the p23 silencing suppressor of *Citrus tristeza virus* confers resistance to the virus in transgenic Mexican lime. *Plant Molecular Biology* 60, 153–165.

Febres, V.J., Ashoulin, L., Mawassi, M., Frank, A., Bar-Joseph, M., *et al.* (1996) The p27 protein is present at one end of *Citrus tristeza virus* particles. *Phytopathology* 86, 1331–1335.

Febres, V.J., Niblett, C.L., Lee, R.F. and Moore, G.A. (2003) Characterization of grapefruit plants (*Citrus paradisi* Macf.) transformed with citrus tristeza closterovirus genes. *Plant Cell Reports* 21, 421–428.

Folimonov, A.S., Folimonova, S.Y., Bar-Joseph, M. and Dawson, W.O. (2007) A stable RNA virus-based vector for citrus trees. *Virology* 368, 205–216.

Folimonova, S.Y., Folimonov, A.S., Tatineni, S. and Dawson, W.O. (2008) *Citrus tristeza virus*: survival at the edge of the movement continuum. *Journal of Virology* 82, 6546–6556.

Folimonova, S.Y., Robertson, C.J., Shilts, T., Folimonov, A.S., Hilf, M.E., *et al.* (2010) Infection with strains of *Citrus tristeza virus* does not exclude superinfection by other strains of the virus. *Journal of Virology* 84, 1314–1325.

Gandía, M., Conesa, A., Ancillo, G., Gadea, J., Forment, J., *et al.* (2007) Transcriptional response of *Citrus aurantiifolia* to infection by *Citrus tristeza virus*. *Virology* 367, 298–306.

Garnsey, S.M. and Cambra, M. (1991) Enzyme-linked immunosorbent assay (ELISA) for citrus pathogens. In: Roistacher, C.N. (ed.) *Graft-Transmissible Diseases of Citrus: Handbook for Detection and Diagnosis*. FAO, Rome, pp. 193–216.

Garnsey, S.M., Gottwald, T.R., Hilf, M.E., Matos, L. and Borbón, J. (2000) Emergence and spread of severe strains of *Citrus tristeza virus* isolates in the Dominican Republic. In: da Graça, J.V., Lee, R.F. and Yokomi, R.K. (eds) *Proceedings of the 14th Conference of the International Organization of Citrus Virologists, Riverside, California,* pp. 57–68.

Garnsey, S.M., Civerolo, E.L., Gumpf, D.J., Paul, C., Hilf, M.E., *et al.* (2005) Biological characterization of an international collection of *Citrus tristeza virus* (CTV) isolates. In: Hilf, M.E., Duran-Vila, N. and Rocha-Peña, M.A. (eds) *Proceedings of the 16th Conference of the International Organization of Citrus Virologists, Riverside, California,* pp. 75–93.

Gmitter, F.G. Jr, Xiao, S.Y., Huang, S., Hu, X.L., Garnsey, S.M., *et al.* (1996) A localized linkage map of the citrus tristeza virus resistance gene region. *Theoretical and Applied Genetics* 92, 688–695.

Gottwald, T., Garnsey, S., Cambra, M., Moreno, P., Irey, M., *et al.* (1996) Differential effects of *Toxoptera citricida* vs. *Aphis gossypii* on temporal increase and spatial patterns of spread of citrus tristeza. In: da Graca, J.V., Moreno, P. and Yokomi, R.K. (eds) *Proceedings of the 13th Conference of the International Organization of Citrus Virologists, Riverside, California,* pp. 120–129.

Gottwald, T.R., Garnsey, S.M. and Borbon, J. (1998) Increase and patterns of spread of *Citrus tristeza virus* infections in Costa Rica and the Dominican Republic in the presence of the brown citrus aphid, *Toxoptera citricida*. *Phytopathology* 88, 621–636.

Gottwald, T.R., Polek, M. and Riley, K. (2002) History, present incidence, and spatial distribution of *Citrus tristeza virus* in the California Central Valley. In: Duran-Vila, N., Milne, R.G. and da Graça, J.V. (eds) *Proceedings of the 15th Conference of the International Organization of Citrus Virologists, Riverside, California,* pp. 83–94.

Gowda, S., Satyanarayana, T., Ayllón, M.A., Albiach-Martí, M.R., Mawassi, M., *et al.* (2001) Characterization of the cis-acting elements controlling subgenomic mRNAs of *Citrus tristeza virus*: production of positive- and negative-stranded 3′-terminal and positive-stranded 5′-terminal RNAs. *Virology* 286, 134–151.

Hajeri, S., Killiny, N., El-Mohtar, C., Dawson, W.O. and Gowda, S. (2014) *Citrus tristeza virus*-based RNAi in citrus plants induces gene silencing in *Diaphorina citri*, a phloem-sap sucking insect vector of citrus greening disease (Huanglongbing). *Journal of Biotechnology* 176, 42–49.

Harper, S.J. (2013) *Citrus tristeza virus*: evolution of complex and varied genotypic groups. *Frontiers in Microbiology* 4, 93.

Harper, S.J., Dawson, T.E. and Pearson, M.N. (2010) Isolates of *Citrus tristeza virus* that overcome *Poncirus trifoliata* resistance comprise a novel strain. *Archives of Virology* 155, 471–480.

Hilf, M.E., Karasev, A.V., Pappu, H.R., Gumpf, D.J., Niblett, C.L., *et al.* (1995) Characterization of *Citrus tristeza virus* subgenomic RNAs in infected tissue. *Virology* 208, 576–582.

Hilf, M.E., Mavrodieva, V.A. and Garnsey, S.M. (2005) Genetic marker analysis of a global collection of iso-lates of *Citrus tristeza virus*: characterization and distribution of CTV genotypes and association with symptoms. *Phytopathology* 95, 909–917.

Karasev, A.V. (2000) Genetic diversity and evolution of Closteroviruses. *Annual Review of Phytopathology* 38, 293–324.

Karasev, A.V., Boyko, V.P., Gowda, S., Nikolaeva, O.V., Hilf, M.E., *et al.* (1995) Complete sequence of the *Citrus tristeza virus* RNA genome. *Virology* 208, 511–520.

Kyriakou, A., Ioannou, N., Gavriel, J., Bar-Joseph, M., Papayiannis, C., *et al.* (1996) Management of *Citrus tristeza virus* in Cyprus. In: da Graça, J.V., Moreno, P. and Yokomi, R.K., (eds) *Proceedings of the 13th Conference of the International Organization of Citrus Virologists, Riverside, California,* pp. 172–178.

Leki, H. and Yamaguchi, A. (1988) Protective interference of mild strains of *Citrus tristeza virus* against a severe strain in Morita navel orange. In: Timmer, L.W., Garnsey, S.M. and Navarro, L. (eds) *Proceedings of the 10th Conference of the International Organization of Citrus Virologists, Riverside, California,* pp. 86–90.

López, C., Cervera, M., Fagoaga, C., Moreno, P., Navarro, L., *et al.* (2010) Accumulation of transgene-derived siRNAs is not sufficient for RNAi-mediated protection against *Citrus tristeza virus* in transgenic Mexican lime. *Molecular Plant Pathology* 11, 33–41.

Lu, R., Folimonov, A., Shintaku, M., Li, W.X., Falk, B.W., *et al.* (2004) Three distinct suppressors of RNA silencing encoded by a 20-kb viral RNA genome. *Proceedings of the National Academy of Sciences USA* 101, 15742–15747.

Mawassi, M., Mietkiewska, E., Gofman, R., Yang, G. and Bar-Joseph, M. (1996) Unusual sequence rela-tionships between two isolates of *Citrus tristeza virus*. *Journal of General Virology* 77, 2359–2364.

Melzer, M.J., Borth, W.B., Sether, D.M., Ferreira, S., Gonsalves, D., *et al.* (2010) Genetic diversity and evi-dence for recent modular recombination in Hawaiian *Citrus tristeza virus*. *Virus Genes* 40, 111–118.

Moreno, P., Ambrós, S., Albiach-Martí, M.R., Guerri, J. and Peña, L. (2008) *Citrus tristeza virus*: a patho-gen that changed the course of the citrus industry. *Molecular Plant Pathology* 9, 251–268.

Muller, G.W. and Garnsey, S.M. (1984) Susceptibility of citrus varieties, species, citrus relatives, and non-Rutaceous plants to slash-cut mechanical inoculation with *Citrus tristeza virus* (CTV). In: Garnsey, S.M., Timmer, L.W. and Dodds, J.A. (eds) *Proceedings of the 9th Conference of the International Organization of Citrus Virologists, Riverside, California,* pp. 33–40.

Narváez, G., Skander, B.S., Ayllón, M.A., Rubio, L., Guerri, J., *et al.* (2000) A new procedure to differentiate *Citrus tristeza virus* isolates by hybridisation with digoxigenin-labelled cDNA probes. *Journal of Viro-logical Methods* 85, 83–92.

Navarro, L., Juárez, J., Pina, J.A. and Ballester, J.F. (1984) The citrus quarantine station in Spain. In: Garnsey, S.M., Timmer, L.W. and Dodds, J.A. (eds) *Proceedings of the 9th Conference of the Inter-national Organization of Citrus Virologists, Riverside, California,* pp. 365–370.

Navarro, L., Juárez, J., Pina, J.A., Ballester, J.F. and Arregui, J.M. (1988) The citrus variety improvement programme in Spain. In: Timmer, L.W., Garnsey S.M. and Navarro, L. (eds) *Proceedings of the 10th Conference of the International Organization of Citrus Virologists, Riverside, California,* pp. 400–406.

Niblett, C.L., Genc, H., Cevik, B., Halbert, S., Brown, L., *et al.* (2000) Progress on strain differentiation of *Citrus tristeza virus* and its application to the epidemiology of citrus tristeza disease. *Virus Research* 71, 97–106.

Olmos, A., Cambra, M., Esteban, O., Gorris, M.T. and Terrada, E. (1999) New device and method for capture, reverse transcription and nested PCR in a single closed-tube. *Nucleic Acids Research* 27, 1564–1565.

Pappu, S.S., Febres, V.J., Pappu, H.R., Lee, R.F. and Niblett, C.L. (1997) Characterization of the 3' prox-imal gene of the citrus tristeza closterovirus genome. *Virus Research* 47, 51–57.

Peña, L., Cervera, M., Fagoaga, C., Romero, J., Juárez, J., *et al.* (2007) Citrus. In: Pua, E.C. and Davey, M.R. (eds) *Biotechnology in Agriculture and Forestry, Transgenic Crops*. Vol. 60. Springer-Verlag, Berlin-Heidelberg, pp. 35–50.

Peng, C.W., Peremyslov, V.V., Mushegian, A.R., Dawson, W.O. and Dolja, V.V. (2001) Functional specializa-tion and evolution of leader proteinases in the family *Closteroviridae*. *Journal of Virology* 75, 12153–12160.

Permar, T.A., Garnsey, S.M., Gumpf, D.J. and Lee, R.F. (1990) A monoclonal-antibody that discriminates strains of *Citrus tristeza virus*. *Phytopathology* 80, 224–228.

Rai, M. (2006) Refinement of the *Citrus tristeza virus* resistance gene (*Ctv*) positional map in *Poncirus trifoliata* and generation of transgenic grapefruit (*Citrus paradisi*) plant lines with candidate resistance genes in this region. *Plant Molecular Biology* 61, 399–414.

Rocha-Peña, M.A., Lee, R.F., Lastra, R., Niblett, C.L., Ochoa-Corona, F.M., *et al.* (1995) *Citrus tristeza virus* and its aphid vector *Toxoptera citricida*: threats to citrus production in the Caribbean and Central and North America. *Plant Disease* 79, 437–445.

Roistacher, C.N. (1991) *Graft-Transmissible Diseases of Citrus: Handbook for Detection and Diagnosis.* FAO, Rome.

Roy, A. and Brlansky, R.H. (2010) Genome analysis of an orange stem pitting *Citrus tristeza virus* isolate reveals a novel recombinant genotype. *Virus Research* 151, 118–130.

Rubio, L., Yeh, H.-H., Tian, T. and Falk, B.W. (2000) A heterogeneous population of defective RNAs is associated with *Lettuce infectious yellows virus*. *Virology* 271, 205–212.

Ruiz-Ruiz, S., Moreno, P., Guerri, J. and Ambrós, S. (2006) The complete nucleotide sequence of a severe stem pitting isolate of *Citrus tristeza virus* from Spain: comparison with isolates from different origins. *Archives of Virology* 151, 387–398.

Ruiz-Ruiz, S., Moreno, P., Guerri, J. and Ambrós, S. (2007) A real-time RT-PCR assay for detection and absolute quantitation of *Citrus tristeza virus* in different plant tissues. *Journal of Virological Methods* 145, 96–105.

Saponari, M., Loconsole, G., Liao, H.H., Jiang, B., Savino, V., *et al.* (2013) Validation of high-throughput real time polymerase chain reaction assays for simultaneous detection of invasive citrus pathogens. *Journal of Virological Methods* 193, 478–486.

Satyanarayana, T., Gowda, S., Boyko, V., Albiach-Martí, M., Mawassi, M., *et al.* (1999) An engineered closterovirus RNA replicon and analysis of heterologous terminal sequences for replication. *Proceedings of the National Academy of Sciences USA* 96, 7433–7438.

Satyanarayana, T., Bar-Joseph, M., Mawassi, M., Albiach, M.R., Ayllón, M.A., *et al.* (2001) Amplification of *Citrus tristeza virus* from a cDNA clone and infection of citrus trees. *Virology* 280, 87–96.

Satyanarayana, T., Gowda, S., Ayllón, M.A., Albiach-Martí, M.R. and Dawson, W.O. (2002a) Mutational analysis of the replication signals in the 3'-nontranslated region of *Citrus tristeza virus*. *Virology* 300, 140–152.

Satyanarayana, T., Gowda, S., Ayllón, M.A., Albiach-Martí, M.R., Rabindran, S., *et al.* (2002b) The p23 protein of *Citrus tristeza virus* controls asymmetrical RNA accumulation. *Journal of Virology* 76, 473–483.

Soler, N., Plomer, M., Fagoaga, C., Moreno, P., Navarro, L., *et al.* (2012) Transformation of Mexican lime with an intron-hairpin construct expressing untranslatable versions of the genes coding for the three silencing suppressors of *Citrus tristeza virus* confers complete resistance to the virus. *Plant Biotechnology Journal* 10, 597–608.

Tatineni, S., Robertson, C.J., Garnsey, S.M. and Dawson, W.O. (2011) A plant virus evolved by acquiring multiple nonconserved genes to extend its host range. *Proceedings of the National Academy of Sciences USA* 108, 17366–17371.

Timmer, L.W., Garnsey, S.M. and Graham, J.H. (2000) *Compendium of citrus diseases*. The American Phytopathological Society (APS Press), St. Paul, Minnesota.

van Vuuren, S.P., Collins, R.P. and da Graça, J.V. (1993) Evaluation of *Citrus tristeza virus* isolates for cross protection of grapefruit in South Africa. *Plant Disease* 77, 24–28.

Wu, G.W., Tang, M., Wang, G.P., Wang, C.X., Liu, Y. *et al.* (2014) The epitope structure of *Citrus tristeza virus* coat protein mapped by recombinant proteins and monoclonal antibodies. *Virology* 448, 238–246.

Yokomi, R.K. and Damsteegt, V.C. (1991) Comparison of *Citrus tristeza virus* transmission efficacy by *Toxoptera citricidus* and *Aphis gossypii*. In: Peters, D.C.J., Webster, A. and Choubler, C.S. (eds) *Proceedings, Aphid-Plant Interactions: Populations to Molecules*. Oklahoma Agricultural Experiment Station, Stillwater, Oklahoma, pp. 229–241.

Yokomi, R., Lastra, R., Stoetzel, M.B., Damsteegt, V., Lee, R., *et al.* (1994) Establishment of the brown citrus aphid (Homoptera, Aphididae) in Central America and the Caribbean Basin and transmission of *Citrus tristeza virus*. *Journal of Economic Entomology* 87, 1078–1085.

Yoshida, T. (1985) Inheritance of susceptibility to *Citrus tristeza virus* in trifoliate orange. *Bulletin of the Fruit Tree Research Station Series B (Okitsu)* 12, 17–26.

Yoshida, T. (1996) Graft compatibility of *Citrus* with plants in the *Aurantioideae* and their susceptibility to *Citrus tristeza virus*. *Plant Disease* 80, 414–417.

15 Rice Tungro

Indranil Dasgupta*

Department of Plant Molecular Biology, University of Delhi, New Delhi, India

15.1 Introduction

Rice tungro bacilliform virus (RTBV) and *Rice tungro spherical virus* (RTSV) are two viruses responsible for the rice tungro disease (RTD). The disease has been known for almost a half a century and has been intensively investigated across various countries in Asia. Today, a large volume of information is available on the viruses, their transmission by insect vectors, their gene functions, the pathological response in rice plants upon infection, and the rice genes that mediate resistance to the viruses. This chapter summarizes what is known about the pathogens and the disease, and discusses the prospects of conventional and biotechnological approaches to controlling RTD – mainly by strengthening the RNA-based defence pathway in rice.

15.2 Disease Symptoms

RTD is characterized by orange–yellow foliar discoloration and stunting of plants to almost half the normal size upon maturity (Figs 15.1 and 15.2). The orange–yellow discoloration varies according to the rice cultivar, but in some instances this symptom is used as the characteristic diagnostic feature for the disease (Fig. 15.3). Panicle exertion is delayed or incomplete, and panicles are short and very often sterile. Tungro, as it is commonly known, has been recorded

over the past 50 years in South and South-East Asia (Rivera and Ou, 1965; Raychaudhuri *et al.*, 1967). It was initially suspected to be caused by nutrient deficiencies, but the viral nature of the disease was demonstrated by Gálvez in the late 1960s (Gálvez, 1967). Gálvez (1967) observed icosahedral particles in affected tissue preparations. However, on closer examination, two distinct particles, icosahedral or bacilliform, were detected. Accordingly, the names RTBV and RTSV were coined for the bacilliform and the icosahedral particles, respectively.

15.3 Distribution

RTD has been recorded in various countries of South-East Asia; seemingly more so in Indonesia, Malaysia and the Philippines than in Cambodia, Laos, Myanmar and Vietnam. In the countries of South Asia, the disease has been reported in Bangladesh and India, but not in Nepal, Pakistan or Sri Lanka. Within India, the disease is restricted to the coastal states bordering the Bay of Bengal (Muralidharan *et al.*, 2003), except for a single report from north-western India (Varma *et al.*, 1999). There is no reason to suspect a non-uniform distribution of the disease in the region, and presumably these reports reflect the available findings of surveys conducted in the different countries. Furthermore, because of its rather uncharacteristic and sometimes misleading symptoms, RTD

*E-mail: indasgup@south.du.ac.in

Fig. 15.1. Symptoms of rice tungro disease (RTD) in the field. Two rows of susceptible varieties alternating with resistant varieties. Symptoms of orange–yellow discoloration and stunting are exhibited by the susceptible varieties.

Fig. 15.2. Symptoms at 24 days post-inoculation, rice variety 'TN-1'. The inoculated plants show stunting and yellowing.

Fig. 15.3. Orange–yellow discoloration in a leaf (left) compared to a healthy leaf (right) of plants growing under greenhouse conditions.

is often overlooked and not recorded unless its occurrence is widespread. Lack of diagnostic tools applicable at the field level also add to the problem.

15.4 Economic Impact

There are no clear statistics regarding economic losses due to RTD. It is generally agreed that in severe cases, losses can reach 100% in susceptible varieties infected at an early growth stage. Herdt (1991) estimated annual losses to rice production attributed to RTD were in excess of US$1 billion worldwide. In India, Muralidharan *et al.* (2003) estimated overall losses to rice yield to be 1% at the national level, but significant losses at the regional level were not reported. More recent estimates put yield losses at 5–10% in South and South-East Asia (Dai and Beachy, 2009). Regardless, sudden and catastrophic epidemics have resulted in famine and death in areas where RTD is prevalent.

15.5 Causative Viruses

Both RTBV and RTSV are detected in RTD-affected plants, indicating that both viruses are jointly responsible for the disease. However, Koch's postulates, which link the causative agent to a disease, have not been rigidly satisfied for both RTBV and RTSV. According to Koch's postulates, if an aetiological agent is isolated from a diseased organism in a pure state and it induces the same disease symptoms on introduction into a healthy organism, the entity can be accepted as the cause of the disease. Proof of disease causation has been challenging with RTBV and RTSV because the viruses are only transmitted by the green leafhoppers *Nephotettix virescens*, *N. cincticeps* and *Resilia dorsalis* (Rivera and Ou, 1967). However, transmission of double-stranded DNA RTBV genome, independent of green leafhoppers (GLH), has been achieved. RTBV DNA has been shown to be infectious when re-introduced into rice plants as a cloned partially duplicated form in a binary plasmid via *Agrobacterium tumefaciens* and by injection into the meristematic region of plants (Dasgupta *et al.*, 1991). The appearance of symptoms of RTD, its detection by electron microscopy and its transmission by GLH to healthy plants satisfied Koch's postulates for RTBV. For RTSV, such experiments have not been reported. Since no other virus has been reported associated with RTD-affected plants, it is generally assumed that only RTBV and RTSV cause RTD.

15.6   Taxonomy

RTBV is a plant pararetrovirus of the family *Caulimoviridae* (unassigned order), in the genus, *Tungrovirus*. Of the 70 recognized families that have not been assigned to an order, the family *Caulimoviridae* includes all members that possess a circular, double-stranded DNA genome of about 8 kb, replicate using reverse transcription of its more than full-length transcript and present genes on only one strand of their DNA. Of the seven genera classified in the family *Caulimoviridae*, the genus *Tungrovirus* has only one species, RTBV, which is transmitted by GLH, whose particles are bullet-shaped. RTSV, on the other hand, is a single-stranded RNA virus that belongs to the order Picornavirales, family *Secoviridae* and genus *Waikavirus*. Two other species within the genus include the *Anthriscus yellows virus* and *Maize chlorotic dwarf virus*.

15.7   Host Range

Cultivated rice (*Oryza sativa*) is the primary host for RTBV and RTSV. Of the various wild rice species tested as potential hosts, *O. australiensis*, *O. barthii*, *O. brachyantha*, *O. eichingeri*, *O. glaberrima*, *O. nivara*, *O. perennis* and *O. punctata* were found to act as hosts for the viruses, although with varying degrees of stunting and incubation periods. Detailed analysis using serological tests were not performed with most of the species examined, however later investigations showed differential susceptibility of some wild rice to the two viruses (Anjaneyulu *et al.*, 1994).

15.8   Transmission

RTBV and RTSV are transmitted in a semi-persistent manner by GLH, but they are not mechanically transmitted (Singh, 1969). Transmission by GLH was shown for the first time in the 1960s (Rivera and Ou, 1965) in the Philippines and soon after in India (Raychaudhuri *et al.*, 1967), indicating the involvement of a similar vector in transmission across various regions. Of note, GLH used in these early studies were referred to as *N. impicticeps*, but they have since been renamed as *N. virescens* (Distant). The related species, *N. nigropictus*, is not as efficient in transmitting the viruses. Likewise, *N. malayanus*, *N. parvus* and *Resilia dorsalis* transmit at lower efficiencies (Rivera and Ou, 1967). Transmission rates were reported dependent on the temperature (high rates are observed at 34°C) and the sex of the insect (females transmit better than males) (Ling and Tiongco, 1975). The time of acquisition for the viruses varies from 5 to 30 minutes and an incubation period has never been observed, which suggests that transmission can be detected immediately following acquisition. The viruses are retained by the vector for up to 5 days (Ling, 1966).

    With the advent of serological methods for detecting RTBV and RTSV, it was revealed that GLH could transmit both RTBV and RTSV, either individually or together from plants having both viruses. Transmission of RBTV was found dependent on the presence of RSTV, but transmission of RTSV was independent (Hibino *et al.*, 1979, 1988). Details of the mechanism of this unique interaction between RTBV, RTSV and GLH have not yet been elucidated. However, the fact that GLH is capable of transmitting RTBV but not RTSV after having fed on anti-RTSV antiserum following acquisition from plants having both RTBV and RTSV, points to the possibility of an RTSV-induced substance, either of plant or viral origin, that acts as a 'helper factor' for RTBV transmission (Hibino and Cabauatan, 1987). The precise nature of this helper factor or its interaction with GLH and RTBV remains to be worked out. Nonetheless, it is clear that RTSV plays the role of a helper virus for insect transmission of RTBV. Individually these viruses exhibit rather mild symptoms. RTSV alone induces mild stunting in some rice cultivars. Infection of rice plants with RTBV, on the other hand, results in severe symptom expression, which is accentuated by co-infection with RTSV. This suggests that the two viruses have co-evolved into a unique disease complex, in which the viruses have developed shared mechanisms that enable the complex to establish systemic infection and co-transmission.

15.9 Molecular Biology

15.9.1 General features

RTD, as indicated earlier, is caused by a complex synergistic interaction between two different viruses, RTBV and RSTV. RTBV contains double-stranded circular DNA, whereas RTSV has a single-stranded positive-sense RNA as the genomic materials. The genome sizes of RTBV DNA and RTSV RNA were determined to be 8 kb and 12 kb, respectively (Jones et al., 1991). Further work based on nucleotide sequence analysis showed that the RTBV DNA has features of caulimoviruses, namely, a coat protein (CP)–protease–reverse transcriptase/RNase H gene arrangement and a reverse transcriptase-mediated replication initiated at a tRNA primer-binding site (Hay et al., 1991). Plants infected with RTBV produce two viral-specific transcripts: a full-length transcript (Hay et al., 1991) and a shorter, spliced transcript responsible for mainly translating the 3' part of the genome. The nucleotide sequences of RTBV DNA suggests the presence of four open reading frames (ORFs): ORF I encoding a potential 24 kDa protein; ORF II, encoding a potential 12 kDa protein; ORF III encoding a polyprotein of 194 kDa; and the last ORF, ORF IV,

encoding a potential 45 kDa protein. The 194 kDa polyprotein has domains resembling a CP, protease and reverse transcriptase/RNase H (Hay et al., 1991). A spliced RTBV transcript was shown to be responsible for the expression of ORF IV (Futterer et al., 1994), after splicing of an intron of about 6.3 kb and an in frame fusion to a short ORF in the RTBV leader sequence. The RTBV transcript is polycistronic, carrying several ORFs on the same transcript. This poses a problem in translation because eukaryotic ribosomes move downstream from the 5' cap end towards the 3' direction, translating ORFs, followed by their disassociation from the transcript. Hence, it was proposed that most RTBV ORFs are translated by a leaky scanning mechanism. In this case, not all ribosomes recognize the first ORF and keep moving along the transcript until they find the next ORF (Futterer et al., 1997). Thus there is reinitiation at downstream cistrons, such that the presence of the upstream ORFs appears not to affect gene expression. This mechanism operates efficiently because of the lack of additional AUGs within about 1 kb region between the start codons of ORFs I and III, the feature also conserved in the closely related badnaviruses. Translation is initiated by a ribosome shunting mechanism.

RTSV RNA sequence analysis revealed features of a picornavirus; namely, the presence of a large polyprotein gene encoding domains suggestive of three CPs, an RNA polymerase, a proteinase and RNA-binding proteins (Shen et al., 1993; Zhang et al., 1993). Additionally, the polyprotein ORF is preceded with an unusually long leader sequence, of about 500 nucleotides, which has several short ORFs and a high propensity to form a stable secondary structure. These structures are known to inhibit 5' end-dependent, scanning-mediated translation initiation on eukaryotic ribosomes. Compelling evidence indicates a ribosome shunt mechanism, with all the cis-acting elements known to drive ribosome shunting in RTBV, is present in RTSV. It is likely that these motifs co-evolved independently in RSTV through adaptation to the rice translational machinery, but there is also the possibility of their horizontal transfer from RTBV to RTSV during co-infection and co-evolution (Pooggin *et al.*, 2012).

15.9.2   Gene functions

Most viral proteins are believed to perform multiple functions, as opposed to their cellular counterparts. The functions of some of the proteins encoded by RTBV and RTSV have been elucidated, and have helped in understanding the way the two viruses bring about infection. The RTBV ORF I and ORF II encode relatively small proteins. The RTBV ORF II was reported to have a non-specific nucleic acid-binding activity, mediated by a basic stretch of amino acids at its C-terminal portion (Jacquot *et al.*, 1997). The ORF III product, based on its sequence homology to other caulimoviruses, was predicted to encode a polyprotein having CP, protease and reverse transcriptase/ribonuclease H domains (Hay *et al.*, 1991; Qu *et al.*, 1991). Experimental evidence for the presence of a 37 kDa CP species was obtained by mass spectral analysis on purified virions (Marmey *et al.*, 1999). Herzog *et al.* (2000) demonstrated that P12 interacts with the CP and it was proposed to have a possible role in the assembly of virions. The ORF III product was subsequently shown to exhibit protease activity, resembling retroviral aspartate proteases, which is responsible for the formation of the CP (Marmey *et al.*, 2005). It has been proposed that the above protease activity is responsible for the maturation of the polyprotein into its constituent active proteins (Laco *et al.*, 2005). Recently, the ORF IV product has been shown to interfere with the spread of cell-to-cell silencing, an anti-viral defence response mounted by plants against invading viruses (Rajeswaran *et al.*, 2014). This function may contribute towards the virulence of the virus.

In RTSV, the genes encoding the three CPs (CP1, CP2 and CP3) were mapped (Zhang *et al.*, 1993) on the viral genome, followed by the demonstration that CP3 was possibly located on the outer surface of the virions (Druka *et al.*, 1996). Subsequently,

putative protease activity was shown in a domain located near the 3′ end of the viral RNA (Thole and Hull, 1998) and the active sites and the substrate specificity were later reported (Thole and Hull, 2002). The protease was further characterized and found to possess self-cleaving properties (Sekiguchi *et al.*, 2005). The remaining genes carried on the RTSV RNA have not yet been functionally characterized.

15.9.3   Promoter activity

Since RTBV and *Cauliflower mosaic virus* (CaMV) belong to the same family, there was considerable interest in studying the RTBV promoter because of the potential likelihood of it being as useful in expressing heterologous genes in plants as the counterpart in CaMV. Unlike the CaMV promoter, the RTBV promoter was reported to be tissue specific, expressing mainly in phloem tissues, and was also weaker than the CaMV promoter when checked in protoplasts. The RTBV promoter, however, expressed better in rice protoplasts compared to tobacco (Bhattacharyya-Pakrasi *et al.*, 1993). Yin and Beachy (1995) described the regions of the promoter necessary for phloem-specific activity, and within it, determined the binding sites of two rice proteins. Although important components of a promoter are generally located upstream to the transcription start site, RTBV promoter was reported to carry elements enhancing transcription both upstream and downstream to the transcription start site (Chen *et al.*, 1996), using protoplast systems. The promoter was further characterized by defining additional upstream elements required for phloem-specific expression in transgenic rice plants and also the rice transcription factors which bind to them and influence the promoter activity (Klöti *et al.*, 1999; He *et al.*, 2000, 2002). Silencing of the RTBV promoter was reported in transgenic plants over several generations due to methylation, which was later seen to spread to adjacent transcribed regions (Klöti *et al.*, 2002). Compared to the isolate from the Philippines, on which most studies have been based, the RTBV from

India was seen to contain a different combination of control elements, and more interestingly, a negative element in the downstream region of the transcription start site (Mathur and Dasgupta, 2007). This negative element was further characterized and was seen to downregulate heterologous promoters, and was active when placed in the opposite orientation and upstream of the promoter (Purkayastha *et al.*, 2010). Thus, multiple elements and interacting proteins have been discovered in rice that control the expression of the RTBV promoter. However, since most studies have been conducted using the isolated promoter, often under transgenic conditions, or in protoplasts under transient expression conditions, it may be difficult to extend the studies to the expression patterns of the promoter during natural infection.

15.9.4 Variability

Variability in viral strains is generally suspected when viral diseases display varying symptoms on the same host. Several such symptomatic variants of RTD were reported from India (Anjaneyulu and John, 1972). However, it is to be noted that the above report appeared even before the involvement of two viruses were discovered. Hence, interpretations of such observations need to be carefully made. Studies on variability were also conducted based on immunological reactions of isolates from various regions and on susceptibility to proteases. RTSV isolates from the Philippines, Thailand, Malaysia and India were reported to be serologically indistinguishable. The Indian isolate could be distinguished by its electrophoretic mobility and proteolytic digestion patterns (Druka *et al.*, 1996). Differences between the RTBV isolates were also documented, mainly between those from the Indian subcontinent and South-East Asia, based on restriction digestion patterns of cloned viral DNA. More interestingly, the genomes of RTBV from the Indian subcontinent were seen to have a 64 base-pair deletion as compared to the others (Fan *et al.*,

1996). Within the Philippines, Villegas *et al.* (1997) reported more variations in RTBV in field isolates, as compared to those collected from plants in the greenhouses of the International Rice Research Institute at Los Baños, Philippines. These observations were based on PCR-restriction fragment length polymorphism studies. Similar observations were also reported with four isolates of RTBV maintained in International Rice Research Institute glasshouses (Cabauatan *et al.*, 1998). Nucleotide sequence comparisons of CP genes of five RTSV isolates from various countries of South and South-East Asia revealed a maximum of 15% difference in the sequences (Zhang *et al.*, 1997). Sequence comparison of four full-length RTBV sequences from South-East Asia revealed scattered nucleotide changes, with the cysteine-rich region of the CP showing the greatest variation. Most changes were as a result of insertions, deletions and recombinations (Cabauatan *et al.*, 1999). Based on comparisons between the CP sequences of a large number of RTSV isolates from the Philippines and Indonesia, Azzam *et al.* (2000) reported on the existence of several genetic groups, which apparently remain isolated from each other. In the Philippines, sequencing of a virulent RTSV strain indicated a difference in sequences in the 5′ leader region of the genome–the difference, however, did not correlate with virulence (Isogai *et al.*, 2000). Complete sequence analysis of two Indian isolates of RTBV indicated that they have both genetically diverged from the South-East Asian isolates, both being highly similar to each other (Nath *et al.*, 2002). A similar study with RTSV indicated that it has not diverged in India from the South-East Asian isolates as much as the RTBV isolates (Verma and Dasgupta, 2007). PCR- restriction fragment length polymorphism analysis of field isolates from India indicated mixed infections of RTBV belonging to different genetic groups (Joshi *et al.*, 2003). Sequence analysis of more recent RTBV and RTSV isolates from India has reinforced the observation that the South Asian and South-East Asian RTBV isolates are genetically distinct, and that within the South Asian isolates there are limited nucleotide changes, both for RTBV and RTSV (Banerjee *et al.*, 2011; Mangrauthia *et al.*, 2012, 2013; Mathur and Dasgupta, 2013). There is evidence of recombination in a southern Indian isolate of RTBV; portions of DNA belonging to other isolates were detected (Sharma *et al.*, 2011).

15.10 Diagnostic Methods

Under field conditions, the early diagnosis of any disease is generally based on the appearance of symptoms. For RTD, orange–yellow discoloration and stunting are used for an early diagnosis, followed by more reliable tools of assessment. Additional support for the presence of RTD is the prevalence of GLH. Transmission of the viruses to a susceptible rice variety, such as 'Taichung Native 1' ('TN1') is also used as supporting evidence. The variety 'TN1' develops clear symptoms of RTD within 2 weeks of GLH-mediated inoculation. For confirmation of RTD, further diagnostic tests are typically recommended. These diagnostic tests are mainly based on two approaches: serological and nucleic acid-based. Serological diagnostic methods depend upon the availability of antibodies against either or both the viruses and can be designed as an agglutination test or ELISA (Takahashi *et al.*, 1991). Commercial kits are available but these tend to be too expensive (and they require equipment) for resource-poor extension services. Research laboratories have examined the possibility of developing cheaper and easier tissue print assays; raising antibodies against the two viruses has posed a challenge to these developments. Immunosensors are also being explored. Immunosensor technologies are based on the combination of specific antigens and antibodies in a solution on a surface support coupled to a signal transducer. These technologies will invariably assist with onsite monitoring and early RTD detection.

Nucleic acid-based diagnostic tests are reported to be at least ten times more sensitive than ELISA for the detection of RTD. These are mainly based on PCR methodology.

PCR-based diagnostic tests have not only increased sensitivity, but are also convenient to use. PCR tests for RTBV, because of its DNA genome, have been reported in several countries and for a variety of samples collected from field conditions, including GLH (Takahashi *et al.*, 1993; Dasgupta *et al.*, 1996). Since RTSV has an RNA genome, PCR-based methods require an additional, prior step of a reverse transcription. Periasamy *et al.* (2006) described a multiplex reverse transcriptase-PCR method, which uses the RTBV transcript and the RTSV RNA to simultaneously detect both the viruses from RNA samples of symptomatic rice plants. More recently, a SYBR Green-based reverse transcriptase-PCR method has been developed for the detection and accurate quantification of RTBV and RTSV in infected plants (Sharma and Dasgupta, 2012). These diagnostic techniques have been very important in monitoring the incidence of RTD under field conditions (Varma *et al.*, 1999), as well as in assessing RTD resistance in various rice plants.

15.11 Management Strategies

15.11.1 Conventional methods

Currently, three broad approaches are possible for the field control of RTD: (i) cultural practices to reduce the build up of the vector; (ii) chemical control of vectors; and (iii) varietal resistance or use of rice genes to control viruses or the vector. In addition, there are several transgenic approaches, which have demonstrated useful levels of resistance under greenhouse conditions.

For the field control of RTD, the most common methods used include the application of a range of cultural practices. Cultural practices for the control RTD are based on the principle of reducing the build up of the vector GLH in fields so as to delay or suppress transmission of the viruses and spread of the disease. GLH populations increase dramatically during the hot humid season in the Indian subcontinent, coinciding with either the onset of monsoon or towards the withdrawal stage of monsoon when the humidity remains high but the rainfall ceases. The population of GLH is usually low during heavy monsoon rains. It is then possible to introduce a break in the infection cycle where no susceptible plants are grown. The other factor to consider is the relationship of age of the plant to susceptibility. Plants are most susceptible between 15 and 30 days post-germination, and beyond this period their susceptibility decreases. One commonly recommended cultural practice to reduce the incidence of RTD has been changing the planting time. Early planting for the monsoon or wet season gives the opportunity for the plants to grow and mature early, and by the time the GLH population rises, the plants are already past the susceptible young age. Similarly, late planting for the dry season in South-East Asia and the kharif season in the Indian subcontinent is recommended to avoid RTD, so that the most susceptible stage of seedlings coincides with a period when the GLH population is low. The other cultural practice recommended is synchronous planting, where a large area is simultaneously planted and subsequently left fallow during the intercropping season. This practice reduces the build up of GLH and prevents the constant availability of young seedlings on which the GLH generally thrives. Other cultural practices include the use of a mixture of varieties, crop rotation and roguing of fields. A mixture of varieties of rice, if used, will have a high chance of reducing the susceptible rice varieties and those which favour GLH multiplication, thereby reducing vector build up in the field. Crop rotation includes the cultivation of non-rice crops, again to reduce GLH build up. Roguing of rice fields involves removal of infected plants during the growing season including stubble following the harvest of the crop, both of which act as inoculum sources for the spread of RTD.

Since RTD is exclusively transmitted by GLH, chemical control of GLH has been applied to the management of RTD, albeit with low efficacy. A variety of general insecticides, such as carbofuran, dicrotophos, parathion and similar compounds, are generally used (Shukla and Anjaneyulu, 1982). These approaches have always been used with caution as the application of insecticides not only represents an increase in the input costs of rice farmers, most of whom are subsistence farmers, but also increases the risk of exposure of farmers and farm workers to hazardous substances and carry-over of these toxic products. Rational and judicious use of insecticides can, however, contribute to increased rice yields in areas where RTD is endemic (Anjaneyulu et al., 1994).

The utilization of genes already present in various rice germplasm is considered by many as the final frontier for durable management of any plant disease. Large-scale screening for new RTD resistance sources has been conducted. The assessment of resistance is, however, complicated by the fact that resistance can be against either or both the viruses or against GLH. Moreover resistance exhibited under controlled conditions is not necessarily expressed under RTD-field conditions. Generally, it has been noted that resistance against GLH is not stable under field conditions, especially under high GLH pressure (Azzam and Chancellor, 2000). Most genes against GLH have not yet been characterized as to their nature or mode of action, but their sources are known and they have been introgressed into popular RTD-susceptible rice varieties. The sources of such resistance genes against GLH include the varieties, 'Pankhari-203', 'ASD-7', 'IR-28' and 'IR-36'. Resistance against RTBV and RTSV has been recognized in a number of rice varieties, and hence there is promise for more durable resistance against the viruses, rather than GLH, under field conditions. The resistance reactions in many of these varieties vary, from total resistance to tolerance, when the viral titres are low and the symptoms are mild. Of these, the resistance genes from the Indonesian variety 'Utri Merah', which confers resistance against RTSV and tolerance against RTBV, have been the most promising (Azzam and Chancellor, 2000). There are no reports of developing durable RTD-resistant varieties for commerce using either resistance genes to GLH or to RTBV or RTSV. Other investigators have studied the relationship of morphological markers and genetic markers, such as simple sequence repeat and inter simple sequence repeat, as tools to select characters contributing to both RTD resistance and to yield (Latif *et al.*, 2013). This holds considerable promise for the future. The utility of such approaches, however, needs to be demonstrated through the appropriate genetic crosses and evaluating RTD resistance at the field level, because the mechanisms of resistance as coded by these genes are still unclear.

An early report suggested that the genetic resistance in 'Utri Merah' against RTBV was polygenic (Shahjahan *et al.*, 1990). The genetic resistance of 'Utri Merah' to RTD has been investigated by genetic means, combining segregation of resistance against RTBV and RTSV in progenies and nearly isogenic lines of 'Utri Merah' with the susceptible variety 'TN-1'. This was combined with the estimation of the levels of virus infection in the progenies. It was concluded from this study that resistance against the two viruses could be genetically separated from each other, indicating for the first time their separate identities (Encabo *et al.*, 2009). Shortly thereafter, Lee *et al.* (2010) reported the characterization of a recessive gene from 'Utri Merah' showing strong association with RTSV resistance. The investigators first mapped the gene, using genetic mapping after crossing of the resistant parent to a susceptible variety, to a region of 200 kb on chromosome 7. Sequence analysis of the region located a gene encoding eukaryotic translation initiation factor (eIF4G) in the above region. Further analysis of this region in several varieties, susceptible and resistant to RTSV, revealed a strong association of resistance with a single nucleotide polymorphism site in the exon 9 of the gene. The finding pointed towards this gene being responsible for RTSV resistance, when mutated or at a particular polymorphic state. It is possible that the eIF4G is required for RTSV replication and certain mutations in the gene affect this function, thereby resulting in resistance. A similar characterization of

RTBV resistance or tolerance from 'Utri Merah' is still awaited.

15.11.2 Non-conventional or transgenic methods

Transgenic approaches towards generating resistance against RTD have been pursued by many groups. Although considerable success has been achieved in this direction, none of the transgenic varieties have yet been tested in the field. One of the first reports of transgenic resistance was against RTSV. It was observed that transgenic rice lines containing the replicase gene in the sense orientation, but incapable of expressing the protein, conferred partial resistance against RTSV, and those containing the same gene in the antisense orientation, exhibited immunity (Huet *et al.*, 1999). The transgenic plants, as expected, inhibited the transmission of RTBV though GLH feeding. Based on the accumulation of the transgene RNA, the resistance was attributed to co-suppression, a RNA-based silencing system in plants active against invading nucleic acids. Simultaneously, Sivamani *et al.* (1999) reported partial resistance against RTSV in rice plants expressing the three CP genes either individually or together. An additive effect in plants expressing more than one CP gene was not observed. Two reports, one using RNA interference with the help of a double-stranded RNA derived from RTBV (Tyagi *et al.*, 2008) and one involving the over-expression of RTBV CP (Ganesan *et al.*, 2009), described resistance in transgenic rice plants against RTBV. In both cases, resistance manifested as a slower build up of the virus in transgenic plants, as compared to untransformed plants. Following GLH-mediated inoculation, RTSV accumulated slower than expected in the transgenic plants and the plants acted as poor sources of inoculum for RTBV transmission (Verma *et al.*, 2012). All the above transgenic lines represent additional sources of resistance available for breeders and eventual integration of these genes into popular, but RTD-susceptible rice varieties. RNA interference-based resistance utilizing the natural defence pathway in plants against viruses is particularly promising because of the absence of any viral protein expression in plants, giving rise to fewer issues related to biosafety of the lines. The RNA interference-based RTBV resistant line has since been used in back crossing programs with several popular rice varieties in India. Enhanced tolerance to RTD was observed (Roy *et al.*, 2012; Jyothsna *et al.*, 2013).

A novel method of obtaining transgenic resistance against RTD has been published by Dai *et al.* (2008). In this report, the investigators used two rice transcription factors, which were earlier shown to be essential for the expression of RTBV promoter function, and over-expressed them in rice plants. These plants, upon inoculation with RTBV, did not exhibit the typical symptoms of infection or symptom expression was very mild. The investigators speculated that the transcription factors, named RF2a and RF2b, not only bound to the RTBV promoter, but were also necessary for the proper expression of a number of rice genes required for the normal growth and development of the rice plant. In support of their speculation, they demonstrated that downregulating the expression of these two transcription factors produced abnormal plants because of the low expression levels of the essential genes whose expression they otherwise activated. Over-expression of RF2a and RF2b compensated for the shortage brought about by the infection of RTBV, and hence, symptom development was masked. This report illustrated an example where over-expression of a plant gene gives rise to virus resistance, in this case, against RTBV.

15.12 Concluding Remarks

RTBV and RTSV are the most important viruses affecting rice production in Asia. RTD can cause extensive damage to the rice crop and thus needs to be managed effectively. The two viruses, though unrelated to each other, display a unique relationship. They interact intimately with the GLH vector, and where RTBV contributes mostly to symptom development, RTSV is responsible for vector transmission. Transmission by GLH is a topic that also needs to be explored. Identifying the elusive helper factor, which is required for the transmission of RTBV, is still a challenge to be met. The genetic variability, especially in RTBV across Asia, should be kept in mind while deploying measures to control RTD across countries in Asia; a functional map of the complete RTBV and RTSV is needed, although more is known of RTBV. Finally, identification and characterization of resistance genes in rice germplasm is essential for the development of rice cultivars with long-lasting RTD resistance. Transgenic technology has demonstrated the possibilities of engineering resistance against both RTBV and RTSV, and can therefore complement traditional breeding efforts.

Acknowledgements

The author is grateful to Dr J. Tarafdar and Mr Gaurav Kumar for the photographs of RTD-affected plants. Research in the author's lab is funded by grants from Department of Biotechnology, Government of India and University of Delhi, under the R&D scheme and DST PURSE.

References

Anjaneyulu, A. and John, V.T. (1972) Strains of rice tungro virus. *Phytopathology* 62, 1116–1119.
Anjaneyulu, A., Satapathy, M.K. and Shukla, V.D. (1994) *Rice Tungro*. Oxford and IBH Publishing Company, New Delhi.
Azzam, O. and Chancellor, T.C.B. (2000) The biology, epidemiology and management of rice tungro disease in Asia. *Plant Disease* 86, 88–100.
Azzam, O., Yambao, M.L., Muhsin, M., McNally, K.L. and Umadhay, K.M. (2000) Genetic diversity of *Rice tungro spherical virus* in tungro-endemic provinces of Philippines and Indonesia. *Archives of Virology* 145, 1183–1197.

Banerjee, A., Roy, S. and Tarafdar, J. (2011) Phylogenetic analysis of *Rice tungro bacilliform virus* ORFs revealed strong correlation between evolution and geographical distribution. *Virus Genes* 43, 398–408.

Bhattacharyya-Pakrasi, M., Peng, J., Elmer, J.S., Laco, G., Shen, P., *et al.* (1993) Specificity of a promoter of *Rice tungro bacilliform virus* for expression in phloem tissues. *The Plant Journal* 4, 71–79.

Cabauatan, P.Q., Arboleda, M. and Azzam, O. (1998) Differentiation of *Rice tungro bacilliform virus* strains by restriction analysis and DNA hybridization. *Journal of Virological Methods* 76, 121–126.

Cabauatan, P.Q., Melcher, U., Ishikawa, K., Omura, T., Hibino, H., *et al.* (1999) Sequence changes in six variants of *Rice tungro bacilliform virus* and their phylogenetic relationships. *Journal of General Virology* 80, 2229–2237.

Chen, G., Rothnie, H.M., He, X., Hohn, T. and Futterer, J. (1996) Efficient transcription from *Rice tungro bacilliform virus* promoter requires elements downstream of the transcription start site. *Journal of Virology* 70, 8411–8421.

Dai, S. and Beachy, R.N. (2009) Genetic engineering of rice to resist rice tungro disease. *In Vitro Cellular and Developmental Biology -Plant* 45, 517–524.

Dai, S., Wei, X., Alfonso, A.A., Pei, L., Duque, U.G., *et al.* (2008) Transgenic rice plants that overexpress transcription factors RF2a and RF2b are tolerant to rice tungro virus replication and disease. *Proceedings of National Academy of Sciences USA* 105, 21012–21016.

Dasgupta, I., Hull, R., Eastop, S., Poggi-Pollini, C., Blakebrough, M., *et al.* (1991) *Rice tungro bacilliform virus* DNA independently infects rice after *Agrobacterium*-mediated transfer. *Journal of General Virology* 72, 1215–1221.

Dasgupta, I., Das, B.K., Nath, P.S., Mukhopadhyay, S., Niazi, F.R., *et al.* (1996) Detection of *Rice tungro bacilliform virus* in field and glasshouse samples from India using polymerase chain reaction. *Journal of Virological Methods* 58, 53–58.

Druka, A., Burns, T., Zhang, S. and Hull, R. (1996) Immunological characterization of *Rice tungro spherical virus* coat proteins and differentiation of isolates from the Philippines and India. *Journal of General Virology* 77, 1975–1983.

Encabo, J.R., Cabauatan, P.Q., Cabunagan, R.C., Satoh, K., Lee, J.H., *et al.* (2009) Suppression of two tungro viruses in rice by separable traits originating from cultivar Utri Merah. *Molecular Plant Microbe Interactions* 22, 1268–1281.

Fan, Z., Dahal, G., Dasgupta, I., Hay, J. and Hull, R. (1996) Variation in the genome of *Rice tungro bacilliform virus*: molecular characterization of six isolates. *Journal of General Virology* 77, 847–854.

Futterer, J., Potrykus, I., Valles-Brau, M.P., Dasgupta, I., Hull, R., *et al.* (1994) Splicing in a plant pararetrovirus. *Virology* 198, 663–670.

Futterer, J., Rothnie, H.M., Hohn, T. and Potrykus, I. (1997) *Rice tungro bacilliform virus* open reading frames II and III are translated from polycistronic pregenomic RNA by leaky scanning. *Journal of Virology* 71, 7984–7989.

Gálvez, G.E. (1967) Purification of virus like particles from rice tungro virus-infected plants. *Virology* 33, 357–359.

Ganesan, U., Suri, S.S., Rajasubramaniam, S., Rajam, M.V. and Dasgupta, I. (2009) Transgenic expression of coat protein gene of *Rice tungro bacilliform virus* in rice reduces the accumulation of viral DNA in inoculated plants. *Virus Genes* 39, 113–119.

Hay, J.M., Jones, M.C., Blakebrough, M.L., Dasgupta, I., Davies, J.W., *et al.* (1991) An analysis of the sequence of an infectious clone of *Rice tungro bacilliform virus*, a plant pararetrovirus. *Nucleic Acids Research* 19, 2615–2621.

He, X., Hohn, T. and Fütterer, J. (2000) Transcriptional activation of the *Rice tungro bacilliform virus* gene is critically dependent upon an activator element located immediately upstream of the TATA box. *Journal of Biological Chemistry* 275, 11799–11808.

He, X., Futterer, J. and Hohn, T. (2002) Contribution of downstream promoter elements to transcriptional regulation of the *Rice tungro bacilliform virus* promoter. *Nucleic Acids Research* 30, 497–506.

Herdt, R.W. (1991) Research priorities for biotechnology. In: Khush, G.S. and Toenissen, G.H. (eds) *Rice Biotechnology*. CAB International, Wallingford, UK, pp 19–54.

Herzog, E., Guerra-Peraza, O. and Hohn, T. (2000) The *Rice tungro bacilliform virus* gene II product interacts with the coat protein domain of the viral gene III polyprotein. *Journal of Virology* 74, 2073–2083.

Hibino, H. and Cabauatan, P.Q. (1987) Infectivity neutralization of rice tungro associated viruses acquired by vector leafhoppers. *Phytopathology* 77, 473–476.

Hibino, H., Saleh, N. and Roechan, M. (1979) Transmission of two kinds of rice tungro associated viruses by insect vector. *Phytopathology* 69, 1266–1268.

Hibino, H., Daquioag, R.D., Cabauatan, P.Q. and Dahal, G. (1988) Resistance to *Rice tungro spherical virus* in rice. *Plant Disease* 72, 843–847.

Huet, H., Mahendra, S., Wang, J., Sivamani, E., Ong, C.A., *et al.* (1999) Near immunity to *Rice tungro spherical virus* achieved in rice by a replicase-mediated resistance strategy. *Phytopathology* 89, 1022–1027.

Isogai, M., Cabauatan, P.Q., Masuta, C., Uyeda, I. and Azzam, O. (2000) Complete nucleotide sequence of the *Rice tungro spherical virus* genome of the highly virulent strain Vt6. *Virus Genes* 20, 79–85.

Jacquot, E., Keller, M. and Yot, P. (1997) A short basic domain supports a nucleic acid binding activity in the *Rice tungro bacilliform virus* open reading frame 2 product. *Virology* 239, 352–359.

Jones, M.C., Gough, K., Dasgupta, I., Subba Rao, B.L., Cliffe, J., *et al.* (1991) Rice tungro disease is caused by a RNA and a DNA virus. *Journal of General Virology* 72, 757–761.

Joshi, R., Kumar, V. and Dasgupta, I. (2003) Detection of molecular variability in rice tungro bacilliform viruses from India using polymerase chain reaction-restriction fragment length polymorphism. *Journal of Virological Methods* 109, 89–93.

Jyothsna, M., Manonmani, S., Rabindran, R., Dasgupta, I. and Robin, S. (2013) Introgression of transgenic resistance for rice tungro disease into mega variety, ASD16 of Tamil Nadu through marker assisted backcross breeding. *Madras Agricultural Journal* 100, 70–74.

Klöti, A., Henrich, C., Bieri, S., He, X., Chen, G., *et al.* (1999) Upstream and downstream sequence elements determine the specificity of the *Rice tungro bacilliform virus* promoter and influence RNA production after transcription initiation. *Plant Molecular Biology* 40, 249–266.

Klöti, A., He, X., Potrykus, I., Hohn, T. and Futterer, J. (2002) Tissue-specific silencing of a transgene in rice. *Proceedings of the National Academy of Sciences USA* 99, 10881–10886.

Laco, G.S., Kent, S.B. and Beachy, R.N. (2005) Analysis of proteolytic processing and activation of the *Rice tungro bacilliform virus* reverse transcriptase. *Virology* 208, 207–214.

Latif, M.A., Rahman, M.M., Ali, M.E., Ashkani, S. and Rafii, M.Y. (2013) Inheritance studies of SSR and ISSR molecular markers and phylogenetic relationship of rice genotypes resistant to tungro virus. *Comptes Rendus Biologies* 336, 125–133.

Lee, J.H., Muhsin, M., Atienza, G.A., Kwak, D.Y., Kim, S.M., *et al.* (2010) Single nucleotide polymorphisms in a gene for translation initiation factor (eIF4G) of rice (*Oryza sativa*) associated with resistance to *Rice tungro spherical virus*. *Molecular Plant-Microbe Interactions* 23, 29–38.

Ling, K.C. (1966) Non-persistence of tungro virus of rice in its leafhopper vector, *Nephotettix impicticeps*. *Phytopathology* 56, 1252–1256.

Ling, K.C. and Tiongco, E.R. (1975) Effect of temperature on the transmission of rice tungro virus by *Nephotettix virescens*. *Philippine Phytopathology* 11, 46–57.

Mangrauthia, S.K., Malathi, P., Agarwal, S., Ramkumar, G., Krishnaveni, D., *et al.* (2012) Genetic variation of coat protein gene among the isolates of *Rice tungro spherical virus* from tungro-endemic states of India. *Virus Genes* 44, 482–487.

Mangrauthia, S.K., Malathi, P., Agarwal, S., Sailaja, P., Singh, J., *et al.* (2013) The molecular diversity and evolution of *Rice tungro bacilliform virus* from Indian perspective. *Virus Genes* 45, 126–138.

Marmey, P., Bothner, B., Jacqout, E., deKochko, A., Ong, C.A., *et al.* (1999) *Rice tungro bacilliform virus* open reading frame 3 encodes a single 37-kDa coat protein. *Virology* 253, 319–326.

Marmey, P., Rojas-Mendoza, A., de Kochko, A., Beachy, R.N. and Fauquet, C.M. (2005) Characterization of the protease domain of *Rice tungro bacilliform virus* responsible for the processing of the capsid protein from the polyprotein. *Virology Journal* 2, 33.

Mathur, S. and Dasgupta, I. (2007) Downstream promoter sequence of an Indian isolate of *Rice tungro bacilliform virus* alters tissue-specific expression in host rice and acts differentially in heterologous system. *Plant Molecular Biology* 65, 259–275.

Mathur, S. and Dasgupta, I. (2013) Further support of genetic conservation in Indian isolates of *Rice tungro bacilliform virus* by sequence analysis of an isolate from North-Western India. *Virus Genes* 46, 387–391.

Muralidharan, K., Krishnaveni, D., Rajarajeswari, N.V.L. and Prasad, A.S.R. (2003) Tungro epidemics and yield losses in paddy fields in India. *Current Science* 85, 1143–1147.

Nath, N., Mathur, S. and Dasgupta, I. (2002) Molecular analysis of two complete *Rice tungro bacilliform virus* sequences from India. *Archives of Virology* 147, 1173–1187.

Periasamy, M., Niazi, F.R. and Malathi, V.G. (2006) Multiplex RT-PCR, a novel technique for the simultaneous detection of the DNA and RNA virus causing the rice tungro disease. *Journal of Virological Methods* 134, 230–236.

Pooggin, M., Rajeswaran, R., Schepetilnikov, M. and Ryabova, L. (2012) Short ORF-Dependent ribosome shunting operates in an RNA picorna-like virus and a DNA pararetrovirus that cause rice tungro disease. *PLoS Pathogen* 8, e1002568.

Purkayastha, A., Sharma, S. and Dasgupta, I. (2010) A negative element in the downstream region of the *Rice tungro bacilliform virus* promoter is orientation- and position-independent and is active with heterologous promoters. *Virus Research* 153, 166–171.

Qu, R., Bhattacharyya, M., Laco, G.S., deKochko, A., Subba Rao, B.L., *et al.* (1991) Characterization of the genome of *Rice tungro bacilliform virus,* comparison with *Commelina yellow mottle virus* and cauli-moviruses. *Virology* 185, 354–364.

Rajeswaran, R., Golyaev, V., Seguin, J., Zvereva, A., Farinelli, L., *et al.* (2014) Interactions of Rice tungro bacilliform pararetrovirus and its protein P4 with plant RNA silencing machinery. *Molecular Plant-Microbe Interactions* 12, 1370–1378.

Raychaudhuri, S.P., Mishra, M.D. and Ghosh, A. (1967) Preliminary note on the transmission of a virus disease resembling tungro of rice in India and other virus-like symptoms. *Plant Disease Reporter* 51, 300–301.

Rivera, C.T. and Ou, S.H. (1965) Leafhopper transmission of tungro disease of rice. *Plant Disease Reporter* 49, 127–131.

Rivera, C.T. and Ou, S.H. (1967) Transmission studies of two strains of rice tungro virus. *Plant Disease Reporter* 51, 877–881.

Roy, S., Banerjee, A., Tarafdar, J., Senapati, B.K. and Dasgupta, I. (2012) Transfer of transgenes for resistance to rice tungro disease into high yielding rice cultivars through gene based marker-assisted selection. *The Journal of Agricultural Science* 150, 610–618.

Sekiguchi, H., Isogai, M., Masuta, C. and Uyeda, I. (2005) 3C-like protease encoded by *Rice tungro spherical virus* is autocatalytically processed. *Archives of Virology* 150, 595–601.

Shahjahan, M., Jalani, B.S., Zakri, A.H., Imbe, T. and Othman, O. (1990) Inheritance of tolerance to *Rice tungro bacilliform virus* (RTBV) in rice (*Oryza sativa*). *Theoretical and Applied Genetics* 80, 513–517.

Sharma, S. and Dasgupta, I. (2012) Development of SYBR Green I based real time PCR assays for quantitative detection of *Rice tungro bacilliform virus* and *Rice tungro spherical virus*. *Journal of Virological Methods* 181, 86–92.

Sharma, S., Rabindran, R., Robin, S. and Dasgupta, I. (2011) Analysis of the complete sequence of *Rice tungro bacilliform virus* from southern India indicates it to be a product of recombination. *Archives of Virology* 156, 2257–2262.

Shen, P., Kaniewska, M., Smith, C. and Beachy, R.N. (1993) Nucleotide sequence and genomic organization of *Rice tungro spherical virus*. *Virology* 193, 621–630.

Shukla, V.D. and Anjaneyulu, A. (1982) Evaluation of systemic insecticides to reduce rice tungro disease incidence in rice nursery. *Indian Phytopathology* 35, 502–504.

Singh, K.G. (1969) Virus-vector relationship in penyakit merah of rice. *Japanese Journal of Phytopathology* 35, 322–324.

Sivamani, E., Huet, H., Shen, P., Ong, C.A., deKochko, A., *et al.* (1999) Rice plant (*Oryza sativa*) containing *Rice tungro spherical virus* (RTSV) coat protein transgenes are resistant to virus infection. *Molecular Breeding* 5, 177–185.

Takahashi, Y., Omura, T., Shohara, K. and Tsuchizaki, T. (1991) Comparison of four serological methods for practical detection of ten viruses of rice in plants and insects. *Plant Disease* 75, 458–461.

Takahashi, Y., Tiongco, E.R., Cabauatan, P.Q., Koganezawa, H., Hibino, H., *et al.* (1993) Detection of *Rice tungro bacilliform virus* by polymerase chain reaction for assessing mild infection of plants and viruliferous vector leafhoppers. *Phytopathology* 83, 655–659.

Thole, V. and Hull, R. (1998) *Rice tungro spherical virus* polyprotein processing, identification of a virus-encoded protease and mutational analysis of putative cleavage sites. *Virology* 247, 106–114.

Thole, V. and Hull, R. (2002) Charaterization of a protein from *Rice tungro spherical virus* with serine proteinase-like activity. *Journal of General Virology* 83, 3179–3186.

Tyagi, H., Rajasubramaniam, S., Rajam, M.V. and Dasgupta, I. (2008) RNA-interference in rice against *Rice tungro bacilliform virus* results in its decreased accumulation in inoculated rice plants. *Transgenic Research* 17, 897–904.

Varma, A., Niazi, F.R., Dasgupta, I., Singh, J., Cheema, S.S., *et al.* (1999) Alarming epidemic of rice tungro disease in North-West India. *Indian Phytopathology* 52, 71–74.

Verma, V. and Dasgupta, I. (2007) Sequence analysis of the complete genomes of two *Rice tungro spherical virus* isolates from India. *Archives of Virology* 152, 645–648.

Verma, V., Sharma, S., Vimla Devi, S., Rajasubramaniam, S. and Dasgupta, I. (2012) Delay in virus accumulation and low virus transmission from transgenic plants expressing *Rice tungro spherical virus* RNA. *Virus Genes* 45, 350–359.

Villegas, L.C., Druka, A., Bajet, N.B. and Hull, R. (1997) Genetic variation of *Rice tungro bacilliform virus* in the Philippines. *Virus Genes* 15, 195–201.

Yin, Y. and Beachy, R.N. (1995) The regulatory regions of *Rice tungro bacilliform virus* promoter and interacting nuclear factors in rice (*Oryza sativa*). *The Plant Journal* 7, 969–980.

Zhang, S., Jones, M.C., Barker, P., Davies, J.W. and Hull, R. (1993) Molecular cloning and sequencing of coat protein-encoding cDNA of *Rice tungro spherical virus* – a plant picornavirus. *Virus Genes* 7, 121–132.

Zhang, S., Lee, G., Davies, J.W. and Hull, R. (1997) Variation on coat protein genes among five geographically different isolates of *Rice tungro spherical virus*. *Archives of Virology* 142, 1873–1879.

16 Sweet Potato Virus Disease

Augustine Gubba* and Benice J. Sivparsad

Department of Plant Pathology, School of Agricultural, Earth and Environmental Sciences, University of KwaZulu-Natal, Pietermaritzburg, South Africa

16.1 Introduction

Sweet potato is ranked as the seventh most important food crop in the world (Woolfe, 1992; FAOSTAT, 2012). Among the major starch staples, it has the largest rates of biomass and nutrient production per unit area per unit time (Woolfe, 1992). Because of its good performance under adverse farming conditions and high carbohydrate and vitamin content, sweet potato has been identified as an ideal starch staple in subsistence economies (Mukasa *et al.*, 2003; Wambugu, 2003; Naylor *et al.*, 2004; Loebenstein *et al.*, 2009).

Virus infection is the main limiting factor in sweet potato production worldwide (Allemann *et al.*, 2004). Moreover, viral diseases rank second after sweet potato weevils as restraining biotic factors and can cause considerable yield reduction of up to 98% (Carroll *et al.*, 2004; Aritua *et al.*, 2006). Sweet potatoes are vegetatively propagated and farmers often use vines from their own field year after year. Thus, virus diseases are inevitably transmitted to the newly planted field, resulting in decreased yields. Moreover, infection in the field often develops by multiple viruses interacting in a synergistic complex, thereby compounding the effect on yields (Loebenstein *et al.*, 2009).

The most devastating viral disease affecting sweet potatoes worldwide is sweet potato virus disease (SPVD) (Kokkinos *et al.*, 2006; Miano *et al.*, 2008). The disease is caused by the synergistic interaction between *Sweet potato feathery mottle virus* (SPFMV) and *Sweet potato chlorotic stunt virus* (SPCSV). SPVD infected plants display chlorosis (Fig 16.1a), vein clearing (Fig. 16.1b), mottling (Fig. 16.1c), severe stunting (Fig. 16.1d) and an almost 99% reduction in tuber yield (Fig 16.2) (Gibson *et al.*, 1998; Karyeija *et al.*, 1998; Tairo *et al.*, 2005). SPVD is currently the most important viral disease complex in East Africa where sweet potato is often the main food staple (Karyeija *et al.*, 1998; Loebenstein *et al.*, 2009). The occurrence of SPVD and other viral co-infections has been reported in almost every sweet potato growing area worldwide.

16.2 Causative Viruses

16.2.1 *Sweet potato feathery mottle virus*

SPFMV (family *Potyviridae*, genus *Potyvirus*) is the most common sweet potato virus, found nearly everywhere sweet potatoes are grown (Moyer and Salazar, 1989; Karyeija *et al.*, 1998; Kreuze and Fuentes, 2009). Some of the synonyms used for SPFMV include russet crack virus, sweet potato virus A, sweet potato ringspot virus, sweet potato leaf cork virus and internal cork virus (Clark and Moyer, 1988; Moyer and Salazar, 1989).

SPFMV has flexuous filamentous particles between 830 nm and 850 nm in length. Particles contain a single-stranded, positive-sense RNA genome of about 10.6 kb and a

*E-mail: gubbaa@ukzn.ac.za

Fig. 16.1. Sweet potato virus disease symptoms observed on field samples from rural farming areas in KwaZulu-Natal, South Africa. (a) Yellowing, mosaic and chlorosis, (b) vein clearing, (c) mottling and purpling on local varieties, and (d) stunting and leaf deformation.

Fig. 16.2. The effect of sweet potato virus disease on storage roots of sweet potato (*Ipomoea batatas* Lam.) plants grown on small-scale farms in KwaZulu-Natal, South Africa.

coat protein (CP) of 38 kDa, both of which are larger than the genome and CP of the average potyvirus (Loebenstein *et al.*, 2003; Kreuze and Fuentes, 2009). Transmission of SPFMV occurs by several species of aphids in a non-persistent manner (Clark and Moyer, 1988; Karyeija *et al.*, 1998). The virus is transmitted by grafting, but not by seed, pollen or by contact between plants (Loebenstein *et al.*, 2003).

SPFMV has a narrower host range than most potyviruses and is mostly limited to the family Convolvulaceae, and especially to the genus *Ipomoea*, although some strains have been reported to infect *Nicotiana benthamiana* Gray and *Chenopodium* spp. (Kreuze and Fuentes, 2009). Symptoms, serology and host range have been used to differentiate SPFMV isolates into two strains: the common (C) strain and the more severe russet crack (RC) strain (Karyeija *et al.*, 1998). However, phylogenetic analyses of the CP sequences have differentiated SPFMV into four strains: common (C), East African (EA), ordinary (O) and russet crack (RC) (Kreuze *et al.*, 2000). Isolates of strains RC, O and EA are closely related and are phylogenetically distant from strain C (Tairo *et al.*, 2005). Strains RC, O and C are distributed worldwide, whereas isolates of the EA strain have been largely restricted to countries in East Africa (Kreuze *et al.*, 2000; Mukasa *et al.*, 2003). However, recent reports have detected isolates of the EA strain outside East Africa in Spain (Valverde *et al.*, 2004a), Vietnam (Ha *et al.*, 2008), Peru (Untiveros *et al.*, 2008) and Easter Island (Rännäli *et al.*, 2009).

Leaf symptoms are generally mild and may consist of the classic irregular chlorotic patterns (feathery mottle), chlorosis of older leaves and vein clearing (Clark and Moyer, 1988; Moyer and Salazar, 1989; Karyeija et al., 1998). Some infected plants may even be symptomless (Moyer and Salazar, 1989; Gibson et al., 1997). In other Ipomoea species, including the indicator plants I. setosa and I. nil, symptoms of vein clearing, mosaic and distortion are more pronounced (Karyeija et al., 1998). Some strains of SPFMV cause necrotic lesions on the root exterior (russet crack disease), whereas another strain produces symptoms on the root interior (internal cork disease) (Clark and Moyer, 1988; Moyer and Salazar, 1989). The main economic loss due to SPFMV is when it acts with SPCSV in the synergistic virus complex, SPVD (Karyeija et al., 1998; Kreuze and Fuentes, 2009). This synergism has been shown to be mediated by the SPCSV-encoded RNase3 protein. RNase3 is a double-stranded RNA-specific class 1 RNA endoribonuclease III that functions as an RNA silencing suppressor and in the digestion of long and short double-stranded RNAs. RNase3 catalytic activity is required for its RNA silencing suppressor activity, thus implicating RNA cleavage in the suppression of RNA silencing suppressor as well as synergistic disease induction. Although the exact mechanism of RNase3 action is not clear, it is evident that it is able to mediate increased susceptibility of sweet potato to a wide range of viruses (Cuellar et al., 2009).

16.2.2 *Sweet potato chlorotic stunt virus*

SPCSV (family *Closteroviridae*, genus *Crinivirus*) is one of the most devastating viruses infecting sweet potato crops worldwide (Winter et al., 1992; Mukasa et al., 2006; Untiveros et al., 2007). In single infections, SPCSV can reduce yields by 50%, causing mild stunting combined with slight yellowing or purpling of older leaves (Kreuze and Fuentes, 2009). However, what makes SPCSV one of the most damaging viruses of sweet potato is its ability to break down the natural resistance of sweet potato to other viruses and mediate severe synergistic viral diseases with other sweet potato viruses (Karyeija et al., 2000; Kreuze et al., 2008). The most common and severe of these diseases is SPVD, and is caused by co-infection with SPFMV (Gibson et al., 1998; Karyeija et al., 1998; Kreuze et al., 2008).

SPCSV is phloem-limited and is transmitted in a semi-persistent manner by whiteflies (Cohen et al., 1992; Loebenstein et al., 2003; Valverde et al., 2004a; Gamarra et al., 2010). Transmission can occur through grafting but not by mechanical inoculation or by contact between plants. Similar to most sweet potato-infecting viruses, the host range of SPCSV is limited mainly to the family Convolvulaceae, and the genus *Ipomoea*, although *Nicotiana* spp. and *Amaranthus palmeri* are susceptible (Brunt et al., 1996; Loebenstein et al., 2003; Kreuze and Fuentes, 2009).

SPCSV has flexuous particles of between 850 nm and 950 nm in length and 12 nm in diameter. The bipartite genome of SPCSV, consisting of RNA1 (9407 nucleotides) and RNA2 (8223 nucleotides), is encapsidated by a 33 kDa major CP (Cohen et al., 1992; Brunt et al., 1996). Based on molecular and serological analyses, SPCSV can be differentiated into EA and West African (WA) serotypes (Alicia et al., 1999; Ishak et al., 2003; Tairo et al., 2005). Outside of West Africa, the WA strain seems to have a wide geographical distribution, occurring in Egypt (Ishak et al., 2003), Israel (Cohen et al., 1992), Australia (Tairo et al., 2005), America (Di Feo et al., 2000; Fenby et al., 2002; Abad et al., 2007), South Africa (Sivparsad and Gubba, 2012) and China (Qin et al., 2013). Isolates of the EA strain have been largely restricted to countries in East Africa, with only one isolate from Peru being reported from outside Africa (Gutiérrez et al., 2003). The first complete genomic sequence of SPCSV was determined by Kreuze et al. (2002). To date, a total of six complete SPCSV genomic sequences can be accessed through GenBank, including two sequences of the SPCSV EA strain and four sequences of the SPCSV-WA strain (Kreuze et al., 2002; Trenado et al., 2009; Cuellar et al., 2011b; Qin et al., 2014).

16.3   Diagnostic Methods

Early detection of the viral disease followed by rapid and accurate identification of the causal viral agent is vital if correct control measures are to be employed. This is particularly true for newly identified sweet potato viruses where novel control strategies may have to be developed alongside characterization of the new virus (Kreuze *et al.*, 2009). The detection and identification of sweet potato viruses is complicated by the frequent occurrence of mixed infections, synergistic complexes and the constant emergence of new viral species and strains (Tairo, 2006; Gutiérrez and Valverde, 2007). Traditional methods such as biological indexing, ELISA, molecular hybridization and PCR are mostly used. However, due to their limited reliability and sensitivity, a combination of methods is required to correctly identify the virus present in a diseased sweet potato plant (Tairo, 2006).

16.3.1   Biological indexing

Biological indexing is one of the oldest methods that employs reactions of plants when infected by viruses. One way is by visual inspection of the material to be tested for characteristic symptoms. The other approach is by indexing for infective virus on indicator hosts and their examination for symptoms (Bos, 1999). In sweet potato, graft inoculation onto *Ipomoea* species is widely used to assay for many sweet potato viruses (Tairo, 2006). *Ipomoea setosa* Ker. is a convenient and nearly universal host for sweet potato viruses. Characteristic symptoms of mosaic, curled or wrinkled leaves and vein clearing on graft-inoculated *I. setosa* is an indication of viral infection (Feng *et al.*, 2000).

Biological indexing has routinely been used to detect virus-infected sweet potato plants in China (Feng *et al.*, 2000). Although indexing is simple and suitable to detect virus-infected sweet potato samples which do not show any obvious symptoms, it requires time, and the production of similar symptoms makes it difficult to distinguish different sweet potato viruses. However, when indexing is combined with serological methods, detection efficiency can be improved (Tairo, 2006; Feng *et al.*, 2011).

16.3.2   Serological tests

Serological methods, based on the specific interaction of an antibody and antigen, are widely used in the diagnosis of plant viral diseases (Bos, 1999). The advantages of such methods are that they are highly specific, rapid and sensitive to small amounts of viral antigen in the plant material (Yao and Hortense, 2005).

A membrane immunobinding assay known as nitrocellulose membrane ELISA has been used with success for the detection of sweet potato viruses (Gutiérrez and Valverde, 2007). This method has been adapted for practical virus detection by the International Potato Center (Peru). To date, a kit containing antiserum for ten sweet potato viruses is available from the International Potato Center, together with a standardized nitrocellulose membrane ELISA protocol. However, the accuracy of the serological detection of sweet potato viruses is hampered by the presence of interfering phenolic substances and inhibitors (Gibb and Padovan, 1993), low concentration and erratic distribution of viruses in infected sweet potato plants (Esbenshade and Moyer, 1982), and the frequency of multiple infections of viruses. Thus, subsequent testing is needed to resolve discrepancies between assays and confirm positive results (Tairo, 2006).

16.3.3   Electron microscopy

Electron microscopy is useful in revealing virus structure and in showing the cytoplasmic effect of viruses in infected tissue (Kado and Agrawal, 1972). Such characteristics are often specific to viral families and are valuable in virus identification. For instance, the long flexuous particles of sweet potato-infecting *Potyviruses* can be directly identified by transmission electron microscopy (TEM) of leaf-dip preparations or in thin sections of infected *I. setosa* plants. Alternatively, these viruses can be identified by the production of characteristic pinwheel cytoplasmic inclusions induced by potyvirus infection (Nome *et al.*, 2006). Immunosorbent electron microscopy (ISEM) has combined the ease of TEM with the specificity of serology. This technique, which focuses on the specific trapping of viral particles on the grid that have been pretreated with antiserum, has increased the sensitivity and specificity of TEM up to 10,000-fold (Bos, 1999). Both TEM and ISEM have been used to detect SPFMV, SPLV and *Tobacco mosaic virus* (TMV) (Yang *et al.*, 1995). However, these methods are not specific in identifying individual viruses and also require expensive equipment. Therefore, TEM and ISEM are often used only to classify and study sweet potato viruses and not for routine diagnosis (Feng *et al.*, 2000).

16.3.4 Molecular techniques

The most recent approaches used for the detection of sweet potato viruses are those based on molecular biology techniques. Techniques such as PCR, quantitative real-time PCR, rolling-circle amplification and deep sequencing of small RNAs have emerged as powerful methods in the identification and characterization of viruses infecting sweet potato (Colinet *et al.*, 1998; Valverde *et al.*, 2004b; Clark *et al.*, 2012). These techniques show great promise and may circumvent the problems associated with serological and biological indexing (Colinet *et al.*, 1998).

PCR is an effective technique for detecting sweet potato viruses such as SPFMV, which are usually irregularly distributed and present at a low titre in the infected plants (Souto *et al.*, 2003). Depending on the choice of primers, PCR assists in the detection of a single species or many members of a group or family of related viruses (Colinet *et al.*, 1998). Sequence information and the knowledge of conserved viral sequences have simplified the design of oligonucleotide primers that enable specific and rapid identification of sweet potato viruses (Colinet *et al.*, 1998; Tairo *et al.*, 2006). The value of PCR for rapid identification and characterization of sweet potato viruses has been demonstrated for *Potyviruses, Criniviruses* and *Geminiviruses* (Colinet and Kummert, 1993; Colinet *et al.*, 1998; Li *et al.*, 2004; Kokkinos and Clark, 2006; Tairo *et al.*, 2006; Opiyo *et al.*, 2010). The restriction fragment length polymorphism (RFLP) technique is a sensitive and simple diagnostic procedure that involves restriction analysis of PCR generated amplicons. Restriction fragment length polymorphism analysis has been used in the rapid identification and differentiation of *Potyvirus* complexes in sweet potato (Colinet *et al.*, 1998; Tairo *et al.*, 2006).

Quantitative real-time PCR has been used in the detection and quantification of SPCSV, SPFMV, *Sweet potato mild mottle virus* (SPMMV), *Sweet potato virus G* (SPVG), *Sweet potato virus 2* (SPV2) and *Sweet potato leaf curl virus* (SPLCV) directly from infected sweet potato plants (Kokkinos and Clark, 2006; Mukasa *et al.*, 2006; McGregor *et al.*, 2009; Perez-Egusquiza *et al.*, 2009). Quantitative real-time PCR has been shown to be more efficient and sensitive in detecting sweet potato viruses than conventional PCR (Kokkinos and Clark, 2006). However, its use in routine virus detection may be restricted by the sequence specificity of the TaqMan probes and primers, as well as the expensive reagents and instruments (Clark *et al.*, 2012).

The rapid evolution of viruses has caused significant problems in the design of PCR primers to detect all viral strains (Zhang and Ling, 2011). To this end, rolling-circle amplification, sometimes combined with restriction fragment length polymorphism, is emerging as a powerful tool for the detection of sweet potato viruses (Paprotka *et al.*, 2010; Clark *et al.*, 2012). Rolling-circle amplification / restriction fragment length polymorphism combined with sequencing has enabled the identification of novel variants, strains and species of sweet potato viruses (Haible *et al.*, 2006; Lozano *et al.*, 2009; Paprotka *et al.*, 2010; Albuquerque *et al.*, 2011).

Novel DNA sequencing techniques, referred to as 'next-generation sequencing', have become available in the last few years and these involve an unbiased approach to plant viral disease diagnosis which requires no prior knowledge of the host or pathogen (Adams *et al.*, 2009). They have been widely used in many projects (e.g., whole genome sequencing, metagenomics, small RNA discovery and RNA sequencing). Their common feature is to provide high speed throughput that can produce an enormous volume of sequences with many possible applications in research and viral diagnostics (Barzon *et al.*, 2011). This interesting strategy, which has led to the discovery of various virus species, exploits the property of invertebrates and plants to respond to infection by processing viral RNA genomes into small RNAs of discrete sizes. A recent study on small RNA libraries sequenced by next-generation sequencing platforms (Wu *et al.*, 2010) showed that viral small silencing RNAs produced by invertebrate animals are overlapping in sequence and can assemble into long contiguous fragments of the invading viral genome. Based on this result, an approach of virus discovery in plants by deep sequencing and assembly of total small RNAs was developed and utilized in the analysis of contigs (i.e. a contiguous length of genomic sequences in which the order of bases is known) assembled from available small RNA libraries (Barzon *et al.*, 2011). The use of deep sequencing of small RNA has been described by Kreuze *et al.* (2009) as a novel means to detect sweet potato viruses. The technique has since been used to determine complete genome sequences of *Sweet potato virus G, Sweet potato virus 2, Sweet potato latent virus* and SPFMV-RC; new sweet potato viruses, badnaviruses, a *Cavemovirus,* a *Solendovirus* and new strains of SPCSV from South America (Cuellar *et al.*, 2011a; Clark *et al.*, 2012) as well as SPFMV-RC and *Sweet potato virus C*, SPCSV-WA, *Sweet potato leaf curl Georgia virus* and *Sweet potato pakakuy virus* strain B (synonym: Sweet potato badnavirus B) from Honduras and Guatemala (Kashif *et al.*, 2012).

16.4   Management Strategies

Control of viral disease in sweet potato is complicated by the frequent occurrence of mixed synergistic viral complexes. In addition, vegetative propagation of infected roots or vines provides a perfect means of perpetuating viruses within the production cycle. Therefore,

effective and durable disease control methods are based on prevention (Clark and Moyer, 1988). However, no single management tool that provides adequate control against the natural viral complexes that infect sweet potato is available.

16.4.1   Cultural methods

A series of cultural methods such as weed control, intercropping and roguing of infected plants have proven effective in minimizing losses due to viral disease in sweet potato (Karyeija *et al.*, 1998; Ndunguru and Aloyce, 2000; Gibson *et al.*, 2004). Weeds play an important role in the incidence and spread of sweet potato viruses as they possibly serve as alternate hosts of insect vectors and viruses. Removal of reservoir weed hosts in a wide area around a crop may relieve the inoculum pressure. The incidence of SPFMV and SPCSV was shown to decrease when weeds, especially wild *Ipomoea* species, were removed in and around sweet potato fields (Karyeija *et al.*, 1998). The use of intercropping to reduce the number of infectious insect vectors attacking the sweet potato crop may help to reduce viral incidence by delaying vector onset and build-up. A sweet potato/maize intercrop was shown to lower SPVD incidence in traditional sweet potato farming systems (Ndunguru and Aloyce, 2000). A recent on-farm site trial in Uganda showed that roguing of diseased cuttings within a month of planting and isolating new crops (15–20 m apart) from diseased crops, can considerably decrease the spread of SPVD to susceptible cultivars (Gibson *et al.*, 2004). However, despite these initial successes, neither cultural control method has been shown to be durable and feasible against the multitude of viral complexes that infect sweet potato.

16.4.2   Distribution of virus-indexed material

At present, the best way to control viral diseases in sweet potato is to supply growers with virus-indexed propagation material

(Cohen *et al.*, 2008; Loebenstein *et al.*, 2009). Such material can be produced by the meristem tip or shoot tip culture techniques, which are based on the propagation of the youngest tissues of the shoot apex that have uneven low virus titres (Wang and Valkonen, 2008). Recently, the combination of meristem tip culture with cryotherapy or thermotherapy was shown to drastically enhance the efficiency of virus elimination in sweet potato (Wang and Valkonen, 2008; Feng *et al.*, 2011; Mashilo *et al.*, 2013). Worldwide, there are many programmes that produce and distribute meristem-derived, virus-free propagation material for sweet potato cultivation (Loebenstein *et al.*, 2009). In South Africa, a sweet potato improvement scheme was implemented in 1971 with the task of establishing virus-free market stock which is supplied to registered sweet potato growers and producers (Joubert *et al.*, 1996). However, in South Africa and in the rest of Africa, such programs are operating on a limited scale, because sweet potatoes are grown mainly as a subsistence crop, and not commercially (Loebenstein *et al.*, 2009).

16.4.3   Resistance

Cultural practices and the distribution of virus-indexed propagation material have only been marginally effective in the management of the multitude of viruses that infect sweet potato. The development of resistant sweet potato varieties is the most promising means of controlling viral disease in the long term (Loebenstein *et al.*, 2009).

Natural resistance

Little success has been reported in the development of sweet potato cultivars with broad virus resistance. All reports so far primarily focus on SPFMV and SPVD (Gibson *et al.*, 1998, 2004; Karuri *et al.*, 2009). Breeding for virus resistance in sweet potato involves the introduction of resistant genes into cultivated varieties without changing any of its desirable characteristics. Wild relatives of sweet potato (e.g. *I. trifida*) may serve as a

source of resistant genes (Agrios, 2005). Breeding programs in Uganda have worked at combining SPVD resistance with desirable agronomical traits such as yield, earliness and acceptable culinary quality in sweet potato cultivars (Karyeija *et al.*, 2000; Mwanga *et al.*, 2002; Loebenstein *et al.*, 2009). Although progress has been made, it remains to be seen if these cultivars will retain their resistance when challenged by differing strains of the SPVD viral components, which may occur in different geographical locations (Loebenstein *et al.*, 2009). In addition to virus variation, conventional breeding is hindered by the amount of time and expense required (Lomonossoff, 1995). Moreover, genetic sources of resistance are scarce and the incorporation of such resistance from the wild diploid species into polyploid sweet potato is a complicated task (Kreuze, 2002).

Engineered resistance

As natural resistance to viruses in sweet potato seems to be of limited use, alternate strategies for obtaining virus resistance through biotechnological means have been attempted (Kreuze, 2002). Most of these strategies are based on the concept of 'pathogen-derived resistance' (PDR), which proposes that pathogen resistance genes may be developed from the pathogen's own genetic material (Sanford and Johnston, 1985). PDR for plant viruses can be mediated by the expression of RNA, or RNA-mediated PDR (Shepherd *et al.*, 2009; Collinge *et al.*, 2010). The post-transcriptional gene silencing process, also known as RNA interference or RNA silencing, is the mechanism of RNA-mediated PDR (Tenllado *et al.*, 2004; Lindbo and Dougherty, 2005; Fuchs and Gonsalves, 2007). Post-transcriptional gene silencing is a specific RNA degradation mechanism of any organism that breaks down aberrant, excess or foreign RNA intracellularly, in a homology-dependent manner, resulting in a resistance phenotype (Dasgupta *et al.*, 2003).

Current genetic engineering efforts have been used to develop transgenic sweet potato with virus resistance using the CP gene of SPFMV and/or SPCSV (Newell *et al.*, 1995; Gama *et al.*, 1996; Otani *et al.*, 1998;

Okada *et al.*, 2002; Kreuze *et al.*, 2008). However, given the multiplicity of viruses occurring under field conditions, this approach has had limited success.

The RNA silencing approach was also used to target SPCSV and SPFMV simultaneously. A construct was designed to produce transcripts that generate a double-stranded RNA structure that was homologous to the polymerase genes of each virus, thus effectively inducing the RNA silencing defence mechanism against both viruses (Kreuze *et al.*, 2008). The study showed that many transgenic lines accumulated only low titres of SPCSV with no symptom development. However, the low titres of SPCSV in transgenic plants were still sufficient to break down the naturally high level of resistance to SPFMV, and SPVD developed.

Recently, Lin *et al.* (2011) demonstrated that multiple virus resistance in transgenic plants can be induced by using viral DNA as a 'silencer' and linking it to DNA segments derived from other viruses. Using the same strategy, Sivparsad and Gubba (2014) attempted to develop transgenic sweet potato cv. 'Blesbok' with broad virus resistance. CP segments of the SPCSV, SPFMV, SPVG and SPMMV infecting sweet potato in Kwa-Zulu Natal, South Africa, were fused to a silencer DNA, the middle half of the nucleocapsid gene of *Tomato spotted wilt virus* (TSWV), and introduced into sweet potato via *Agrobacterium*-mediated transformation. Transgenic sweet potato plants showing varying levels of resistance to SPFMV and SPVG were obtained by post-transcriptional gene silencing. The effectiveness of the resistance displayed by transgenic sweet potato plants developed in this study against multiple virus infection in the field is unknown and remains to be tested.

16.5 Concluding Remarks

The growing importance of sweet potato as a staple food in developing countries despite its global status as a low economic valued 'orphan' crop, attests to its vital historical role in meeting the food and income needs of the world's poorest and fastest-growing populations. However, present studies report an increase in the incidence of SPVD as well as the identification of additional viruses that can form synergistic relationships with SPCSV and SPFMV and hence potentially contribute to a new SPVD complex. This upsurge of viral incidence, along with the diversity and emergence of new viruses justifies the need for continuously updating information on virus identity and distribution. Such information can be obtained by conducting regular field surveys in previously surveyed virus endemic locations. A continuous updated record of viral prevalence and distribution will be crucial in understanding the dynamics involved in SPVD development, thus laying the foundation for the development of sustainable control/management strategies for the disease.

References

Abad, J.A., Parks, E.J., New, S.L., Fuentes, S., Jesper, W., *et al.* (2007) First report of *Sweet potato chlorotic stunt virus*, a component of sweet potato virus disease, in North Carolina. *Plant Disease* 91, 327.

Adams, I.P., Glover, R.H., Monger, W.A., Mumford, R., Jackeviciene, E., *et al.* (2009) Next-generation sequencing and metagenomic analysis: a universal diagnostic tool in plant virology. *Molecular Plant Pathology* 10, 537–545.

Agrios, G.N. (2005) *Plant Pathology*. Academic Press, San Diego, California, pp. 165–169, 314–315, 724–752.

Albuquerque, L.C., Inoue-Nagata, A.K., Pinheiro, B., Ribeiro, S.D., Resende, R.O., *et al.* (2011) A novel monopartite begomovirus infecting sweet potato in Brazil. *Archives of Virology* 156, 1291–1294.

Alicia, T., Fenby, N.S., Gibson, R.W., Adipala, E., Vetten, H.J., *et al.* (1999) Occurrence of two serotypes of *Sweet potato chlorotic stunt virus* in East Africa and their associated differences in coat protein and HSP70 homologue gene sequences. *Plant Pathology* 48, 718–726.

Allemann, J., Laurie, S.M., Thiart, S. and Vorster, H.T. (2004) Sustainable production of root and tuber crops (potato, sweet potato and indigenous potato, cassava) in Southern Africa. *South African Journal of Botany* 70, 60–66.

Aritua, V., Bau, B., Vetten, H.J., Adipala, E. and Gibson, R.W. (2006) Incidence of five viruses infecting sweet potatoes in Uganda; the first evidence of Sweet potato caulimo-like virus in Africa. *Plant Pathology* 56, 324–331.

Barzon, L., Lavezzo, E., Militello, V., Toppo, S. and Palù, G. (2011) Applications of next-generation sequencing technologies to diagnostic virology. *International Journal of Molecular Sciences* 12, 7861–7884.

Bos, L. (1999) *Plant Viruses, Unique and Intriguing Pathogens – A Textbook of Plant Virology*. Backhuys Publishers, Leiden, The Netherlands.

Brunt, A., Crabtree, K., Dallwitz, M., Gibbs, A. and Watson, L. (1996) *Viruses of Plants: Descriptions and Lists from VIDE Database*. CAB International, Wallingford, UK, pp. 1216–1229.

Carroll, H.W., Villordon, A.Q., Clark, C.A., La Bonte, D.R. and Hoy, M.W. (2004) Studies on Beauragard sweet potato clones naturally infected with viruses. *International Journal of Pest Management* 50, 101–106.

Clark, C.A. and Moyer, J.W. (1988) *Compendium of Sweet Potato Diseases*. APS Press, St. Paul, Minnesota, pp. 49–53.

Clark, C.A., Davis, J.A., Abad, J.A., Cuellar, W.J., Fuentes, S., *et al.* (2012) Sweet potato viruses: 15 years on understanding and managing complex diseases. *Plant Disease* 96, 168–185.

Cohen, J., Franck, A., Vetten, H.J., Lesemann, D.E. and Loebenstein, G. (1992) Purification and properties of *Closterovirus*-like particles associated with a whitefly-transmitted disease of sweet potato. *Annals of Applied Biology* 121, 257–268.

Cohen, J., Gal-On, A. and Loebenstein, G. (2008) Viral diseases of sweet potato and their control. *Phytoparasitica* 36, 124–125.

Colinet, D. and Kummert, J. (1993) Identification of a *Sweet potato feathery mottle virus* from China (SPFMV-CH) by the polymerase chain reaction with degenerate primers. *Journal of Virological Methods* 45, 149–159.

Colinet, D., Nguyen, M., Kummer, J. and Lepoivre, P. (1998) Differentiation among potyviruses infecting sweet potato based on genus- and virus-specific reverse transcription polymerase chain reaction. *Plant Disease* 82, 223–229.

Collinge, D.B., Jorgensen, H.J.L., Lund, O.S. and Lyngkjaer, M.J. (2010) Engineering pathogen resistance in crop plants: current trends and future prospects. *Annual Review of Phytopathology* 48, 269–291.

Cuellar, W.J., Kreuze, J.F., Rajamäki, M.-L., Cruzado, K.R., Untiveros, M., *et al.* (2009) Elimination of antiviral defense by viral RNase III. *Proceedings of the National Academy of Sciences USA* 106, 354–358.

Cuellar, W.J., De Souza, J., Barrantes, I., Fuentes, S. and Kreuze, J.F. (2011a) Distinct cavemoviruses interact synergistically with *Sweet potato chlorotic stunt virus* (genus *Crinivirus*) in cultivated sweet potato. *Journal of General Virology* 92, 1233–1243.

Cuellar, W.J., Cruzado, K.R., Fuentes, S., Untiveros, M., Soto, M., *et al.* (2011b) Sequence characterization of a Peruvian isolate of *Sweet potato chlorotic stunt virus*: further variability and a model for p22 acquisition. *Virus Research* 157, 111–115.

Dasgupta, I., Malathi, V.G. and Mukherjee, S.K. (2003) Genetic engineering for virus resistance. *Current Science* 84, 341–354.

Di Feo, L., Nome, S.F., Biderbost, E., Fuentes, S. and Salazar, L.F. (2000) Etiology of sweet potato chlorotic dwarf disease in Argentina. *Plant Disease* 84, 223–229.

Esbenshade, P.R. and Moyer, J.W. (1982) Indexing system for *Sweet potato feathery mottle virus* in sweet potato using enzyme-linked immunosorbent assay. *Plant Disease* 66, 911–913.

FAOSTAT (2012) *FAO statistics*. Food and Agricultural Organization of the United Nations, Rome (accessed 2 February 2015).

Fenby, N., Foster, G., Gibson, R.W. and Seal, S. (2002) Partial sequence of HSP70 homologue gene shows diversity between West African and East African isolates of *Sweet potato chlorotic stunt virus*. *Tropical Agriculture* 79, 26–30.

Feng, G., Yifo, G. and Pinbo, Z. (2000) Production and deployment of virus-free sweet potato in China. *Crop Protection* 19, 105–111.

Feng, C., Yin, Z., Ma, Y., Zhang, Z., Chen, L., *et al.* (2011) Cryopreservation of sweet potato (*Ipomoea batatas*) and its pathogen eradication by cryotherapy. *Biotechnology Advances* 29, 84–93.

Fuchs, M. and Gonsalves, D. (2007) Safety of virus-resistant transgenic plants two decades after their introduction: lessons from realistic field risk assessment studies. *Annual Review of Phytopathology* 45, 173–202.

Gama, M.I.C.S., Leite, R.P., Cordeiro, A.R. and Cantliffe, D.J. (1996) Transgenic sweet potato plants obtained by *Agrobacterium tumefaciens*-mediated transformation. *Plant Cell Tissue Organ Culture* 46, 237–244.

Gamarra, H.A., Fuentes, S., Morales, F.J., Glover, R., Malumphy, C., *et al.* (2010) *Bemisia afer* sensu lato, a vector of *Sweet potato chlorotic stunt virus. Plant Disease* 94, 510–514.

Gibb, K.S. and Padovan, A.C. (1993) Detection of sweet potato feathery mottle potyvirus in sweet potato grown in northern Australia using an efficient and simple assay. *International Journal of Pest Management* 39, 223–228.

Gibson, R.W., Mwanga, R.O.M., Kasule, S., Mpembe, I. and Carey, E.E. (1997) Apparent absence of viruses in most symptomless field-grown sweet potato in Uganda. *Annals of Applied Biology* 130, 481–490.

Gibson, R.W., Kaitisha, G.C., Randrianaivoriavony, J.M. and Vetten, H.T. (1998) Identification of the East African strain of *Sweet potato chlorotic stunt virus* (SPCSV) as a major component of sweet potato virus disease (SPVD) in southern Africa. *Plant Disease* 82, 1063.

Gibson, R.W., Aritua, V., Byamukama, E., Mpembe, I. and Kayongo, J. (2004) Control strategies for sweet potato virus disease in Africa. *Virus Research* 100, 115–122.

Gutiérrez, D.L. and Valverde, R.A. (2007) Detection of sweet potato viruses by NCM-ELISA. *HortScience* 42, 450–451.

Gutiérrez, D.L., Fuentes, S. and Salazar, L.F. (2003) Sweet potato virus disease (SPVD): distribution, incidence, and effect on sweet potato yield in Peru. *Plant Disease* 87, 297–302.

Ha, C., Revill, P., Harding, R.M., Vu, M. and Dale, J.L. (2008) Identification and sequence analysis of potyviruses infecting crops in Vietnam. *Archives of Virology* 153, 45–60.

Haible, D., Kober, S. and Jeske, H. (2006) Rolling circle amplification revolutionizes diagnosis and genomics of geminiviruses. *Journal of Virological Methods* 135, 9–16.

Ishak, J.A., Kreuze, J.F., Johansson, A., Mukasa, S.B., Tairo, F., *et al.* (2003) Some molecular characteristics of three viruses from sweet potato virus disease (SPVD)-affected sweet potato plants in Egypt. *Archives of Virology* 148, 2449–2460.

Joubert, T.G., Ia, G., Klesser, P. and Nel, D.D. (1996) The value of virus-indexed sweet potato material. In: *Vegetable Production in South Africa*. ARC Roodeplaat, Pretoria, pp. 1–2.

Kado, C.I. and Agrawal, H.O. (1972) *Principles and Techniques in Plant Virology*. Van Nostrand Reinhold Company, New York.

Karuri, H.W., Ateka, E.M., Amata, R., Nyende, A.B. and Muigai, A.W.T. (2009) Characterization of Kenyan sweet potato genotypes for SPVD resistance and high dry matter content using simple sequence repeat markers. *African Journal of Biotechnology* 8, 2169–2175.

Karyeija, R.F., Gibson, R.W. and Valkonen, J.P.T. (1998) The significance of *Sweet potato feathery mottle virus* in subsistence sweet potato production in Africa. *Plant Disease* 82, 4–15.

Karyeija, R.F., Kreuze, J.F., Gibson, R.W. and Valkonen, J.P.T. (2000) Two serotypes of *Sweet potato feathery mottle virus* in Uganda and their interaction with resistant sweet potato cultivars. *Phytopathology* 90, 1250–1255.

Kashif, M., Pietilä, S., Artola, K., Jones, R.A.C., Tugume, A.K., *et al.* (2012) Detection of viruses in sweet potato from Honduras and Guatemala augmented by deep-sequencing of small-RNAs. *Plant Disease* 96, 1430–1437.

Kokkinos, C.D. and Clark, C.A. (2006) Real-time PCR assays for detection and quantification of sweet potato viruses. *Plant Disease* 90, 783–788.

Kokkinos, C.D., Clark, C.A., McGregor, C.E. and LaBonte, D.R. (2006) The effect of sweet potato virus disease and its viral components on gene expression levels in sweet potato. *Journal of the American Society of Horticultural Science* 131, 657–666.

Kreuze, J. (2002) Molecular studies on the sweet potato virus disease and its two causal agents. Doctoral thesis, Swedish University of Agricultural Sciences, Uppsala, Sweden, pp. 24–33.

Kreuze, J. and Fuentes, S. (2009) Sweet potato viruses. In: Mahy, B.W.J. and Van Regenmortel, M.H.V. (eds) *Encyclopedia of Virology*. Academic Press, Amsterdam, The Netherlands, pp. 659–669.

Kreuze, J.F., Karyeija, R.F., Gibson, R.W. and Valkonen, J.P.T. (2000) Comparisons of coat protein gene sequences show that East African isolates of *Sweet potato feathery mottle virus* form a genetically distinct group. *Archives of Virology* 145, 567–574.

Kreuze, J.F., Savenkov, E. and Valkonen, J.P.T. (2002) Complete genome sequence and analyses of the subgenomic RNAs of *Sweet potato chlorotic stunt virus* reveal several new features for the genus *Crinivirus. Journal of Virology* 76, 9260–9270.

Kreuze, J.F., Salmolski, K.I., Lazaro, M.U., Chuquiyuri, W.J.C., Morgan, G.L., *et al.* (2008) RNA silencing-mediated resistance to a *Crinivirus* (*Closteroviridae*) in cultivated sweet potato (*Ipomoea batatas* L.) and development of sweet potato virus disease following co-infection with a potyvirus. *Molecular Plant Pathology* 9, 589–598.

Kreuze, K.F., Perez, A., Untiveros, M., Quispe, D., Fuentes, S., *et al*. (2009) Complete viral genome sequence and discovery of novel viruses by deep sequencing of small RNAs: a generic method for diagnosis, discovery and sequencing of viruses. *Virology* 388, 1–7.

Li, R., Salih, S. and Hurtt, S. (2004) Detection of Geminiviruses in sweet potato by polymerase chain reaction. *Plant Disease* 88, 1347–1351.

Lin, C.Y., Ku, H.M., Tsai, W.S., Green, S.K. and Jan, F.J. (2011) Resistance to a DNA and a RNA virus in transgenic plants by using a single chimeric transgene construct. *Transgenic Research* 20, 261–270.

Lindbo, J.A. and Dougherty, W.G. (2005) Plant pathology and RNAi: a brief history. *Annual Review of Phytopathology* 43, 191–204.

Loebenstein, G., Fuentes, S., Cohen, J. and Fuentes, S. (2003) Sweet potato. In: Loebenstein, G. and Thottappilly, G. (eds) *Virus and Virus-Like Diseases in Plants of Developing Countries*. Kluwer Academic Publishers, Dordrecht, The Netherlands, pp. 223–246.

Loebenstein, G., Thottappilly, G., Fuentes, S. and Cohen, J. (2009) Viruses and phytoplasma diseases. In: Loebenstein, G. and Thottappilly, G. (eds) *The Sweet Potato*. Springer-Verlag New York Inc., New York, p. 114.

Lomonossoff, G.P. (1995) Pathogen-derived resistance to plant viruses. *Annual Review of Phytopathology* 33, 323–343.

Lozano, G., Trenado, H.P., Valverde, R.A. and Navas-Castillo, J. (2009) Novel begomovirus species of recombinant nature in sweet potato (*Ipomoea batatas*) and *Ipomoea indica*: taxonomic and phylogenetic implications. *Journal of General Virology* 90, 2550–2562.

Mashilo, J., Niekerk, R. and Shanahan, P. (2013) Combined thermotherapy and meristem-tip culture for efficient elimination of feathery mottle virus in sweet potato (*Ipomoea batatas* L.). *Acta Horticulturae* 1007, 719–725.

McGregor, C.E., Miano, D., LaBonte, D., Hoy, M. and Clark, C. (2009) The effect of the sequence of infection of the causal agents of sweet potato virus disease on symptom severity and individual virus titres in sweet potato cv. Beauregard. *Journal of Phytopathology* 157, 514–517.

Miano, D.W., LaBonte, D.R. and Clark, C.A. (2008) Identification of molecular markers associated with sweet potato resistance to sweet potato virus disease in Kenya. *Euphytica* 160, 15–24.

Moyer, J.W. and Salazar, L.F. (1989) Virus and virus-like diseases of sweet potato. *Plant Disease* 73, 451–455.

Mukasa, S.B., Rubaihayo, P.R. and Valkonen, J.P.T. (2003) Incidence of viruses and virus like diseases of sweet potato in Uganda. *Plant Disease* 87, 329–335.

Mukasa, S.B., Rubaihayo, P.R. and Valkonen, J.P.T. (2006) Interactions between a crinivirus, an ipomovirus and a potyvirus in coinfected sweet potato plants. *Plant Pathology* 55, 458–467.

Mwanga, R.O.M., Yencho, G.C. and Moyer, J.W. (2002) Diallel analysis of sweetpotatoes for resistance to sweetpotato virus disease. *Euphytica* 128, 237–249.

Naylor, R.L., Falcon, W.P., Goodman, R.M., Jahn, M.M., Sengooba, T., *et al*. (2004) Biotechnology in the developing world: a case for increased investments in orphan crops. *Food Policy* 29, 15–44.

Newell, C.A., Lowe, J.M., Merryweather, A., Rooke, L.M. and Hamilton, W.D.O. (1995) Transformation of sweet potato (*Ipomoea batatas* Lam.) with *Agrobacterium tumefaciens* and regeneration of plants expressing cowpea trypsin inhibitor and snowdrop lectin. *Plant Science* 107, 215–227.

Ndunguru, J. and Aloyce, R.C. (2000) Incidence of sweet potato virus disease in sweet potato grown under different traditional cropping systems in the Lake Victoria zone of Tanzania. *African Potato Association Conference Proceedings* 5, 405–408.

Nome, C.F., Laguna, I.G. and Nome, S.F. (2006) Cytological alterations produced by *Sweet potato mild speckling virus*. *Journal of Phytopathology* 154, 504–507.

Okada, Y., Nishiguchi, M., Saito, A., Kimura, T., Mori, M., *et al*. (2002) Inheritance and stability of the virus-resistant gene in the progeny of transgenic sweet potato. *Plant Breeding* 121, 249–253.

Opiyo, S.A., Ateka, E.M., Owuor, P.O., Manguro, L.O.A. and Miano, D.W. (2010) Development of a multiplex PCR technique for the simultaneous detection of *Sweet potato feathery mottle virus* and *Sweet potato chlorotic stunt virus*. *Journal of Plant Pathology* 92, 363–366.

Otani, M., Shimada, T., Kimura, T. and Saito, A. (1998) Transgenic plant production from embryogenic callus of sweet potato [*Ipomoea batatas* (L.) Lam] using *Agrobacterium tumefaciens*. *Plant Biotechnology* 15, 11–16.

Paprotka, T., Bioteux, L.S., Fonseca, M.E.N., Resende, R.O., Jeske, H., *et al*. (2010) Genomic diversity of sweet potato geminiviruses in a Brazilian germplasm bank. *Virus Research* 149, 224–233.

Perez-Egusquiza, Z., Ward, L.I. and Glover, G.R.G. (2009) Detection of *Sweet potato virus 2* in sweet potato in New Zealand. *Plant Disease* 93, 427.

Qin, Y., Zhang, Z., Qiao, Q., Zhang, D., Tian, Y., *et al*. (2013) Molecular variability of *Sweet potato chlorotic stunt virus* (SPCSV) and five potyviruses infecting sweet potato in China. *Archives of Virology* 158, 491–495.

Qin, Y., Wang, L., Zhang, Z., Qiao, Q., Zhang, D., *et al.* (2014) Complete genomic sequence and comparative analysis of the genome segments of *Sweet potato chlorotic stunt virus* in China. *PLoS One* 9, e106323.

Rännäli, M., Czekaj, V., Jones, R.A.C., Fletcher, J.D., Davis, R.I., *et al.* (2009) Molecular characterization of *Sweet potato feathery mottle virus* (SPFMV) isolates from Easter Island, French Polynesia, New Zealand, Southern Africa. *Plant Disease* 93, 933–939.

Sanford, J.C. and Johnston, S.A. (1985) The concept of pathogen-derived resistance – deriving resistance genes from the parasite's own genome. *Journal of Theoretical Biology* 113, 395–405.

Shepherd, D.N., Martin, D.P. and Thomson, J.A. (2009) Transgenic strategies for developing crops resistant to geminiviruses. *Plant Science* 176, 1–11.

Sivparsad, B.J. and Gubba, A. (2012) Molecular resolution of the genetic variability of major viruses infecting sweet potato (*Ipomoea batatas* L.) in the province of Kwazulu-Natal in the Republic of South Africa. *Crop Protection* 41, 49–56.

Sivparsad, B.J. and Gubba, A. (2014) Development of transgenic sweet potato (*Ipomoea batatas* L.) with multiple virus resistance in South Africa (SA). *Transgenic Research* 23, 377–388.

Souto, E.R., Sim, J., Chen, J., Valverde, R.A. and Clark, C.A. (2003) Properties of strains of *Sweet potato feathery mottle virus* and two newly recognized potyviruses infecting sweet potato in the United States. *Plant Disease* 87, 1226–1232.

Tairo, F. (2006) Molecular resolution of genetic variability of major sweet potato viruses and improved diagnosis of potyviruses co-infecting sweet potato. MSc. thesis, Swedish University of Agricultural Sciences, Uppsala, Sweden, pp. 9–36.

Tairo, F., Mukasa, S.R., Jones, R.A.C., Kullaya, A., Rubaihayo, P.R., *et al.* (2005) Unraveling the genetic diversity of the three main viruses involved in sweet potato virus disease (SPVD), and its practical implications. *Molecular Plant Pathology* 6, 199–211.

Tairo, F., Jones, R.A.C. and Valkonen, J.P.T. (2006) Potyvirus complexes in sweetpotato: occurrence in Australia, serological and molecular resolution, and analysis of the *Sweet potato virus 2* (SPV2) component. *Plant Disease* 90, 1120–1128.

Tenllado, F., Llave, C. and Díaz-Ruíz, J.R. (2004) RNA interference as a new biotechnology tool for the control of virus diseases in plants. *Virus Research* 102, 85–96.

Trenado, H.P., Franco, A.O. and Navas-Castillo, J. (2009) Sweet potato chlorotic stunt virus strain West African segment RNA1, complete sequence. Available at: http://www.ncbi.nlm.nih.gov/nuccore/FJ807784.1?report=gbwithparts&log$=seqview (accessed 7 September 2015).

Untiveros, M., Fuentes, S. and Salazar, L.F. (2007) Synergistic interactions between *Sweet potato chlorotic stunt virus* (Crinivirus) with carla-, cucumo-, ipomo- and potyviruses infecting sweet potato. *Plant Disease* 91, 669–676.

Untiveros, M., Fuentes, S. and Kreuze, J. (2008) Molecular variability of *Sweet potato feathery mottle virus* and other potyviruses infecting sweet potato in Peru. *Archives of Virology* 153, 473–483.

Valverde, R.A., Kokkinos, C.D. and Clark, C.A. (2004a) Whitefly transmission of sweet potato viruses. *Virus Research* 100, 123–128.

Valverde, R.A., Kokkinos, C.D. and Clark, C.A. (2004b) *Sweet potato leaf curl virus*: detection by molecular hybridization. *Phytopathology* 94, S105.

Wambugu, F.M. (2003) Development and transfer of genetically modified virus-resistant sweet potato for subsistence farmers in Kenya. *Nutritional Reviews* 61, s110–s113.

Wang, Q.C. and Valkonen, J.P.T. (2008) Elimination of two viruses which interact synergistically from sweetpotato by shoot tip culture and cryotherapy. *Journal of Virological Methods* 154, 135–145.

Winter, S., Purac, A., Leggett, F., Frison, E.A., Rossel, H.W., *et al.* (1992) Partial characterization and molecular cloning of a closterovirus from sweet potato infected with the sweet potato virus disease complex from Nigeria. *Phytopathology* 82, 869–875.

Woolfe, J.A. (1992) *Sweet Potato – An Untapped Food Resource*. Cambridge University Press, New York.

Wu, Q., Luo, Y., Lu, R., Lau, N., Lai, E.C., *et al.* (2010) Virus discovery by deep sequencing and assembly of virus-derived small silencing RNAs. *Proceeding of the National Academy of Sciences USA* 107, 1606–1611.

Yang, C.L., Xing, J.Y. and Wu, J.Y. (1995) The types of main sweetpotato viruses in China and productive program of virus-free plantlets. *Jiangsu Agricultural Science* 6, 35–36.

Yao, D.J.E. and Hortense, A.D. (2005) Detection and distribution of *Sweet potato feathery mottle virus* in sweetpotato using membrane immunobinding assay. *African Journal of Biotechnology* 4, 717–723.

Zhang, S.C. and Ling, K.-S. (2011) Genetic diversity of sweet potato begomoviruses in the United States and identification of a recombinant between *Sweet potato leaf curl virus* and *Sweet potato leaf curl Georgia virus*. *Archives of Virology* 156, 955–968.

17 Mealybug Wilt Disease

Cherie Gambley[1]* and John Thomas[2]

[1]Department of Agriculture and Fisheries, Stanthorpe, Queensland, Australia; [2]Queensland Alliance for Agriculture and Food Innovation, The University of Queensland, Brisbane, Queensland, Australia

17.1 Introduction

Mealybug wilt disease (MWD) is a serious field disease of pineapples worldwide that was first described in Hawaii in 1910 (German et al., 1992). Depending on the age of the plant at the onset of the disease, reductions in fruit yields range from 30% to 55% in Hawaii (Sether and Hu, 2002a). The disease is often referred to as isolated wilt as it typically occurs in secluded patches within the crop or along the edges (Sether et al., 2010) as shown in Fig. 17.1. MWD is thought to be caused by a complex involving viruses, mealybugs and ants. The viruses are transmitted by mealybugs, which in turn are tended by ants. Although a number of distinct viruses have been associated with the disease, the identity of the causal agent(s) has not been determined unequivocally.

17.2 Disease Symptoms

MWD of pineapple progresses in four stages: (i) the leaves redden; (ii) the leaves change from red to pink; (iii) the leaf margins roll downwards; and (iv) the leaves lose turgidity, droop and the plant wilts (Carter, 1945a). In this final stage of wilt, the leaves are almost completely dry. Before leaf symptoms become obvious, root elongation stops and the roots collapse (Carter, 1962, 1963). Figure 17.2 shows typical disease symptoms.

Disease symptoms can vary in different clonal plant lines, depending on the natural levels of anthocyanin in the leaves. Those lines with low anthocyanin develop yellowing symptoms rather than the typical reddening symptom associated with smooth cayenne clones (Carter, 1963). Disease development and incidence is affected by plant age at the onset of mealybug infestation, with younger plants displaying symptoms earlier and at a greater incidence than older plants (Carter, 1945a). Therefore, the impact of MWD on yield is greater for crops which are infested early in the plant cycle.

MWD-affected pineapple plants may 'recover' and produce new, symptomless leaves if subsequently kept mealybug-free, but such plants remain a positive source for disease spread (Carter, 1945a). Recovery is usually associated with the appearance of new roots (Carter, 1962, 1963). As symptoms can recur if reinfested with mealybugs from an alternative disease source, plants that have recovered are not immune from MWD (Carter, 1945a; Ito, 1959).

17.3 History of Research into the Aetiology of Mealybug Wilt Disease

Since MWD was first described in Hawaii in 1910, it has been recorded in all major pineapple-growing regions worldwide, including Africa, the Americas, the islands of the Caribbean and Pacific, and South and South-East Asia (Carter, 1942; Singh and Sastry, 1974; Rohrbach, 1983; Lim, 1985; Borroto et al.,

*E-mail: cherie.gambley@daf.qld.gov.au

Fig. 17.1. A commercial pineapple field affected by mealybug wilt disease. The disease often occurs as isolated patches within or along the edges of the crop. The yellowing plants in the foreground of the photograph exhibit typical symptoms.

Fig. 17.2. Symptoms of mealybug wilt disease in plants of the commercial cultivar Smooth Cayenne clone F180 (right) compared to a healthy plant (left). Disease symptoms include leaf tip dieback, downward rolling of the leaf margins and discoloration of leaves (Gambley *et al.*, 2008b).

1998; Hughes and Samita, 1998; Nickel *et al.*, 2000; Shen *et al.*, 2009; Hernandez-Rodriguez *et al.*, 2014). Early hypotheses attributed symptoms of MWD to a reaction to phytotoxins released by mealybugs when feeding (Illingworth, 1931; Carter, 1933a,b, 1945a,b; Carter and Collins, 1947). However, the identification of plants as either positive or negative sources for MWD led to the proposal of a 'latent transmissible factor' as the causal agent for the disease (Ito, 1959). Insect

transmission of the disease was observed from positive source plants, including their vegetative progeny, and transmission occurred irrespective of symptoms in either the mother or progeny disease source plants. This was further explored by Carter (1963), who confirmed that the disease was caused by a transmissible factor, probably an unidentified virus. Carter (1963) also suggested that mealybugs, in addition to being vectors of the pathogen, had a further role in MWD development by rendering plants more susceptible to the disease through the stress caused by their feeding. Subsequently, viruses from two families, the *Closteroviridae* and the *Caulimoviridae*, have been detected in pineapple MWD-affected plants (German *et al.*, 1992; Wakman, 1994; Wakman *et al.*, 1995; Hu *et al.*, 1996, 1997; Thomson *et al.*, 1996; Sether *et al.*, 2001, 2005, 2009, 2012; Gambley *et al.*, 2008a,b).

In 1989, filamentous virions and double-stranded RNA (a virus replicative intermediate) typical of a *Closteroviridae* species were reported from pineapple plants (German *et al.*, 1992). Polyclonal and monoclonal antibodies have been developed to Australian and Hawaiian isolates of the virus(es), named pineapple closterovirus (PCV) (Ullman *et al.*, 1989; Wakman *et al.*, 1995; Hu *et al.*, 1996), and serological assays were used to detect the virus(es) in pineapples from Australia, France, Malaysia and Taiwan (Wakman, 1994). PCV was also detected in *A. comosus* var. *bracteatus* (syn. *Ananas bracteatus*) (Wakman, 1994; Hu *et al.*, 1997), *A. comosus* var. *ananassoides* (syn. *Ananas ananassoides*) (Hu *et al.*, 1997) and in mealybugs collected from symptomatic, but not asymptomatic plants (Hu *et al.*, 1996). At least two serotypes of PCV were detected in Australia and Hawaii using polyclonal (Wakman *et al.*, 1995) and monoclonal antibodies (Hu *et al.*, 1996), respectively.

What was thought to be a new badnavirus species, and named Pineapple bacilliform virus, was also detected in pineapples in Australia and Hawaii (Wakman *et al.*, 1995; Sether and Hu, 2002b). Using PCR assays thought to be specific, positive reactions were obtained from all Australian commercial clones irrespective of MWD symptoms (Thomson *et al.*, 1996). However, the sequence attributed to Pineapple bacilliform virus was subsequently shown to be that of a retrotransposon present in *Ananas* spp. and not that of a virus (Gambley *et al.*, 2008a). The bacilliform particles seen in these plants were most likely those of one of the subsequently detected badnaviruses, such as *Pineapple bacilliform CO virus* (PBCOV) (Gambley *et al.*, 2008a).

Research on the aetiology of MWD and molecular characterization of the viruses infecting pineapple continues to advance. Additional viruses have been discovered and full genome sequences of some viruses are now available, including at least two ampeloviruses (Melzer *et al.*, 2001, 2008; Sether *et al.*, 2009) and one badnavirus (Wu *et al.*, 2010; Sether *et al.*, 2012). In total, there are two badnaviruses and five ampeloviruses described from pineapple.

17.4   Ampeloviruses Infecting Pineapple

Ampeloviruses (family *Closteroviridae*) are plant-infecting viruses with flexuous or filamentous particles 12 nm in diameter; they have a linear, positive sense, single-stranded RNA genome and are typically transmitted by mealybugs (Martelli *et al.*, 2005).

A total of five distinct ampeloviruses have been reported to infect pineapple (Sether *et al.*, 2001, 2005, 2009; Gambley *et al.*, 2008b). The viruses were named *Pineapple mealybug wilt-associated virus* 1 (PMWaV-1), PMWaV-2, PMWaV-3, PM-WaV-4 and PMWaV-5. Of these, PMWaV-1 and PMWaV-2 are synonymous to the two serotypes of PCV previously reported from Hawaii (Hu *et al.*, 1996; Sether *et al.*, 2001). Table 17.1 details the known distribution

Table 17.1. Global detections of viruses in commercial crops of pineapple.

Virus species (Acronym)	Detections	References
Pineapple mealybug wilt-associated virus 1 (PMWaV-1)	Australia, Brazil, China, Costa Rica, Cuba, Guyana, Honduras, India, Indonesia, Kenya, Martinique, Philippines, Sri Lanka, Taiwan, Thailand, USA (Hawaii)	Sether *et al.*, 2001, 2009; Hernandez-Rodriguez *et al.*, 2014; Gambley *et al.*, 2008b
Pineapple mealybug wilt-associated virus 2 (PMWaV-2)	Australia, Brazil, Costa Rica, Cuba, Honduras, Indonesia, Kenya, Malaysia, Philippines, Sri Lanka, Taiwan, USA (Hawaii, Puerto Rico)	Sether *et al.*, 2001, 2009; Hernandez-Rodriguez *et al.*, 2014; Gambley *et al.*, 2008b
Pineapple mealybug wilt-associated virus 3 (PMWaV-3)	Australia, Cuba Taiwan, USA (Hawaii)	Sether *et al.*, 2005; Gambley *et al.*, 2008b; Shen *et al.*, 2009; Hernandez-Rodriguez *et al.*, 2014
Pineapple mealybug wilt-associated virus 4 (PMWaV-4)	Hawaii	Sether *et al.*, 2009
Pineapple mealybug wilt-associated virus 5 (PMWaV-5)	Australia	Gambley *et al.*, 2008b
Pineapple bacilliform CO virus (PBCOV)	Australia, China, Hawaii	Gambley *et al.*, 2008a; Wu *et al.*, 2010; Sether *et al.*, 2012
Pineapple bacilliform ER virus (PBERV)	Australia	Gambley *et al.*, 2008a
Ananas metavirus[a] (syn. Pineapple bacilliform virus; PBV)	Worldwide	Gambley *et al.*, 2008a; Sether *et al.*, 2012
Endogenous pineapple pararetrovirus 1[a] (ePPRV-1)	Worldwide	Gambley *et al.*, 2008a; Sether *et al.*, 2012

[a]Ananas metavirus and endogenous Pineapple pararetrovirus 1 are endogenous elements present in the genome of pineapple.

of these viruses in commercial crops worldwide.

The distribution of four ampeloviruses in various *Ananas* species and varieties from germplasm accessions maintained at the Maroochy Research Station, Department of Agriculture, Fisheries and Forestry, Queensland, Australia, the Centre de Coopération Internationale en Recherche Agronomique pour le Développement (CIRAD), Martinique and the United States Department of Agriculture-Agricultural Research Service National Clonal Germplasm Repository, Hilo, Hawaii, USA is shown in Table 17.2 (Sether *et al.*, 2001, 2005; Gambley *et al.*, 2008a,b). *Bromelia pinguin*, used as a hedgerow plant in Cuba, was found to harbour PMWaV-2 (Hernandez-Rodriguez *et al.*, 2014).

The involvement of these viruses in MWD is discussed below.

17.5 Badnaviruses and Retroelements in Pineapple

Badnaviruses (family *Caulimoviridae*) are plant-infecting pararetroviruses with bacil-liform virions and a double-stranded DNA genome of about 7–8 kbp (Hull *et al.*, 2005). The majority of badnaviruses are pathogens of tropical and subtropical plant species, including banana, sugarcane, yam, cacao, black pepper and taro, and where vectors are known, these viruses are mostly transmitted by mealybugs (de Silva *et al.*, 2002; Loebenstein and Thottappilly, 2003; Yang *et al.*, 2003).

Two distinct badnaviruses infecting *Ananas* spp. are known. PBCOV and PBERV have been detected by molecular sequence analyses and electron microscopy (Gambley *et al.*, 2008a). In the same study, both viruses were also shown to be transmissible in pineapple by the pink pineapple mealybug, *Dysmicoccus brevipes*. In Hawaii, PBCOV was transmitted by the grey pineapple mealybug, *D. neobrevipes* (Sether *et al.*, 2012). The full genome of two isolates of PB-COV has been sequenced (Wu *et al.*, 2010; Sether *et al.*, 2012).

During the 2008 study of retroelements in pineapple, a third unique caulimovirid sequence was detected from pineapple, but was not derived from encapsidated DNA. It is thought to be present as an endogenous

Table 17.2. Detection of pineapple mealybug wilt-associated viruses 1 to 5 (PMWaV-1 to PMWaV-5), *Pineapple bacilliform CO virus* (PBCOV) and *Pineapple bacilliform ER virus* (PBERV) in pineapple germplasm accessions from Australia, Martinique and USA and from field samples of *Bromelia pinguin* from Cuba.

Species	PMWaV-1	PMWaV-2	PMWaV-3	PMWaV-5	PBCOV	PBERV
A. comosus var. comosus	+	+	+	+	+	–
A. comosus var. ananassoides	+	+	+	+	+	–
A. comosus var. bracteatus	+	–	+	–	NT	NT
A. comosus var. erectifolius	+	+	+	–	–	+
A. comosus var. parguazensis	+	–	–	–	NT	NT
A. comosus var. comosus × *A. macrodontes* (syn. *P. sagenarius*)	–	–	–	NT	NT	NT
Bromelia pinguin	NT	+	NT	NT	NT	NT

+, virus detected by reverse transcriptase PCR or tissue blot immunoassay; –, virus not detected in the samples examined; NT, not tested.

form, and as such, was named endogenous Pineapple pararetrovirus-1 (ePPRV-1) (Gambley *et al.*, 2008a). In phylogenetic analyses, the ePPRV-1 sequence clusters with caulimovirids, but does not appear to belong to any of the existing genera within the *Caulimoviridae* family. In addition, the identity of a further reverse transcribing element of pineapple was clarified. The partial sequence published by Thomson *et al.* (1996) for Pineapple bacilliform virus was shown not to be viral, but instead from a previously undescribed Ty3-gypsy retrotransposon, subsequently named Ananas metavirus (Gambley *et al.*, 2008a).

Additional sequence variants of PBCOV, endogenous pararetroviral sequences and *Metaviridae*-like sequences were reported from Hawaiian pineapples (Sether *et al.*, 2012).

Table 17.1 details the known distribution of these viruses and retroelements worldwide, and Table 17.2 their detection in various plant species.

Indexing accessions in pineapple germplasm collections by electron microscopy and specific PCR has highlighted the possible existence of further viruses. Bacilliform-shaped virions were observed by immunosorbent electron microscopy from four accession samples, all of which were negative for PBCOV and PBERV by PCR indexing (Gambley *et al.*, 2008a).

17.6 Mealybugs and Ants

There are three major mealybug species associated with MWD, namely, *D. brevipes*, *D. neobrevipes* and *Pseudococcus longispinus* (Carter, 1933a; German *et al.*, 1992). *D. brevipes* and *D. neobrevipes* were initially classified as one species, *Pseudococcus brevipes*, but distinct differences in morphology and reproduction led to the reclassification into the current two species (Carter, 1946; Rohrbach *et al.*, 1988). *D. brevipes* and *D. neobrevipes* are similar in their life histories. Their average lifespan is 90 to 94 days, during which they progress through three instar stages (crawlers), a process which takes about 34 days to complete (Ito, 1938). Following this they enter pre-larvaposition,

then larvaposition periods each around 25 days in duration, and finally, a post-larvaposition of about 5 days (Ito, 1938). Although similar in their life histories, *D. neobrevipes* produces more crawlers than *D. brevipes* (Ito, 1938). Alternative hosts of *Dysmicoccus* spp. include *Agave sisalana* (sisal), *Chloris radiata* (finger grass), *Eragrostis tenella* (lovegrass), *Musa* spp. (banana) *Megathyrsus maximus* (syn. *P. maximum*) (Guinea grass), *Panicum barbinode*, *P. repens* (Wainaku grass), *Paspalum* sp., *Portulaca oleracea*, *Saccharum officinarum* (sugarcane), *Sorghum halepense* (Johnson grass) and *Melinis repens* (syn. *Tricholaena rosea*) (Carter, 1933b; Sether *et al.*, 2001).

An association between mealybugs and MWD was first reported in 1933 (Carter, 1933a) and since then *D. brevipes* and *D. neobrevipes* have been shown to transmit PMWaV-1, PMWaV-2 and PMWaV-3 (Sether *et al.*, 1998, 2001, 2005). Similarly, Australian isolates of PMWaV-1, PMWaV-2 and PMWaV-3 were transmitted by local colonies of *D. brevipes* in glasshouse studies (Gambley *et al.*, 2008b). Third instar female nymphs of *D. neobrevipes* transmitted PMWaV-1 at the highest efficiency, whereas post-larvapositional females were unable to acquire the virus (Sether *et al.*, 1998). Transmission efficiencies were also greater when groups of 10, 20 or 40 *D. neobrevipes* were used for inoculation, as opposed to individuals or groups of 5 (Sether *et al.*, 1998). The transmission efficiency of *D. brevipes* at various life stages has not been investigated.

Both badnaviruses, PBCOV and PBERV, were transmitted by *D. brevipes* and PBCOV also by *Planococcus citri* from virus-infected pineapple plants to uninfected pineapple plants in glasshouse transmission experiments; however, no disease symptoms were observed in any of the plants (Gambley *et al.*, 2008a). Similarly, PBCOV was transmitted by *D. neobrevipes* (Sether *et al.*, 2012).

The presence of ants in pineapple crops is not a problem *per se*, but their symbiotic association with mealybug colonies has implications for MWD development. Ants provide the mealybugs protection from predators and parasites and in return they consume the mealybug-secreted honeydew (Rohrbach *et al.*, 1988). The ants also disperse mealybugs into and around the crop, and a positive correlation has been established between ant numbers and the percentage of mealybug-infested pineapple plants (Rohrbach *et al.*, 1988).

The ant species of importance in Hawaiian pineapple crops are *Pheidole megacephala* (syn. *Formica megacephala*, big-headed ant), *Linepithema humile* (syn. *Iridomyrmex humilis*) (Argentine ant) and *Solenopsis geminata* (tropical fire ant) (Rohrbach *et al.*, 1988). The natural enemies of *D. brevipes* and *D. neobrevipes* include the parasitoid *Anagyrus ananatis* (Hymenoptera: Encyrtidae) and the predators *Nephus bilucernarius* (Coleoptera: Coccinellidae) and *Lobodiplosis pseudococci* (Diptera: Cecidomyiidae) (González-Hernández *et al.*, 1999a,b). These species were introduced into Hawaii as biocontrol agents and had a significant effect on *D. brevipes* populations, but only in the absence of the ant species *P. megacephala* (González-Hernández *et al.*, 1999a,b). There is little information available on ants associated with pineapple mealybugs from other countries, but a symbiotic relationship with local ant species is presumed.

17.7   Summary of the Aetiology of Mealybug Wilt Disease

A range of different viruses are known to infect pineapple. Of these, several ampeloviruses, namely, PMWaV-1, PMWaV-2 and PMWaV-3 have all been shown to be associated with the disease in some situations (Sether and Hu, 2002b; Gambley *et al.*, 2008b). To date, none of the badnaviruses or retroviral sequences present in pineapple is known to be associated with the disease.

Investigations into the aetiology of MWD in Hawaii have focussed on PMWaV-1, PMWaV-2 and PMWaV-3, with only PMWaV-2 reported as a critical component of the disease, and only when in combination with a mealybug infestation (Sether and Hu, 2002b). Pot trials and field studies indicated that the presence of either PMWaV-2 or mealybugs alone was insufficient to induce symptoms The presence of PMWaV-1 or PMWaV-3 alone or in combination with mealybugs was not associated

with MWD symptoms in Hawaiian studies (Sether and Hu, 2001; Sether *et al.*, 2005). In pineapple samples collected from around the world and indexed for PMWaV-1 and PMWaV-2, only PMWaV-1 was commonly detected irrespective of symptoms. PMWaV-2 was detected from all symptomatic samples and in a small proportion of those from asymptomatic plants (Sether *et al.*, 2001). However, direct comparisons between symptomatic and asymptomatic plants from the same country and cultivar were mostly not available.

In surveys of four MWD-affected crops in Australia, no single virus was clearly associated with the disease at all survey sites (Gambley *et al.*, 2008b). A statistically significant association ($P<0.001$) between the presence of PMWaV-2 and symptoms was observed at one survey site (site 3), but the virus was at a low incidence at the remaining three survey sites. By contrast, although PMWaV-1 and PMWaV-3 were equally distributed between symptomless and MWD-affected plants at site 3, there was a significant ($P<0.001$) association between each of these two viruses and MWD at sites 1 and 4. At site 2, there was a significant ($P<0.001$) association only between PMWaV-3 and MWD. PMWaV-1 was the most commonly found of the four viruses, and conversely, PMWaV-5 was only occasionally found. PMWaV-4 was not included in the study.

The association between the various PMWaVs and MWD in other countries is largely unknown as thorough investigations of the aetiology of the disease have not been done.

Although several ampelovirus species were transmitted by mealybugs in various controlled experiments (Ito, 1959; Sether *et al.*, 2001, 2012; Gambley *et al.*, 2008a,b), subsequent symptom development in newly infected plants was inconsistent. It is believed there is an important but as yet unidentified environmental factor which contributes to development of MWD symptoms. Carter (1945a) first noted the unreliable development of MWD symptoms, especially under glasshouse conditions, and proposed that the quality and intensity of sunlight were factors in symptom development. Transmission studies in Hawaii, demonstrating an association between PMWaV-2 and MWD symptoms used plants grown outside in containers and field grown plants (Sether and Hu, 2002a).

Although mealybug feeding contributes to disease symptoms (Sether et al., 2001), Ito in 1959 demonstrated that this only occurred when mealybugs were from an alternative source than that used for primary infections. MWD is not normally lethal and the plants usually recover from disease symptoms. However, the recovered plants are not immune to MWD and can succumb to the disease again, but only if challenged with mealybugs from an alternative source of MWD-affected plants (Ito, 1959; Carter, 1963). Ito (1959) proposed that recovered plants were cross-protected from strains of the same virus, and that introduction of a heterologous virus resulted in the re-expression of disease symptoms. Given the variability in association of single virus species with disease symptoms, the putative cross-protection proposed by Ito (1959) would seem to have a feasible role in MWD. Therefore, the causal agent(s) could be several of the multiple sequence variants of ampelovirus species and the reintroduction of new viruses, strains or sequence variants responsible for the reappearance of disease symptoms.

17.8   Disease Management

MWD is traditionally managed by minimizing mealybug activity within the crop as this subsequently affects both the spread of viruses and also the level of stress on plants induced by their feeding. Control of mealybugs and ants is achieved through a combination of crop hygiene to remove alternative insect hosts such as weeds and grasses and application of pesticides to reduce insect levels within crops. Additionally, physical barriers and the absence of ant corridors, for example, concrete drains within the crop, assist in preventing or reducing ant movement into and around the crop (Rohrbach et al., 1988).

Removal of MWD-affected plants from commercial plantations and avoidance of MWD-affected planting material are useful in reducing virus spread (Sether and Hu, 2001; Sether et al., 2001). Furthermore, studies on the spatial and temporal incidences of MWD in field plantings show a lower disease incidence within the block as compared to block edges and it is recommended to source planting material from the internal areas (Sether et al., 2010). No alternative hosts of the pineapple ampeloviruses and badnaviruses are known outside the genus Ananas (Sether et al., 2001), implying that the common occurrence of MWD-affected plants along the edge of crops is due to the influx of mealybugs rather than external sources of virus infection.

In Australian studies, there was a difference in the incidence of virus in different pineapple cultivars. The incidence of viruses in a crop of a newly planted hybrid was less than 50% compared to 100% in a crop of a very old cultivar (Gambley et al., 2008b). This demonstrates an accumulation of viruses in the vegetative planting material over time and that virus-free planting material schemes may have merit for control of MWD, though there are no such schemes currently in use.

17.9   Concluding Remarks

As is common with vegetatively propagated crops, pineapples accumulate and become chronically infected with a number of viruses. In the last 20 years, significant advances have been achieved in identifying these viruses, and gaining a better understanding of MWD, building on the excellent earlier work of Carter and Ito in Hawaii. However, the interactions between the viruses, mealybugs and environmental factors are complicated, and the conditions required for the expression of MWD have only been partially elucidated at this time. The possible role of gene silencing, the identity of the additional ampelovirus(es) and badnavirus(es) that have been detected but not characterized, and the interaction between these disease-inducing factors are fertile areas for future research.

References

Borroto, E.G., Cintra, M., Gonzalez, J. and Borroto, C. (1998) First report of a closterovirus-like particle associated with pineapple plants (*Ananas comosus* cv. Smooth Cayenne) affected with pineapple mealybug wilt in Cuba. *Plant Disease* 82, 263.

Carter, W. (1933a) The pineapple mealybug, *Pseudococcus brevipes*, and wilt of pineapples. *Phytopathology* 23, 207–242.

Carter, W. (1933b) The spotting of pineapple leaves caused by *Pseudococcus brevipes*, the pineapple mealybug. *Phytopathology* 23, 243–259.

Carter, W. (1942) The geographical distribution of mealybug wilt with notes on some other insect pests of pineapple. *Journal of Economic Entomology* 35, 10–15.

Carter, W. (1945a) Some etiological aspects of mealybug wilt. *Phytopathology* 35, 305–315.

Carter, W. (1945b) The influence of plant nutrition on susceptibility of pineapple plants to mealybug wilt. *Phytopathology* 35, 316–323.

Carter, W. (1946) Insect notes from South America with special reference to *Pseudococcus brevipes* and mealybug wilt. *Journal of Economic Entomology* 42, 761–766.

Carter, W. (1962) The systemic phytotoxemias: mealybug wilt of pineapple. In: Carter, W. (ed.) *Insects in Relation to Plant Disease*. Wiley-VCH, New York, pp. 238–265.

Carter, W. (1963) Mealybug wilt of pineapple: a reappraisal. *Annals of the New York Academy of Sciences* 105, 741–764.

Carter, W. and Collins, J.L. (1947) Resistance to mealybug wilt of pineapple with special reference to a cayenne-queen hybrid. *Phytopathology* 37, 332–348.

de Silva, D.P.P., Jones, P. and Shaw, M.W. (2002) Identification and transmission of *Piper yellow mottle virus* and *Cucumber mosaic virus* infecting black pepper (*Piper nigrum*) in Sri Lanka. *Plant Pathology* 51, 537–545.

Gambley, C., Geering, A.D.W., Steele, V. and Thomas, J.E. (2008a) Identification of viral and non-viral reverse transcribing elements in pineapple (*Ananas comosus*), including members of two new badnavirus species. *Archives of Virology* 153, 1599–1604.

Gambley, C., Steele, V., Geering, A.D.W. and Thomas, J.E. (2008b) The genetic diversity of ampeloviruses in Australian pineapples and their association with mealybug wilt disease. *Australasian Plant Pathology* 37, 95–105.

German, T.L., Ullman, D.E. and Gunashinghe, U.B. (1992) Mealybug wilt of pineapple. In: Harris, K.F. (ed.) *Advances in Disease Vector Research*. Springer-Verlag, New York, pp. 242–259.

González-Hernández, H., Reimer, N. and Johnson, M. (1999a) Survey of the natural enemies of *Dysmicoccus* mealybugs on pineapple in Hawaii. *BioControl* 44, 47–58.

González-Hernández, H., Johnson, M. and. Reimer, N. (1999b) Impact of *Pheidole megacephala* (F.) (Hymenoptera: Formicidae) on the biological control of *Dysmicoccus brevipes* (Cockerell) (Homoptera: Pseudococcidae). *Biological Control* 15, 145–152.

Hernandez-Rodriguez, L., Ramos-Gonzalez, P.L., Garcia-Garcia, G., Zamora, V., Peralta-Martin, A.M., *et al.* (2014) Geographic distribution of mealybug wilt disease of pineapple and genetic diversity of viruses infecting pineapple in Cuba. *Crop Protection* 65, 43–50.

Hu, J.S., Sether, D.M. and Ullman, D.E. (1996) Detection of pineapple closterovirus in pineapple plants and mealybugs using monoclonal antibodies. *Plant Pathology* 45, 829–836.

Hu, J.S., Sether, D.M., Ullman, D.E. and Lockhart, B.E.L. (1997) Mealybug wilt of pineapple: pineapple viruses and two-step heat treatment of pineapple crowns. *Acta Horticulturae* 425, 485–492.

Hughes, G. and Samita, S. (1998) Analysis of patterns of pineapple mealybug wilt disease in Sri Lanka. *Plant Disease* 82, 885–890.

Hull, R., Geering, A.D.W., Lockhart, B.E. and Schoelz, J.E. (2005) Family *Caulimoviridae*. In: Fauquet, C., Mayo, M., Maniloff, J., Desselberger, U. and Ball, L. (eds) *Virus Taxonomy: Eighth Report of the International Committee on Taxonomy of Viruses*. Elsevier Academic Press, San Diego, California, pp. 385–437.

Illingworth, J.F. (1931) Preliminary reports on evidence that mealybugs are an important factor in mealybug wilt. *Journal of Economic Entomology* 24, 877–889.

Ito, K. (1938) Studies on the life history of the pineapple mealybug, *Pseudococcus brevipes* (Ckll.). *Journal of Economic Entomology* 31, 291–298.

Ito, K. (1959) Terminal mottle as a symptomatological aspect of mealybug wilt with evidence supporting the hypothesis of a virus etiology of the disease. *Pineapple Research Institute Research Report* 62, 1–37.

Lim, W.H. (1985) *Diseases and Disorders of Pineapples in Peninsular Malaysia*. Malaysian Agricultural Research and Development Institute Report 97, Kuala Lumpur, Malaysia.

Loebenstein, G. and Thottappilly, G. (2003) *Virus and Virus-Like Disease of Major Crops in Developing Countries*. Kluwer Academic Publishers, London.

Martelli, G.P., Agranovsky, A.A., Bar-Joseph, M., Boscia, D., Candresse, T., *et al.* (2005) Family *Closteroviridae*. In: Fauquet, C., Mayo, M., Maniloff, J., Desselberger, U. and Ball, L. (eds) *Virus Taxonomy Eighth Report of the International Committee on Taxonomy of Viruses*. Elsevier Academic Press, San Diego, California, pp. 1077–1087.

Melzer, M.J., Karasev, A.V., Sether, D.M. and Hu, J.S. (2001) Nucleotide sequence, genome organization and phylogenetic analysis of *Pineapple mealybug wilt-associated virus-2*. *Journal of General Virology* 82, 1–7.

Melzer, M., Sether, D., Karasev, A., Borth, W. and Hu, J. (2008) Complete nucleotide sequence and genome organization of *Pineapple mealybug wilt-associated virus-1*. *Archives of Virology* 153, 707–714.

Nickel, O., Chagas, C.M. and Vasconcelos, A.P. (2000) Association of pineapple mealybug wilt with closterovirus-like particles and dsRNA in Bahia, Brazil. *Fitopatologia Brasileira* 36, 200–202.

Rohrbach, K.G. (1983) Pineapple diseases and pests and their potential for spread. In: Sing, K.G. (ed.) *Exotic Plant Quarantine Pests and Procedures for Introduction of Plant Materials*. ASEAN Plant Quarantine Centre and Training Institute, Sardang, Selangor, Malaysia, pp. 145–171.

Rohrbach, K.G., Beardsley, J.W., German, T.L., Reimer, N.J. and Sanford, W.G. (1988) Mealybug wilt, mealybugs and ants on pineapple. *Plant Disease* 72, 558–565.

Sether, D.M., Hu, J.S. (2001) The impact of *Pineapple mealybug wilt-associated virus-1* and reduced irrigation on pineapple yield. *Australasian Plant Pathology* 30, 31–36.

Sether, D.M. and Hu, J.S. (2002a) Yield impact and spread of *Pineapple mealybug wilt associated virus-2* and mealybug wilt of pineapple in Hawaii. *Plant Disease* 86, 867–874.

Sether, D.M., and Hu, J.S. (2002b) Closterovirus infection and mealybug exposure are necessary for the development of mealybug wilt of pineapple disease. *Phytopathology* 92, 928–935.

Sether, D.M., Ullman, D.E. and Hu, J.S. (1998) Transmission of pineapple mealybug wilt-associated virus by two species of mealybug (*Dysmicoccus* spp.). *Phytopathology* 88, 1224–1230.

Sether, D.M., Karasev, A.V., Okumura, C., Arakawa, C., Zee, F., *et al.* (2001) Differentiation, distribution and elimination of two different pineapple mealybug wilt-associated viruses found in pineapple. *Plant Disease* 85, 856–864.

Sether, D.M., Melzer, M.J., Busto, J., Zee, F. and Hu, J.S. (2005) Diversity and mealybug transmissibility of ampeloviruses in pineapple. *Plant Disease* 89, 450–456.

Sether, D.M., Melzer, M.J., Borth, W. and Hu, J.S. (2009) Genome organization and phylogenetic relationship of *Pineapple mealybug wilt associated virus-3* with family *Closteroviridae* members. *Virus Genes* 38, 414–420.

Sether, D.M., Borth, W., Melzer, M.J. and Hu, J.S. (2010) Spatial and temporal incidence of pineapple mealybug wilt-associated viruses in pineapple planting blocks. *Plant Disease* 94, 196–200.

Sether, D.M., Melzer, M.J., Borth, W. and Hu, J.S. (2012) *Pineapple bacilliform CO virus*: diversity, detection, distribution and transmission. *Plant Disease* 96, 1798–1804.

Shen, B.N., Zheng, Y.X., Chen, W.H., Chang, T.Y., Ku, H.-M., *et al.* (2009) Occurrence and molecular characterization of three pineapple mealybug wilt associated viruses in pineapple in Taiwan. *Plant Disease* 93, 196.

Singh, S.J. and Sastry, S.M. (1974) Wilt of pineapple – a new virus disease in India. *Indian Phytopathology* 27, 300–303.

Thomson, K., Dietzgen, R., Thomas, J.E. and Teakle, D.S. (1996) Detection of pineapple bacilliform virus using the polymerase chain reaction. *Annals of Applied Biology* 129, 57–69.

Ullman, D.E., German, T.L., Gunashinge, U.B. and Ebesu, R.H. (1989) Serology of a closterovirus-like particle associated with mealybug wilt of pineapple. *Phytopathology* 79, 1341–1345.

Wakman, W. (1994) Two clostero-like viruses and a bacilliform virus in pineapple plants in Australia. PhD Thesis, Department of Microbiology, The University of Queensland, St. Lucia, Brisbane, Australia.

Wakman, W., Teakle, D.S., Thomas, J.E. and Dietzgen, R.G. (1995) Presence of a clostero-like virus and a bacilliform virus in pineapple plants in Australia. *Australian Journal of Agricultural Research* 46, 947–958.

Wu, L., Ruan, X., Shen, W., Tan, Y. and Li, H. (2010) Sequencing and analysis of the complete genomic sequence of *Pineapple bacilliform comosus virus*. *Scientia Agricultura Sinica* 43, 1969–1976.

Yang, I.C., Hafner, G.J., Revill, P.A., Dale, J.L. and Harding, R.M. (2003) Sequence diversity of South Pacific isolates of *Taro bacilliform virus* and the development of a PCR-based diagnostic test. *Archives of Virology* 148, 1957–1968.

18 Viruses Affecting Tropical and Subtropical Crops: Future Perspectives

Gustavo Fermin[1]* and Paula Tennant[2]

[1]Instituto Jardín Botánico de Mérida, Faculty of Sciences, Universidad de Los Andes, Mérida, Venezuela; [2]Department of Life Sciences, The University of the West Indies, Mona Campus, Jamaica

Fifteen chapters spanned a range of highly divergent taxonomic groups of plant viruses, their effects on host phenotypes and implications for the management of virus diseases in tropical and subtropical agriculture. Although the diseases covered can be grouped under two categories, namely major (or traditional) virus diseases versus minor diseases of less economic significance and/or limited geographic distribution, all are considered equally important with respect to maintaining the sanitary status and food security of a region. Some of the diseases covered in this book are initiated by infections with a single virus pathogen that is transmitted by only one or two vector species. The aetiology is less conclusive for others. At the extreme, there are diseases elicited by a complex of different viruses or by a complex involving a number of different viruses along with different groups of insects. Papaya ringspot is an example of a disease caused by a virus with a narrow host range, infecting only papaya, its wild relatives and members of two or so other plant families under natural conditions. Conversely, the aetiological agents of tomato leaf curl and tomato spotted wilt can infect and cause disease in hundreds of crop species and possibly its vector, in the case of the latter. Several species of ampeloviruses and badnaviruses infect pineapple.

Over the past few decades, there has been mounting interest in the increasing number of viruses causing disease epidemics. All too often outbreaks have seriously stretched local resources but more importantly, the capacity to identify and control emerging diseases remains limited in poorer regions where many of these diseases have originated. It can be expected that new virus disease outbreaks will occur in the future due to the expansion of international trade that contributes to the distribution of viruses and virus strains into regions where conditions suitable to their establishment are encountered. In addition, new cultural practices may enhance the probabilities of a new virus, a new host plant or a new vector becoming established. Climate change adds another layer of complexity. Climate change could affect the interactions between the host and the pathogen resulting in changes in host resistance including shifts in geographical distribution of the host and pathogen. Rises in temperature, for example, are expected to introduce major changes in the distribution of major pests, and hence, the viruses they carry.

The major issue facing food production is the need for economic benefits through increased production of saleable products at low production costs. There is also the need for environmental benefits through the

---

*E-mail: fermin@ula.ve

reduced use of pesticides and a better use of water and fertilizers. Finally, there is the need to meet consumer demands for convenience foods, nutritional and health benefits and improved flavour. New cultivars will invariably have an important role to play in addressing these issues. In other words, 'smart' crop varieties that yield more with fewer inputs will be pivotal to comply with producer and consumer demands. Plant virus disease resistance, notably built-in resistance against pests and diseases, is essential to the generation of these 'smart' crop varieties. Unfortunately, sources of resistant genes are presently not available for every cultivated crop.

Recent advances in sequencing strategies, deeper knowledge of vector dynamics and the discovery of new viruses, and even the furthering of our knowledge of virus biology and virus–vector, vector–host and virus–host interactions, will allow for the development of new disease-resistant varieties and strategies that mitigate the damages and losses inflicted by plant diseases of viral aetiology. All investment in research focused on plant pathogens, including plant viruses, is of direct application to food production. Emphasis on research and education, particularly in universities, thus ought to be reflected in national budget allocation. But education and research should also be accompanied by ways of involving producers and consumers alike. By having farmers and other end-users involved in the development of new crop varieties, feedback mechanisms are enhanced as is, more importantly, the relevance of the breeding activities to the needs of the producers.

Integrated disease management must be put in force and make use of all available methods, both classical and modern. Engineered resistance developed beyond pathogen derived resistance (i.e. traditional transgenic crops) will be an important option in crop improvement if based on an understanding and manipulation of virus recognition and response driven by major, dominant R resistance genes (Harris et al., 2013). Besides R genes, the use of recessive resistance genes, targeted manipulation of RNA interference pathways and plant hormone-mediated resistance will add to the arsenal of weapons against disease (Nicaise, 2014). Efforts to address the constraints that impede the use of the technology and the cultivation of transgenic varieties have been significant. The recent discovery of sweet potato plants naturally transformed by *Agrobacterium tumefaciens* (Kyndt et al., 2015) will surely help to facilitate the adoption and acceptance of transgenic crops improved for human consumption.

To be effectively implemented, integrated strategies of disease management must also amalgamate all the essential participants: government, researchers, producers and consumers – from national to local community levels – in a way that plant-protection policies can be dispassionately proposed, implemented and adopted. To be effective at national and international levels, quarantine measures based on effective methods of diagnostics (mostly molecular in nature) must also be adopted. Germplasm movement can be deterred if clear rules are established and virus or vector dissemination to new areas is strictly avoided. Such measures should take into consideration both vertical versus horizontal transmission of viruses. Seed certification programmes, focused on reliable diagnosis and (secure) quick distribution thereafter, guarantee easy access to virus-free planting material.

Widening our knowledge of the plant virosphere will allow us not only to uncover and describe new agents of diseases, but also viruses that do not cause harm, but probably are helpful to our understanding of the ecology of plant viruses, and hence ways to alleviate the damage they cause (McDiarmid et al., 2013). On the other hand, this and other technical advances, will surely aid in understanding virus variability and the evolutionary forces that shape virus populations and explain the emergence of more virulent strains, the true molecular and physiological nature of the host range of viruses, reservoirs, as well as the sanitary status of cultivated hosts. To be useful, however, clear rules on nomenclature and their fast application have to be implemented in a way that adoption will translate into clear communication among researchers and regulators. Bioinformatic tools have to be exploited to their full capacity to facilitate these new developments; new viruses and their isolates, new hosts, availability of host resistant varieties, certification and quarantine measures, all 'omics' research related to viruses, their hosts and vectors, virus epidemiology, etc. Finally, disease forecasting, along with the deployment of plant tolerant and resistant varieties in those areas more afflicted with virus diseases, will benefit not only farmers and their nationals, but also the global community.

Dissecting the molecular biology of viruses, and eventually their control, will provide new avenues into the use of plant viruses as raw materials for the development of vectors useful for the production of novel goods, particularly nutraceuticals and plant vaccines (Rybicki, 2014), and in engineering resistance mechanisms against viruses and other pathogens and pests (Ambrós *et al.*, 2013; Hajeri *et al.*, 2014). The application of biofarming or plant molecular farming that was anticipated in the 1990s has been long in coming. It is speculated that biofarmed viral vaccines will be approved more widely in the near future, at least for animal use. It is interesting that the very traits employed by viruses to establish infection and induce disease in their plant hosts are now being manipulated for the production of plant-derived biologics that are safe, efficacious and offer solutions to the challenge of providing inexpensive medicines without cold chain requirements, or for poorer regions, lacking an established medical infrastructure, where many of these viruses originated.

References

Ambrós, S., Ruiz-Ruiz, S., Peña, L. and Moreno, P. (2013) A genetic system for *Citrus tristeza virus* using the non-natural host *Nicotiana benthamiana*: an update. *Frontiers in Microbiology* 4, 165.

Hajeri, S., Killiny, N., El-Mohtar, C., Dawson, W.O. and Gowda, S. (2014) *Citrus tristeza virus*-based RNAi in citrus plants induces gene silencing in *Diaphorina citri*, a phloem-sap sucking insect vector of citrus greening disease (Huanglongbing). *Journal of Biotechnology* 176, 42–49.

Harris, C.J., Slootweg, E.J., Goverse, A. and Baulcombe, D.C. (2013) Stepwise artificial evolution of a plant disease resistance gene. *Proceedings of the National Academy of Sciences USA* 110, 21189–21194.

Kyndt, T., Quispe, D., Zhai, H., Jarret, R., Ghislain, M., et al. (2015) The genome of cultivated sweet potato contains *Agrobacterium* T-DNAs with expressed genes: an example of a naturally transgenic food crop. *Proceedings of the National Academy of Sciences USA* 112, 5844–5849.

McDiarmid, R., Rodoni, B., Melcher, U., Ochoa-Corona, F. and Roossinck, M. (2013) Biosecurity implications of new technology and discovery in plant virus research. *PLOS Pathogens* 9, e1003337.

Nicaise, V. (2014) Crop immunity against viruses: outcomes and future challenges. *Frontiers in Plant Science* 5, 660.

Rybicki, E.P. (2014) Plant-based vaccines against viruses. *Virology Journal* 11, 205.

Index

Page numbers in **bold** type refer to figures and tables.